Fundamentals of Multiaccess Optical Fiber Networks

For a complete listing of the *Artech House Optoelectronics Library*,
turn to the back of this book.

Fundamentals of Multiaccess Optical Fiber Networks

Denis J. G. Mestdagh

Artech House
Boston • London

Library of Congress Cataloging-in-Publication Data
Mestdagh, Denis J. G.
Fundamentals of multiaccess optical fiber networks / Denis J. G. Mestdagh.
Includes bibliographical references and index.
ISBN 0-89006-666-3
1. Optical communications. 2. Electronic data processing – distributed processing. I. Title.
TK5103.59.M37 1995 94-29879
621.382'75–dc20 CIP

British Library Cataloguing in Publication Data
Mestdagh, Denis J. G.
Fundamentals of Multiaccess Optical Fiber Networks
I. Title
621.38275

ISBN 0-89006-666-3

685 Canton Street
Norwood, MA 02062

International Standard Book Number: 0-89006-666-3
Library of Congress Catalog Card Number: 94-29879

10 9 8 7 6 5 4 3 2 1

To the memory of my father

Contents

PART II: MULTIACCESS NETWORKS

Preface

During the last few years, research on optical-fiber communication technology has led to the advent of a new generation of fiber systems, which can be referred to as *multiaccess optical-fiber networks*. In these networks, the huge bandwidth available within the fibers is shared among all connected nodes. Even though the operating speed of node transceivers is limited by the speed of electronics, an aggregate network capacity much in excess of the individual node's bandwidth can be achieved. This is accomplished by taking advantage of the beneficial properties offered by optics, such as the possibility to perform passive optical multiplexing/demultiplexing using distinct wavelengths, subcarriers, or short pulse-code sequences.

The intense activity in this field has resulted in a number of advances that will find applications in many domains of optical-fiber communication engineering.

Although much work has already been done, most of the material is available only in the form of research papers. This book is intended to provide a comprehensive account of various multiaccess optical-fiber networks with focus on single-mode fiber-based systems.

The book is aimed at scientists and research engineers who will start working or who anticipate future involvement in the field of optical-fiber networks. The implications of various design options on system performance are emphasized throughout the text. It is also useful for technical managers who want an up-to-date overview of advances in this area.

An attempt is made to make the book self-contained so that graduate students with some knowledge of opto-electronics should be able to follow the text easily. Furthermore, each chapter refers to many relevant papers that can be consulted for further study.

I would like to take the opportunity to thank Julie Lancashire for inviting me to write this book and Kate Hawes for her encouragement all along the writing process.

I would also like to thank the reviewers of the book who took time and effort to provide helpful comments. Their suggestions have helped to improve the clarity of the text. They were Patrick Vankwikelberge, Ingrid Van de Voorde, Yves Gigase, Bernard Biran, Gert Van der Plas, Chris Sierens, Paul Spruyt, Tony Van Regemorter, Bruno Vande Sompele, Terry Defruyt from Alcatel Bell Telephone, Ivan Andonovic from the University of Strathclyde, Jean-Michel Gabriagues from Alcatel Alsthom Recherche, and Rick Gross from GTE Labs.

Last but not least, I am very grateful to my wife Brigitte for her patience and comprehension during my many fits of abstraction when I was thinking about this fascinating field of fiber-optic networks.

Chapter 1
Introduction

During the last 15 years or so, rapid advances have been realized in optical-fiber communication technology. Reduction of single-mode fiber losses, progress in high-sensitivity optical receivers, development of high-speed semiconductor laser diodes, and the advent of optical amplifiers have led to a steady increase in demonstrated transmission capacity characterized by the highest possible bit rate over the longest unrepeatered distance.

These advances have had an important impact in the field of telecommunications and it is now universally recognized that optical fibers will be widely used in many application domains of communication engineering in the near future [Miller 1988, Brackett 1989, Karol 1993].

They are already being used in several distinct fields of applications such as submarine transmission systems, telephone networks, computer networks, ships and aircrafts, and more recently cable television networks.

Though impressive capacities have been achieved, today's systems utilize only a small fraction of the potential transmission capacity offered by optical fibers. The main reason is that opto-electronic components at the input and output ends of the fiber are not capable of operating at speeds commensurate with the fiber's bandwidth. It should be desirable to share the available fiber bandwidth between several communication nodes by allowing *multiple access* to the same resource.

Therefore, efforts are now turned to a new generation of fiber systems that may be called *multiaccess optical-fiber networks* [Cheung, 1990]. In these networks the entire path between several end nodes, which share the same resource, is passive (except when optical amplification is applied) and optical: no opto-electronic nor electro-optic conversion is performed except at the ends of a connection.

These new networks make it possible to circumvent the electronic bottlenecks by taking advantage of the beneficial properties offered by optics. For example, high-speed signal multiplexing/demultiplexing might be performed optically, allowing the electronics to operate at much lower bit rates than the aggregate throughput of the network.

Examples of multiaccess optical-fiber network architectures are illustrated in Figure 1.1. The nodes are connected together by using passive optical components such as star couplers, taps, wavelength division multiplexers (WDMs), or a combination of these devices.

Multiple access to the same resource can be accomplished by multiplexing several communication nodes in either the spectral domain or the time domain. In the spectral domain, one can distinguish wavelength division multiple access (WDMA) and subcarrier multiple access (SCMA). In the time domain, two implementation approaches can be distinguished: time division multiple access (TDMA) and code division multiple access (CDMA).

A brief description of these four multiple access techniques, which will be our focus throughout the text, follows.

WDMA [Brackett, 1990]

The use of many wavelengths is a natural exploitation of the enormous optical-fiber bandwidth, and WDMA effectively uses this bandwidth by assigning a unique wavelength to each accessing node. As an example, consider Figure 1.1(a) in which N lasers tuned to a fixed and distinct wavelength are interconnected with N tunable receivers through a star coupler. By tuning one of the receivers to one of the N laser wavelengths, a connection through the network can be set up. Multiple connections can be present at the same time, so long as they are on different wavelengths. This architecture also supports multicast connections since more than one receiver can be tuned to the same source wavelength at the same time. Of course, a contention-resolution protocol is required to prevent collisions that could occur when different sources simultaneously attempt to reach the same destination. This protocol together with the means for communicating the tuning information to the output port receivers are important design issues in determining the performances of these networks. WDMA is not restricted to the simple previous example, and several possibilities exist depending on whether the input lasers, the output receivers, or both, are made tunable.

SCMA [Darcie, 1987]

The data from each node is used to modulate (ASK, PSK, or FSK) a dedicated microwave subcarrier frequency, which is then used to modulate an optical carrier. The name *subcarrier* refers to the fact that the primary carrier frequency is the optical signal (10^{14} Hz), while the microwave signal (at 10^8 to 10^{10} Hz) is the modulated subcarrier. At the node receiver, a photodetector receives the sum of all transmitted subcarrier channels. The intended channel for the node is then selected and demodulated using conventional microwave techniques, similar to those used in radiofrequency (RF) radio tuners. The functional features of SCMA systems are close to those obtained with WDMA except that the multiple access is done in the microwave frequency domain supported by optical channels.

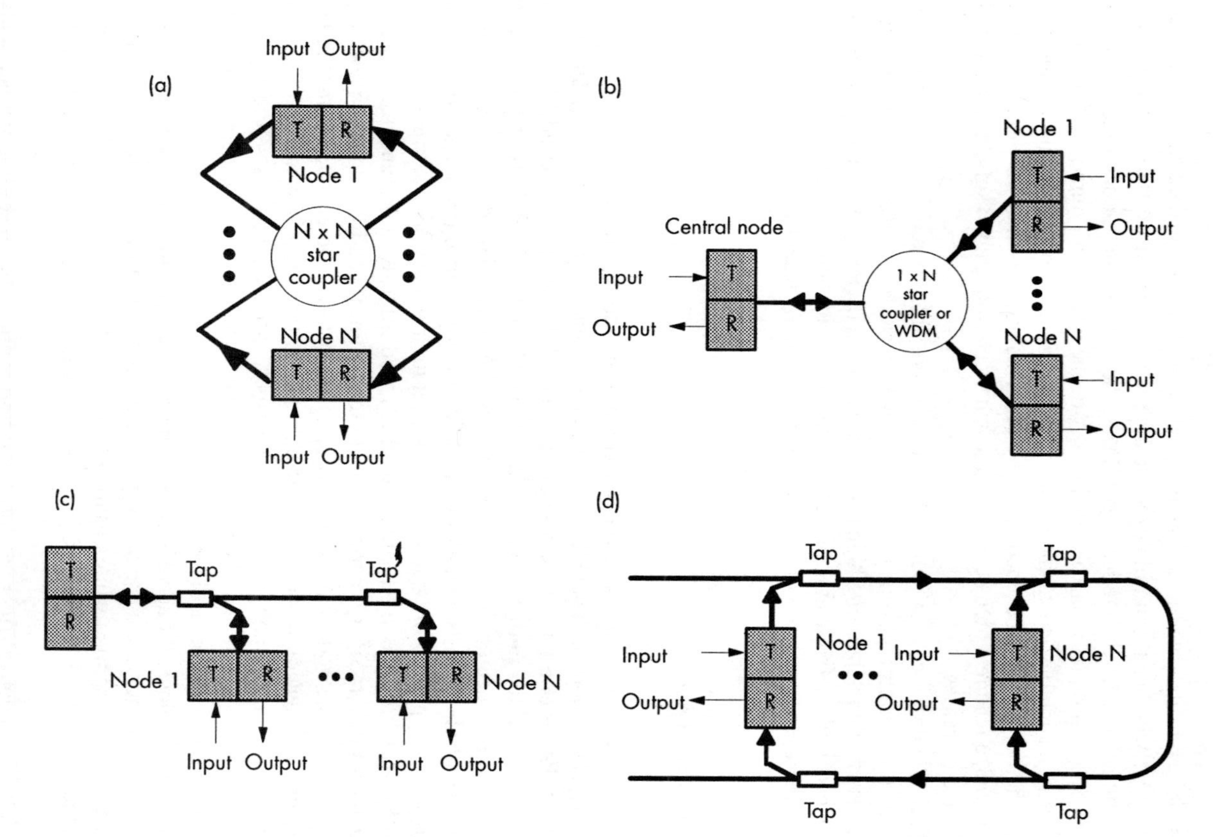

Figure 1.1 Examples of multiaccess optical-fiber networks: (a) $N \times N$ star, (b) $1 \times N$ star, (c) single-bus, and (d) folded-bus.

TDMA [Stern 1987, Hinton 1989]

Each node sends its data in a synchronized way. Two implementation options exist depending on whether the multiplexed data stream is created by bit-interleaving or by packet-multiaccess. In bit-interleaving TDMA, the N accessing nodes send the sampled data with bit duration less than or equal to $1/N$ of the original bit duration. These bits are combined in an interleaved fashion, with preassigned time position, into a time period equal to the duration of the uncompressed bit. This requires that all the nodes be bit-synchronized, a feature which becomes increasingly complicated to solve when the node's data bit rate increases.

Packet multiaccess, by contrast, makes it possible to relax this synchronization requirement by permitting some unused time regions, called guard time, between adjacent data packets while maintaining high transport network efficiency. In some systems, nodes are provided with a fixed assignment in time slots. In others, nodes may access the resource in a dynamic fashion controlled by the medium access protocol.

CDMA [Salehi, 1989]

This type of multiple access method is accomplished through the use of specific optical codes to represent both the data bits and the nodes. At each node, optical encoders encode each bit of data into a unique very high rate optical sequence (composed of very short optical pulses), which is then coupled into the single-mode fiber. This sequence, called either the address code or signature sequence, is recognizable by the intended receiver in the presence of all other data signals at the same time, provided that the set of sequences is orthogonal (or pseudo-orthogonal). The recognition is done by a correlation process; that is, by matching the input signal to the stored address code, and to a threshold comparator for data recovery. For example, to send information from node j to node k, the address code for the node k is impressed upon the data by the encoder at the jth node. Furthermore, the generation of ultrashort light pulses (the present world record is 6 f/s) and the ability to encode and decode these extremely fast optical pulses would suggest the possibility of ultrahigh-speed CDMA systems having capacities of orders of magnitude above those that could be achieved with TDMA.

Since technological tradeoffs exist for each of these multiple access techniques, one of the main tasks for the system designer is to find the most suitable technique (or a combination) to be applied while ensuring compatibility with the network specifications, such as the architecture, the application and the corresponding requirements, the cost, and so on.

Multiaccess optical-fiber networks will find applications in several distinct fields of optical-fiber communication engineering. Among these fields (to name but the most important envisaged up to now) are:

1. Local area networks (e.g., computer networks);
2. Metropolitan area networks;
3. Public telecommunication networks;
4. Photonic switching (including optical interconnection).

In some of these application fields the very high capacity feature offered by multiaccess optical-fiber networks is exploited. In others, the possibility of overall system cost reduction provided by these new networks becomes the prime concern.

These issues along with the adopted technical solutions will be addressed throughout the text.

Overview

This book is composed of two parts comprising nine chapters. Each chapter contains a detailed reference section for readers who wish to study a particular topic in greater detail. The references are listed in chronological order. This made it easier to include recent significant publications during the writing process, resulting, so the author hopes, in a book text that is as up to date as possible.

Broadly viewed, Chapters 2 to 5 of Part I provide the background material needed for understanding the various technological aspects of multiaccess optical-fiber networks. Except where explicitly indicated otherwise, single-mode fibers are assumed throughout the book. Part II (from Chapters 6 to 10) deals with multiaccess network architectures and performances. Chapter 6 is devoted to passive optical devices and network topologies. Chapters 7 to 10 focus on multiaccess techniques; namely, WDMA, SCMA, TDMA and CDMA. A brief outline of each chapter follows to provide an overview of the text.

Chapter 2 is devoted to the description of semiconductor light sources ranging from the light-emitting diodes (LEDs) to laser diodes (LDs) of several types; Fabry-Perot LDs, distributed feedback LDs and wavelength tunable LDs. The chapter begins with a brief description of the fundamental physical mechanisms responsible for the generation of light within such devices. The relevant characteristics for multiaccess optical-fiber networks, such as the modulation bandwidth, the linearity, the output optical spectrum, and the wavelength tunability as well as other specific properties, are then described for each type of semiconductor light source.

Chapter 3 provides the basic mathematics needed for understanding light propagation in optical fibers. Starting with Maxwell's equations, the wave propagation equation is derived and used to discuss the single-mode condition. Attenuation and dispersion characteristics of single-mode fibers are then described and dispersion-induced pulse broadening is derived using the Gaussian pulse envelope approximation. Since nonlinear effects can give rise to limiting constraints in multiaccess optical-fiber networks, they are presented as well.

Chapter 4 focuses on optical detection techniques. It starts with a brief description of the physical processes used for light detection within semiconductor photodetectors (i.e., PIN and avalanche photodiodes). The performances of direct detection receivers are then described and the dominant noise sources are outlined. Coherent lightwave detection techniques, which may play an important role in future multiaccess optical networks, are described with emphasis on their improved system performance compared to direct detection.

Chapter 5 considers the current status of optical amplifiers with focus on semiconductor and doped-fiber devices.

Chapter 6 is devoted to passive optical devices and multiaccess optical-fiber network topologies. The network losses corresponding to the distinct topologies are derived and some techniques that make it possible to overcome the network power budget limitation are discussed.

Chapter 7 focuses on the WDMA technique. The basic concepts behind WDMA networks are introduced first, with focus on the so-called single-hop WDMA systems. The chapter then describes the different types of wavelength tunable optical filters, with particular emphasis on their applicability within direct detection WDMA systems. Design and performance issues of single-hop WDMA networks are then discussed and some relevant examples of architectures that have been proposed so far are described. The chapter ends with a description of the so-called multihop WDMA networks and two experimental systems are reviewed.

Chapter 8 considers the SCMA technique, which takes advantage of conventional microwave components to provide high-capacity fiber networks. Design issues for the laser (or LED) transmitters and the receivers are discussed. The limitations of the network capacity due to optical noise sources, such as shot noise, electronic noise, laser-intensity noise, optical beat interference noise, nonlinear distortion, and clipping effects are presented. Recent experimental results are also reviewed.

Chapter 9 considers the TDMA technique, with focus on packet-based TDMA. Although TDMA does not really take advantage of the beneficial properties offered by the optical domain, this technique is currently applied in several commercially available systems. Implementation issues such as synchronization and optical power leveling are described, and examples of TDMA systems are discussed with regard to their performances and potential applications.

Chapter 10 is devoted to CDMA, which can be viewed as a spread spectrum approach to TDMA. The chapter introduces a specific class of codes; namely, optical orthogonal codes, that are suitable for CDMA fiber-optic networks. The techniques and the components required to implement CDMA nodes are discussed. Finally, a novel CDMA technique based upon spectral encoding of ultrashort light pulses, referred to as coherent optical CDMA, is presented.

REFERENCES

[Darcie, 1987] Darcie, T. E. "Subcarrier multiplexing for multiple access lightwave networks," *IEEE J. Light Technology*, Vol. LT-5, No. 8, Aug. 1987, pp. 1103–1110.

[Stern, 1987] Stern, J. R., et al. "Passive optical local networks for telephony applications and beyond," *Electronic Letters*, Vol. 23, No. 24, Nov. 1987, pp. 1255–1257.

[Miller, 1988] Miller, S. E., and I. P. Kaminow. *Optical Fiber Telecommunications II*, eds., Cambridge, MA: Academic Press, 1988.

[Hinton, 1989] Hinton, H. S. "Photonic time-division switching systems," *IEEE Circuits and Devices Magazine*, July 1989, pp. 39–43.

[Salehi, 1989] Salehi, J. A., et al. "Code division multiple-access techniques in optical fiber networks—Parts I and II," *IEEE Trans. on Communication*, Vol. 37, No. 8, Aug. 1989, pp. 824–842.

[Brackett, 1989] Brackett, C. A., ed. "Special Issue on Lightwave Systems and Components," *IEEE Communication Magazine*, Vol. 27, No. 10, Oct. 1989.

[Cheung, 1990] Cheung, N. K., K. Nosu, and G. Winzer, eds. "Special Issue on Dense Wavelength Multiplexing for High Capacity and Multiple Access Communications Systems," *IEEE J. Selected Areas of Communication*, Vol. JSAC-8, No. 6, Aug. 1990.

[Brackett, 1990] Brackett, C. A. "Dense wavelength division multiplexing networks: principles and applications," *IEEE J. Selected Areas of Communication*, Vol. JSAC-8, Aug. 1990, pp. 948–964.

[Karol, 1993] Karol, M. J., G. Hill, C. Lin, and K. Nosu, eds. "Special Issue on Broad-Band Optical Networks," *IEEE J. Light. Technology*, Vol. LT-11, No. 5/6, May/June 1993.

Part I
Background

Chapter 2
Semiconductor Light Sources

Light sources for optical-fiber communication systems are almost exclusively semiconductor light-emitting diodes (LEDs) or semiconductor laser diodes (LDs). The fundamental difference between these two light sources is that light from LEDs is produced by spontaneous emission and is thus incoherent, while LDs operate in the stimulated emission regime, making them sources of coherent radiation. This difference has important consequences for the performance of fiber systems and the choice of the type of light source is generally dictated by a compromise between the required system performance and its cost of implementation.

Although the first demonstration of laser action in semiconductors was reported as early as 1962 [Hall 1962, Nathan 1962], it was only after the 1970s that semiconductor light sources became practical for use in fiber communication systems. Since then, device characteristics such as efficiency, modulation bandwidth, and reliability have continuously been improved.

This chapter is devoted to semiconductor light sources, with emphasis on their characteristics relevant to the main subject of the book. The basic concepts needed for an understanding of light generation within semiconductors are introduced in Section 2.1. Section 2.2 will discuss the characteristics of LEDs from the standpoint of their applicability in single-mode fiber systems. Section 2.3 is devoted to laser diodes. Sections 2.3.1 and 2.3.2 will introduce the concepts of optical gain and laser threshold, which are common to any type of laser diode. The Fabry-Perot LD is discussed in Section 2.3.3, while Section 2.3.4 deals with the basic laser cavity structure for single longitudinal mode operation; namely, the distributed feedback (DFB) laser diode. Subsequently, Section 2.3.5 is devoted to advanced devices that are wavelength tunable. These wavelength tunable laser sources are currently being investigated in R&D laboratories and may become of major importance in future multiaccess optical-fiber networks, as will become apparent throughout this book. Finally, Section 2.4 summarizes the main results of the chapter.

Semiconductor light sources have been extensively discussed in the literature on optical communications; a few review articles and books are listed in the references [Kressel 1977, Senior 1985, Agrawal 1986, Kobayashi 1988, Tien Pei Lee 1989, Koch 1990].

2.1 ABSORPTION AND EMISSION OF LIGHT

2.1.1 The Two-Level Atomic Gas System

Prior to a discussion of light generation within semiconductor light sources it is instructive to consider the interaction of light in a two-level atomic gas system. Figure 2.1 illustrates such a system where the energy level E_1 corresponds to the ground state and E_2 corresponds to the excited state.

In thermal equilibrium, the populations of the two energy levels are distributed according to the Boltzmann statistics, which gives:

$$N_2/N_1 = e^{-E_\Delta k/T} \tag{2.1}$$

where N_1 and N_2 are the atomic densities in the ground and excited states, respectively, $E_\Delta = E_2 - E_1$, k is the Boltzmann's constant, and T is the absolute temperature.

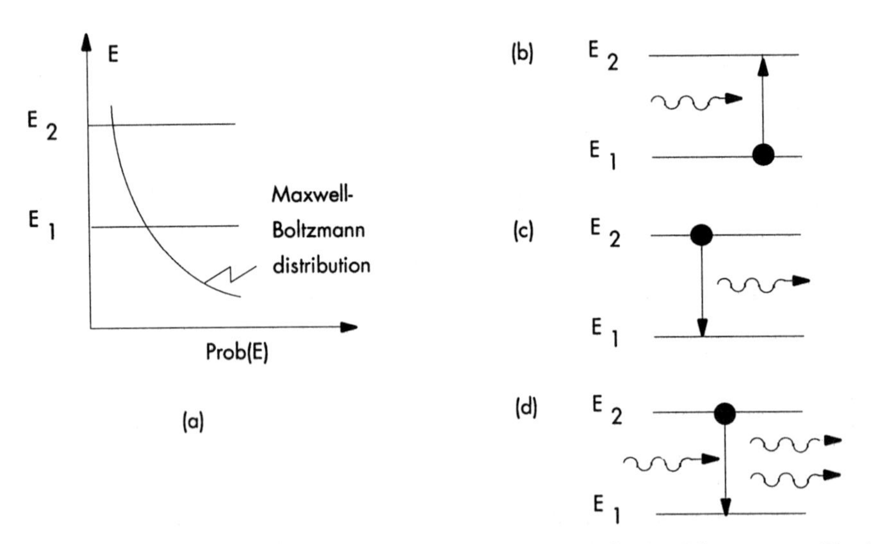

Figure 2.1 (a) Energy levels and probability of their occupancy for a two-level atomic gas system. The three fundamental processes of (b) (stimulated) absorption, (c) spontaneous emission, and (d) stimulated emission between the two energy levels.

If a photon with energy $h\nu = E_\Delta$ (where h is the Planck's constant and ν the frequency associated with the photon) is incident on the atomic system, it may be absorbed by an electron in the ground state E_1, which then rises to the excited state E_2. This process is sometimes referred to as stimulated absorption to distinguish it from thermal excitation. The rate of this process is proportional to the density of electrons in the ground state, N_1, and to the spectral density, $\varrho_{ph}(\nu)$, of the radiation energy at the transition frequency ν. It can thus be written as

$$R_{abs}(\nu) = B_{12} \cdot N_1 \cdot \varrho_{ph}(\nu) \tag{2.2}$$

where B_{12} is a constant of proportionality known as the Einstein coefficient of absorption.

Alternatively, the emission of a photon with frequency $\nu = E_\Delta/h$ may occur if an electron is initially in the excited state E_2 and makes a transition to the ground state E_1. This light emission mechanism can occur through two fundamental processes, known as *spontaneous emission* and *stimulated emission.*

In the spontaneous emission process, the light is emitted in an entirely random manner; that is, the emitted photons have no phase relationship among them and propagate in random directions. This process is at the heart of LED operation (Section 2.2). The random nature of the spontaneous emission process results in *incoherent* radiation. The rate of spontaneous emission is simply proportional to the density of electrons in the excited state, N_2, and can thus be written as:

$$R_{spon}(\nu) = A \cdot N_2 \tag{2.3}$$

where A is the Einstein coefficient of spontaneous emission and is equal to the reciprocal of the spontaneous lifetime of electrons at energy level E_2.

In contrast with spontaneous emission, stimulated emission provides *coherent* radiation and is at the heart of light generation within laser sources. Stimulated emission is initiated by an existing photon at frequency ν, which interacts with an electron in the excited state E_2, causing it to return to the ground state E_1 with the creation of a second photon. This second photon matches the original photon in its essential characteristics, such as the energy (or equivalently its frequency), polarization, and direction of propagation.

The rate of stimulated emission can be obtained in the same way as the rate of stimulated absorption and the result is

$$R_{stim}(\nu) = B_{21} \cdot N_2 \cdot \varrho_{ph}(\nu) \tag{2.4}$$

where B_{21} is the Einstein coefficient of stimulated emission.

For a system in thermal equilibrium, the rates of emission and absorption must be equal. Therefore, using (2.2), (2.3), and (2.4), the following equation must be satisfied:

$$A \cdot N_2 + B_{21} \cdot N_2 \cdot \varrho_{\mathrm{ph}}(\nu) = B_{12} \cdot N_1 \cdot \varrho_{\mathrm{ph}}(\nu) \tag{2.5}$$

Solving (2.5) in $\varrho_{\mathrm{ph}}(\nu)$ and using (2.1) with $E_\Delta = h\,\nu$, we obtain

$$\varrho_{\mathrm{ph}}(\nu) = \frac{A/B_{21}}{(B_{12}/B_{21}) \cdot e^{h\nu/kT} - 1} \tag{2.6}$$

Since the atomic system under consideration is in thermal equilibrium, it should produce a radiation spectral density identical to the blackbody radiation spectral density given by Planck's formula [Jenkins, 1981]

$$\varrho_{\mathrm{ph}}(\nu) = \frac{8\pi h\nu^3/c^3}{e^{h\nu/kT} - 1} \tag{2.7}$$

A direct comparison of (2.6) and (2.7) provides the Einstein relations

$$B_{12} = B_{21} = B \text{ and } A = \frac{8\pi h\nu^3}{c^3} \cdot B \tag{2.8}$$

Two important conclusions can be derived from the above results. First, in thermal equilibrium the probability of stimulated absorption and stimulated emission is equal ($B_{12} = B_{21}$). However, the ratio of the rate of stimulated absorption to that of stimulated emission is given by (2.1). Thus, for radiation in the visible or near-infrared regions ($h\nu \approx 1eV$), absorption always dominates by far over stimulated emission at room temperature ($kT \approx 25meV$). Therefore, at room temperature, the thermal equilibrium involves essentially the interplay between stimulated absorption and spontaneous emission.

Second, to produce coherent light we need to satisfy the condition $R_{\mathrm{stim}} > R_{\mathrm{abs}}$ which can be achieved only when $N_2 > N_1$ [(2.2) and (2.4)]. This condition is referred to as *population inversion* and can never be realized when the system is in thermal equilibrium, as can be seen from (2.1). In other words, to produce coherent light, the system must be out of thermal equilibrium to achieve population inversion. This can be achieved by pumping the system with an external energy source.

2.1.2 Semiconductors

The same conclusions for population inversion hold for semiconductors. However, many differences exist between semiconductors and the extremely simplified two-level atomic gas system that has just been considered.

Indeed, the atoms in a semiconductor are sufficiently close to each other for various interactions to occur between them. These interactions result in important changes in the distribution of energy levels. In semiconductors, energy bands are formed because when atoms are brought close together, the fundamental Pauli's exclusion principle can no longer be neglected.

For our concerns, only the two highest energy bands, called the valence band and the conduction band, are of crucial importance, as is the forbidden energy gap between them, which we shall refer to as E_g. Since electrons (and holes) obey Pauli's principle, the probability $P(E)$ that an electron occupies a particular energy level E when the system is in thermal equilibrium at an absolute temperature T is given by the Fermi-Dirac distribution

$$P(E) = \frac{1}{1 + e^{(E-E_F)/kT}} \tag{2.9}$$

where E_F is known as the Fermi level.

The Fermi level is a mathematical parameter that gives an indication of the distribution of carriers (electrons or holes) in the semiconductor, and its physical interpretation will become clear further on.

Equation (2.9) gives the probability to find an electron at a certain energy level E. To obtain the carrier concentration at that energy level, we need to know the carrier density as a function of energy within the permitted energy band. For the conduction band, this function, $Z_c(E)$, defined as the density of states or the number of permitted energy states per unit energy per unit volume, is found to be [Kressel, 1977]

$$Z_c(E) = \mu_c \cdot \sqrt{E - E_c} \tag{2.10}$$

where μ_c is a constant that depends on the effective mass of the electrons and E_c is the energy level of the bottom of the conduction band.

Therefore, the electron concentration, N, in the conduction band of the semiconductor is given by summing the products of the density of states and the occupancy probability over the conduction band; that is,

$$N = \int_{E_c}^{\infty} Z_c(E)P(E)dE \tag{2.11}$$

Similarly, for holes in the valence band, energy is measured from the top of the valence band E_v and

$$Z_v(E) = \mu_v \cdot \sqrt{E_v - E} \tag{2.12}$$

The concentration of holes in the valence band is then given by

$$P = \int_{-\infty}^{E_v} Z_v(E)[1 - P(E)]dE \tag{2.13}$$

This is illustrated in Figure 2.2 for three distinct semiconductor types. Figure 2.2(a) shows the situation of an intrinsic semiconductor (i.e., a semiconductor containing no impurities or lattice defects). The valence and conduction bands are separated by the forbidden energy bandgap E_g which depends on the semiconductor material. The Fermi level is located at the center of the bandgap, indicating that there is a small probability of electrons occupying energy levels at the bottom of the conduction band and a corresponding number of holes occupying energy levels at the top of the valence band.

This situation is drastically changed when the semiconductor material is doped with impurity atoms, which create either more free electrons (donor impurities, in which case the semiconductor is said to be of the *n*-type) or holes (acceptor impurities, in which case the semiconductor is said to be of the *p*-type).

Figure 2.2(b) shows the situation in *n*-type semiconductors. The Fermi level is raised to a position above the center of the bandgap. The thermally excited electrons from the donor atoms are raised into the conduction band to create an excess of negative charge carriers and the majority carriers of the material are electrons.

In *p*-type semiconductors, as shown in Figure 2.2(c), the Fermi level is lowered below the center of the bandgap. The thermally excited electrons are raised from the valence band to the acceptor impurity levels, leaving an excess of positive charge carriers in the valence band and the majority carriers are holes.

We clearly see how the Fermi level characterizes the type of semiconductor in thermal equilibrium. In *n*-type (*p*-type) semiconductors, it moves towards the conduction (valence) band as the dopant concentration increases. For heavily doped semiconductors, the Fermi level can even lie within the conduction (*n*-type) or the valence band (*p*-type).

2.1.3 The *p-n* Junction

The basic structure of a semiconductor light source is obtained by adjoining *p* and *n*-type semiconductors, as shown in Figure 2.3(a). This is called a *p-n* junction.

At the vicinity of the junction, a number of important phenomena occur. The abundant holes at the left side (*p*-type) will tend to diffuse towards the right side of the junction to neutralize the majority electrons of the *n*-type semiconductor. A thin depletion region is formed at the junction through carrier recombination, which effectively leaves it free of mobile charge carriers. At the *p*-side, within a distance comprised between *0* and $-x_p$ (see Figure 2.3 for the definition of x_p and x_n), the crystal contains fewer holes than in the farthest region of the junction, but also more electrons. The electrical neutrality is thus destroyed and a negatively charged region is created. Symmetrically, at the *n*-side, within a distance comprised between *0* and x_n, a positively charged region is created. Thus, at either side of the junction there exists a double repartition of charge that establishes a potential barrier between the *p* and *n*-type regions (Figure 2.3(b)). At the vicinity of the junction plane, a positive charge is attracted towards the *p*-side because of the electric field that is created across the depletion region. This field is called the built-in

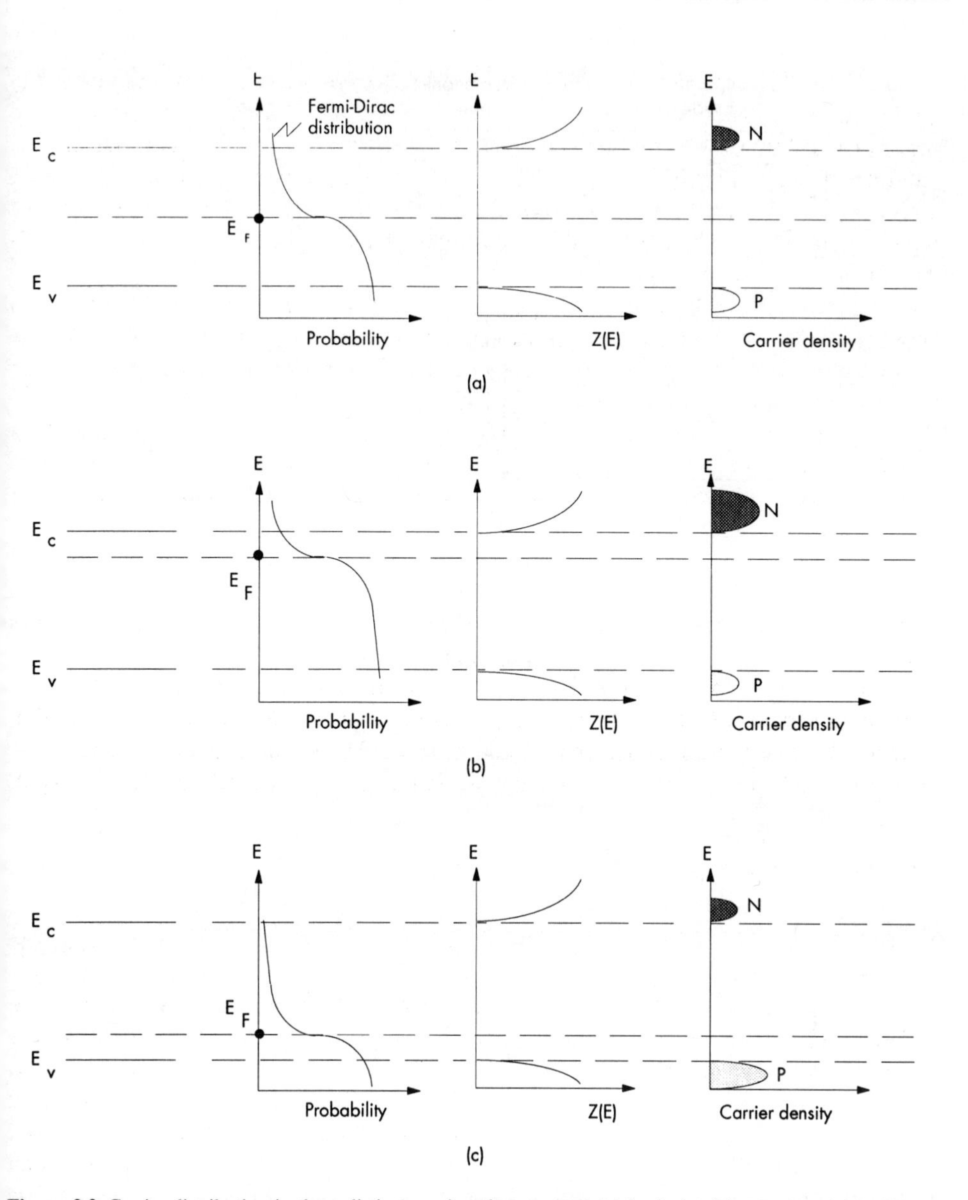

Figure 2.2 Carrier distribution in three distinct semiconductor types: (a) intrinsic, (b) *n*-type, and (c) *p*-type.

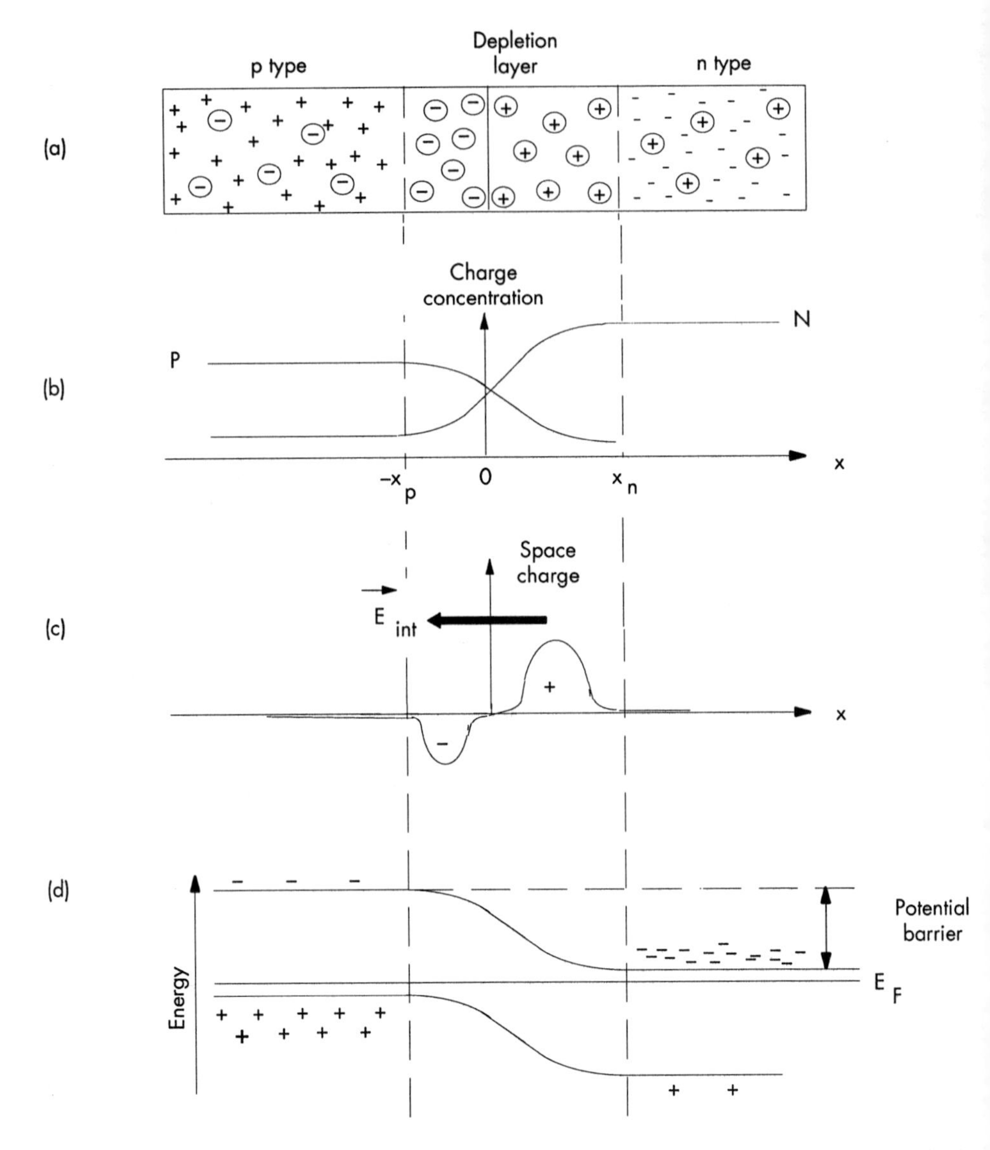

Figure 2.3 (a) The *p-n* homojunction formed by adjoining a *p*-doped and a *n*-doped semiconductor, (b) the carrier concentration, (c) the space charge and built-in electric field $\vec{E}_{int}$, (d) the energy band diagram in thermal equilibrium.

electric field $\vec{E}_{int}$ and it is directed from n towards p (Figure 2.3(c)). It restricts the further interdiffusion of majority carriers from their respective regions. In the absence of an externally applied voltage, no current flows through the junction (except the leaky current).

The width of the depletion region and thus the magnitude of the potential barrier depends upon the doping concentration in the p and n-type regions and any external applied voltage.

When an external negative voltage is applied to the p-type region with respect to the n-type region (Figure 2.4(a)), the external electric field ($\vec{E}_{ext}$) has the same direction as $\vec{E}_{int}$ and thus reinforces its action. The width of the depletion region and the resulting potential barrier are increased, preventing even more the current flow through the junction. In fact, there is always a small current flow, called the leaky current, due to the presence of minority carriers in both p- and n-type regions, for which the passage from one region to the other is favored by the applied field. Moreover, electrons and holes always appear in the depletion region because of thermal agitation. These carriers are immediately separated by the electric field, electrons going towards the n-type region and holes towards the p-type region. This generation of carriers in the depletion region can also be obtained by photo-electric effect when the junction is illuminated with light: this mode of operation is basically that of semiconductor photodiodes and will be discussed Chapter 4.

Conversely, when an external positive voltage V is applied to the p-type region with respect to the n-type region, the generated external electric field $\vec{E}_{ext}$ is opposed to the built-in electric field $\vec{E}_{int}$, resulting in a reduction of the potential barrier. The p-n junction is then said to be forward-biased. Electrons from the n-type region and holes from the p-type region can flow more easily across the junction into the opposite type region. As a result, an electric current I builds up, which increases exponentially with the applied voltage V according to the well-known expression [Kressel, 1977]

$$I = I_s \cdot \left[\exp\left(\frac{eV}{kT}\right) - 1\right] \tag{2.14}$$

where I_s is the saturation current, which depends on the diffusion coefficients associated with electrons and holes.

As seen in Figure 2.4(b), when the p-n junction is forward-biased, free electrons and holes are simultaneously present within the depletion region. These carriers can recombine together and generate light as a result of the energy released from the recombination.

The rates of emission and absorption of light can be obtained following an analysis similar to that for the two-level atomic system. The total spontaneous emission rate at a frequency ν is obtained by summing over all possible transitions between the conduction and valence bands such that $h\nu = E_2 - E_1$, where E_2 is the energy state occupied by an

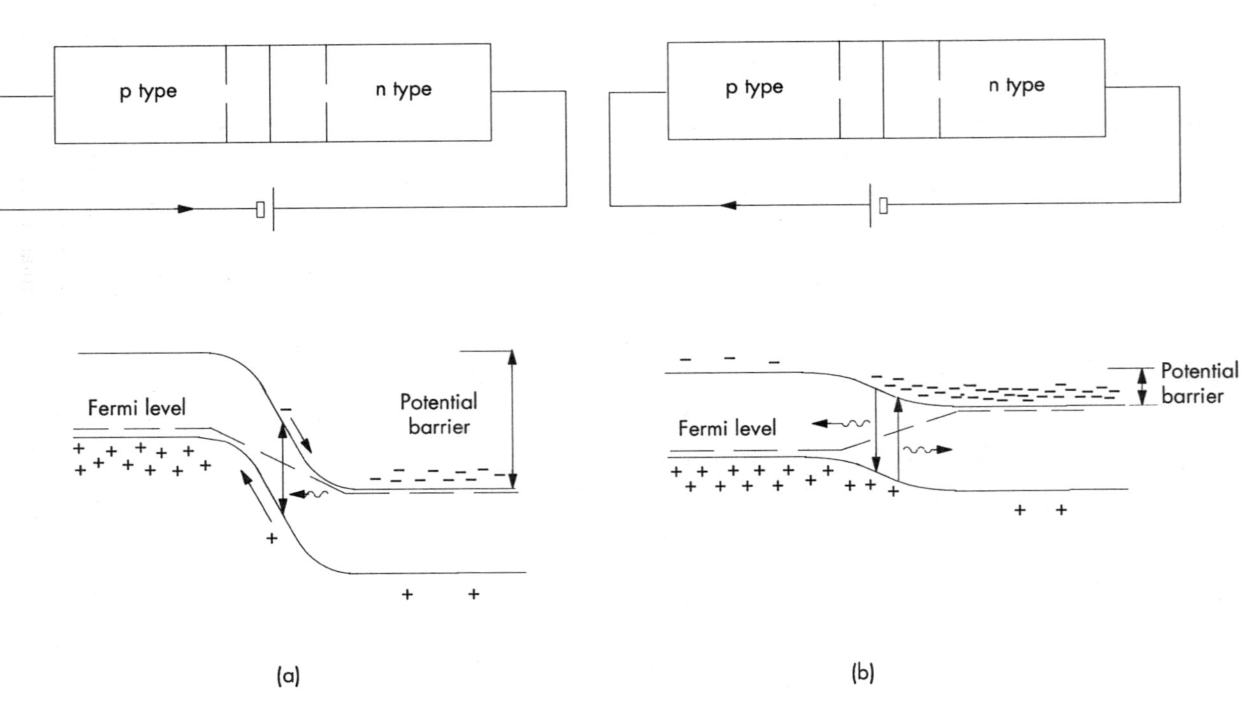

Figure 2.4 Energy-band diagram of a *p-n* homojunction (a) under reverse-bias (photodiode mode) and (b) under forward-bias (laser mode).

electron in the conduction band and E_1 is an empty energy state (i.e., occupied by a hole) in the valence band.

The probability that an electron in the conduction band occupies the energy state E_2 is given by the Fermi-Dirac distribution as

$$P_c(E_2) = \frac{1}{1 + e^{(E_2 - E_{Fc})/kT}} \quad (2.15)$$

where E_{Fc} is referred to as the quasi-Fermi level of the conduction band.

Similarly, the probability that the energy state E_1 in the valence band is occupied by a hole is given by

$$1 - P_v(E_1) = 1 - \frac{1}{1 + e^{(E_1 - E_{Fv})/kT}} \quad (2.16)$$

where E_{FV} is the quasi-Fermi level of the valence band.

Therefore, the total spontaneous emission rate at a frequency ν is given as

$$R_{\mathrm{spon}}(v) = \int_{E_c}^{\infty} A(E_1, E_2)P_C(E_2)[1 - P_V(E_1)]Z_c(hv)dE_2 \quad (2.17)$$

And the total stimulated absorption and emission rates at frequency ν are respectively given by

$$R_{\mathrm{abs}}(v) = \int_{E_c}^{\infty} B(E_1, E_2)P_V(E_1)[1 - P_C(E_2)]Z_c(hv)\varrho_{\mathrm{ph}}(v)dE_2 \quad (2.18)$$

$$R_{\mathrm{stim}}(v) = \int_{E_c}^{\infty} B(E_1, E_2)P_C(E_2)[1 - P_V(E_1)]Z_c(hv)\varrho_{\mathrm{ph}}(v)dE_2 \quad (2.19)$$

By comparing the integrands in (2.18) and (2.19), the condition $R_{stim} > R_{abs}$ is readily obtained and leads to $P_c(E_2) > P_V(E_1)$. Using (2.15) and (2.16), this condition yields

$$E_g < hv = E_2 - E_1 < E_{\mathrm{FC}} - E_{\mathrm{FV}} \quad (2.20)$$

The condition (2.20) known as the Bernard-Durafour condition expresses the fact that in order to obtain population inversion, the quasi-Fermi levels must be separated by a value larger than the minimum value E_g. This can be obtained at the *p-n* junction by heavy doping of both *p-* and *n*-type regions.

Furthermore, it can also be seen from (2.20) that the frequency ν of the stimulated light is greater than E_g/h. However, this is only true for direct bandgap semiconductors in which the minimum energy point E_c in the conduction band coincides with the maximum energy point of the valence band (*E-k* diagram). Typical direct bandgap semiconductors are GaAs, InP, InGaAsP, and AlGaAs.

By contrast, in indirect bandgap semiconductors such as Si and Ge, the position of E_c in the *E-k* diagram differs from that of E_v. As a consequence, electron-hole recombination cannot occur without a change of momentum. In that case, a transition occurs with lattice vibrations (phonons) and thus heat. The efficiency of light generation in indirect bandgap semiconductors is much smaller than for direct bandgap semiconductors because of the requirement of phonon-assisted transition. This is why semiconductor light sources are exclusively made of direct bandgap semiconductor materials.

2.1.4 Heterojunctions

Early semiconductor light sources were *p-n* homojunctions formed with the same semiconductor material on both sides of the junction. A problem with homojunctions is that it is difficult to realize high carrier densities in the active region, resulting in a low efficiency for light generation.

This carrier-confinement problem has been solved with the so-called heterojunctions, which are formed with distinct semiconductor materials having different bandgaps. In double-heterojunctions (DH), as shown in Figure 2.5, a thin semiconductor layer is sandwiched between *p* and *n*-type semiconductor layers having larger bandgaps. The bandgap difference helps to confine electrons and holes into the sandwiched layer, which acts as the active layer for light generation. Moreover, since the bandgap of the active layer is smaller than that of the materials surrounding it, its refractive index is slightly larger. As a consequence, the active layer also acts as a dielectric waveguide for the light generated inside it (Chapter 3). Therefore, the double-heterostructure leads to the confinement of both injected carriers and generated light. It is the heterojunction that has made semiconductor light sources practical for use in optical-fiber communication systems.

To reduce the formation of lattice defects at the interfaces between the distinct semiconductor materials, double heterostructures with lattice-matched materials are fabricated using special epitaxial growth techniques. In particular, to form ternary or quaternary compounds, a fraction of the lattice sites (e.g., GaAs) is replaced by another element. An important compound is $In_{1-x}Ga_xAs_yP_{1-y}$, in which x and y represent the mixing fractions of the corresponding elements. With a suitable choice of x and y, the light source made of such a compound (often referred to as InGaAsP light sources) can be designed to operate in the wavelength range 1.0–1.65 μm, and is thus particularly suitable for optical-fiber communication systems (Chapter 3).

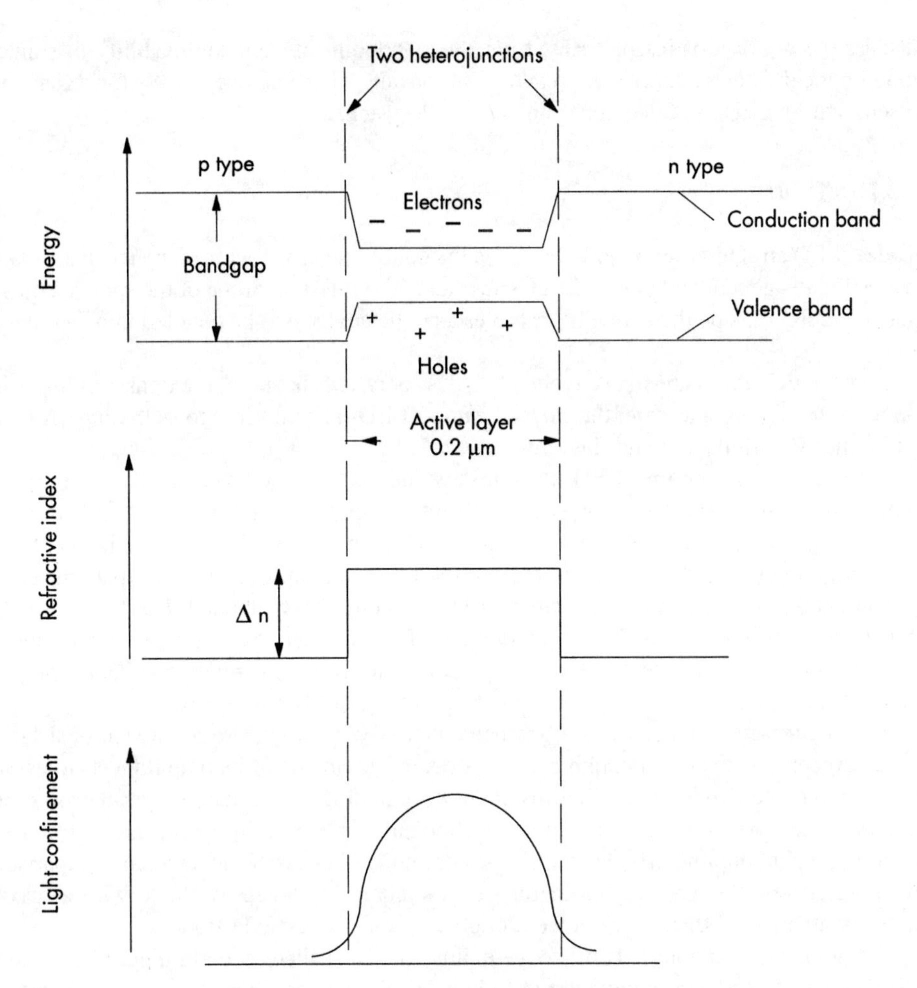

Figure 2.5 The double heterojunction allowing simultaneous confinement of charged carriers and generated light inside the active layer.

2.2 LIGHT-EMITTING DIODES (LEDs)

While LEDs have been superseded by LDs in most single-mode optical-fiber communication systems, their specific features have been exploited in applications where LDs are not necessarily required. Attractive LED features include high reliability, less temperature dependency, simpler drive circuitry (no threshold), and low cost. This section will briefly discuss the characteristics of LEDs from the standpoint of their applicability in single-mode optical-fiber systems. The reader can consult [Tien Pei Lee, 1988] for a detailed discussion of LEDs in telecommunication fiber systems.

2.2.1 LED Structures

Today, LED structures are mainly based on the double-heterostructure. The mechanism by which light is generated from an LED is spontaneous emission. Some of the spontaneously generated light inside the active layer can escape the device and be coupled into an optical fiber.

Although there are many types of LEDs, only two have been extensively used in optical-fiber communication: the *surface-emitter* LED (also referred to as the *Burrus*-type LED after its originator) and the *edge-emitter* LED.

In the surface-emitter LED, the emissive area is restricted to a small active region within the device. The light power coupled into a single-mode fiber is dependent upon many factors, including total internal reflection at the interface, the distance/alignment and the medium between the emissive area and the fiber end, and the emission pattern. With proper coupling design, surface-emitter LEDs can couple only about 1% of the internally generated light power into a single-mode fiber. This corresponds to a maximum of about −23 dBm (50 μW) injected power as a typical value for commercially available components (see Appendix A for the correspondence between dBm and mW).

An edge-emitter LED has a geometry that is very similar to a conventional laser diode except that an antireflection coating is deposited on one of its output facets to avoid laser action (see Section 2.3). The resulting waveguiding in the active layer narrows the output beam divergence and increases the radiance at the emitting facet. As a consequence, higher coupling efficiency into a single-mode fiber can be achieved in comparison with surface-emitter LEDs. Commercial edge-emitter LEDs are available with a maximum of about −13 dBm light power coupled into a single-mode fiber.

Elongated versions of the edge-emitter LED, called superluminescent diodes (SLDs), have also been manufactured to increase the power coupled into a single-mode fiber with typical values in the range −5 dBm to −3 dBm.

2.2.2 Light-Current Curve

An important external characteristic of any semiconductor light source is the output light power versus injection current curve, or L-I curve. Figure 2.6 shows the L-I curves at several temperatures for a typical 1.3-μm edge-emitter LED. It is seen that the output power increases as the injection current increases from zero (i.e., there is no threshold). The gradient dL/dI (or the linearity of the device) remains constant over a limited current range, usually for $I \leq 100$mA.

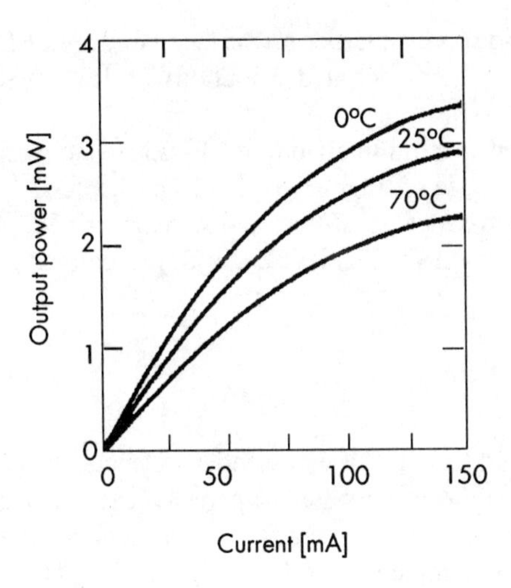

Figure 2.6 Light-current curves at several temperatures for a typical 1.3-μm edge-emitter LED. [After Agrawal, 1992.]

For higher current, dL/dI decreases due to the temperature increase in the active region, which leads to less radiative recombinations. For some applications, like analog transmission, it is sometimes necessary to use predistortion linearization driving circuitry to ensure a linear L-I curve over a larger current range than usual.

The decrease of radiative recombination rates at high temperatures is also the reason why LEDs emit less power at high operating temperatures, as can be seen in Figure 2.6.

2.2.3 Spectral Distribution

The spectral distribution of LEDs can be deduced from $R_{spon}(\nu)$ given by (2.17). Although $R_{spon}(\nu)$ is generally calculated numerically, an approximated analytical expression can be obtained as [Saleh, 1991]

$$R_{spon}(\nu) = A_0 \cdot \sqrt{h\nu - E_g} \cdot e^{-(h\nu - E_g)/kT} \tag{2.21}$$

where A_0 is a constant that depends on various material parameters.

The full-width at half maximum (FWHM) of $R_{spon}(\nu)$ is easily obtained from (2.21) and is given by $\Delta\lambda = 1.8kT\lambda^2/hc$, where we used the relation $\Delta\lambda = (\lambda^2/c)\Delta\nu$ (see Appendix B).

Therefore, $\Delta\lambda$ increases as λ^2. For InGaAsP LEDs emitting at 1.3 μm, $\Delta\lambda$ ranges between 60 nm and 80 nm, while for LEDs emitting at 1.55 μm, $\Delta\lambda$ is about 80 nm to 100 nm. The spectral linewidth is also larger for surface-emitting LEDs than for edge-emitting LEDs. Figure 2.7 shows the spectral distributions of 1.3-μm surface and edge-emitter LEDs.

Although the large spectral linewidth of LEDs leads to performance degradations in optical-fiber systems because of fiber dispersion (see Chapter 3), it can be advantageously exploited in the so-called *spectrum slicing* and *subcarrier multiple access* (SCMA) techniques that will be discussed in Chapters 7 and 8, respectively.

2.2.4 Modulation Bandwidth

The modulation bandwidth of LEDs is essentially dependent upon two parameters: the recombination lifetime of the carriers and the parasitic capacitance of the packaged device.

Assuming negligible parasitic capacitance, the fundamental limit of LED modulation response can be determined by solving the rate equation for the carrier density [Agrawal, 1986]. This leads to

$$P(f) = \frac{P(f=0)}{\sqrt{1 + (2\pi f \tau_c)^2}} \tag{2.22}$$

where $P(f)$ is the amplitude of the modulated optical output power at the modulation frequency f, and τ_c is the carrier lifetime.

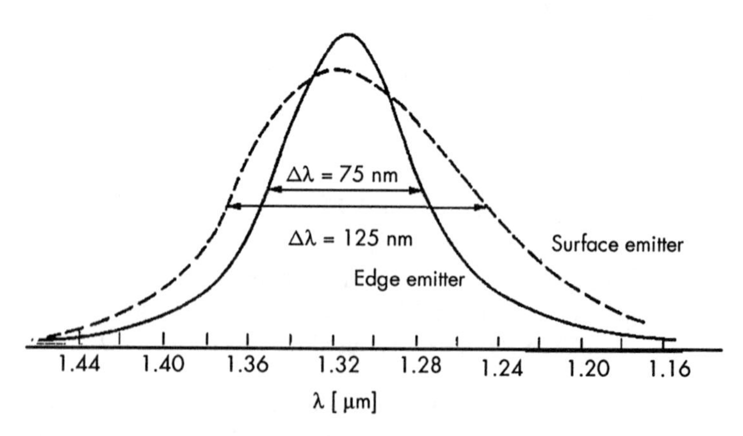

Figure 2.7 Spectral output curve for 1.3-μm InGaAsP surface and edge-emitting LEDs. [After Tien Pei Lee, 1988.]

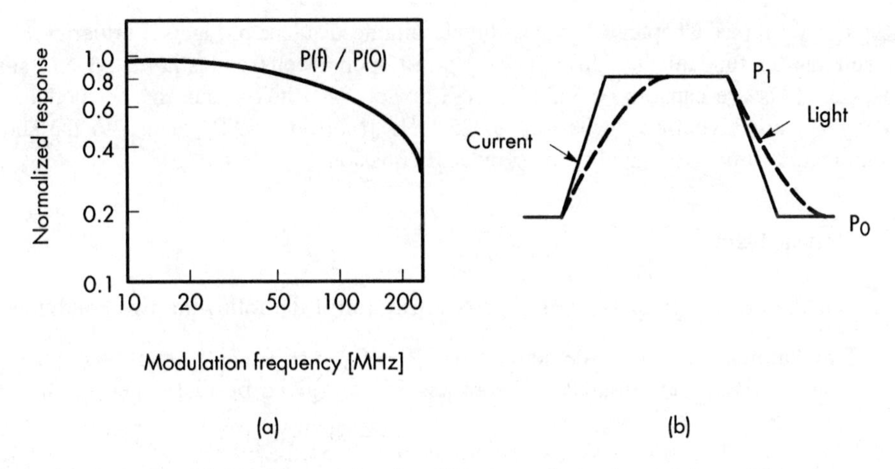

Figure 2.8 Typical modulation characteristics of an LED: (a) small-signal modulation and (b) pulse response. [After Dixon, 1991.]

The 3-dB bandwidth, f_{3dB}, is defined as the modulation frequency at which $P(f)$ is reduced by 3 dB (reduction by a factor 2) with respect to $P(0)$. From (2.22), we obtain

$$f_{3dB} = \frac{\sqrt{3}}{2\pi\tau_c} \tag{2.23}$$

Typically, τ_c is in the range 2 to 5 ns for InGaAsP LEDs and thus f_{3dB} ranges between 50 MHz and 140 MHz.

Therefore, although research LEDs have achieved modulation speeds of up to 1 Gbps, commercial LEDs are mainly restricted to applications where the modulation rates do not exceed 200 Mbps.

2.3 LASER DIODES (LDs)

Unlike an LED, an LD is a semiconductor device that generates coherent light through the stimulated emission process.

Compared to incoherent light emitted by LEDs, the coherent feature of light emitted by LDs provides many advantages for optical-fiber communication systems. A relatively higher directivity of the output beam permits higher coupling efficiency (≈ 50%) into single-mode fibers. A narrower spectral width makes it possible to push the fiber dispersion limit (see Chapter 3) towards higher bit rate-distance products. Furthermore, as a result of the fundamental difference between spontaneous and stimulated emission processes, LDs are capable of emitting high powers ($\approx$ 10mW) and can be modulated directly at very high frequencies (up to 25 GHz [Morton, 1992]) thanks to the short recombination time associated with stimulated emission.

2.3.1 Optical Gain

LDs, like any other type of laser, must necessarily fulfill the following two conditions:

1. Population inversion in order to realize $R_{stim} - R_{abs} > 0$, where the rates of stimulated emission R_{stim} and stimulated absorption R_{abs} are given by (2.19) and (2.18), respectively;
2. Optical feedback to convert an amplifier into an oscillator.

As discussed in Section 2.1, population inversion can be achieved by carrier injection into the active layer of the DH semiconductor device. However, the LD does not emit coherent light (i.e., population inversion is not realized) until the injected carrier density N exceeds a critical value, known as the transparency value N_T. For an injected carrier density larger than N_T, the condition $R_{abs} - R_{stim} > 0$ is satisfied and the active layer exhibits optical gain: an optical signal propagating inside the LD is then amplified by a factor exp(gz), where g is the gain coefficient proportional to ($R_{stim} - R_{abs}$) and z is the propagation length.

Figure 2.9 shows the optical gain per unit length, g, of a 1.3$-\mu$m InGaAsP active layer at different values of N. For $N = 1\times10^{18}$ cm^{-3}, the active layer is opaque to light of the bandgap wavelength and absorbs light rather than emitting it. As N increases, population inversion occurs, and g becomes positive. The wavelength range over which $g > 0$ also increases with N. It has been observed that the peak value of the gain, g_{peak}, varies almost linearly with N so that the peak gain above threshold can be empirically approximated by [Agrawal, 1986]

$$g_{peak}(N) = A \cdot (N - N_T) \tag{2.24}$$

where N_T is the carrier density at which the active layer becomes transparent and $A = dg/dN$ is the differential gain. Both parameters N_T and A are material-dependent. For InGaAsP DH lasers, typical values are $N_T \approx 0.9 - 1.5\times10^{18}$cm^{-3} and $A \approx 1.2 - 2.5 \times 10^{-16}$cm^2.

Smaller values of N_T together with larger values of A are attractive because they would allow the LD to operate at low injected carrier density while providing the same amount of optical power. Quantum well semiconductor lasers are particular devices that make it possible to achieve this by ''quantum confinement'' of the carriers in very thin active layers.

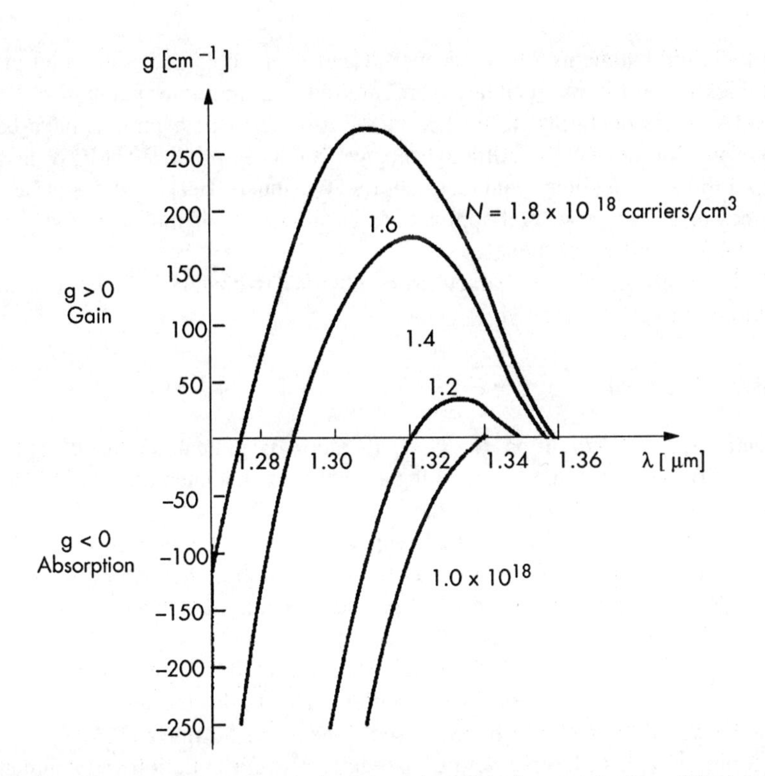

Figure 2.9 Calculated gain coefficient versus λ of the emitted light for several values of the injected carrier density *N*. [After Green, 1993.]

In a single-quantum well (SQW) laser, the width of the active layer is of the order of 10 nm, that is about 10 times thinner than the width of the active layer in DH bulk lasers. This very thin active layer leads to drastic changes in the electronic and optical properties of the laser diode [Arakawa, 1986].

The density of states, *Z(E)*, changes from the parabolic dependency [(2.10) and (2.12)] to a step-like dependency. The implications for laser characteristics are significant:

1. Smaller value of N_T, resulting in a lower threshold current density (see Section 2.3.2);
2. Larger differential gain *A* by about a factor 2;
3. Lower temperature dependency of the threshold current (see Section 2.3.2);
4. Higher modulation speed;
5. Narrower linewidth and reduced chirp (see Section 2.3.4.2);
6. Larger wavelength tunability (see Section 2.3.5).

Further improvements of device performance are obtained with multiple active layers of thickness ≈ 10 nm. Such lasers are referred to as multiple-quantum well (MQW) lasers. MQW lasers operating in the 1.3- or 1.5-μm wavelength regions have been successfully made for some time. Although not yet routinely produced, MQW lasers have been shown to have excellent manufacturability [Weisbuch, 1991] and might become the light sources of choice in several applications because of their multiple advantages over conventional DH bulk laser diodes.

In what follows, device characteristics obtained with DH bulk lasers and MQW lasers will be explicitly indicated.

2.3.2 Laser Threshold

The second necessary condition for laser operation is optical feedback. This can be achieved in different ways and the resulting output light characteristics will be dependent upon the technique used to accomplish this feedback.

For instance, suppose that the amplifying semiconductor medium is placed between two partially reflecting mirrors with power reflection coefficients R_1 and R_2, as depicted in Figure 2.10. The optical cavity formed provides positive feedback of the photons by reflection at the mirrors placed at either end of the cavity. Hence, the optical wave generated inside the active layer is fed back many times whilst receiving amplification as it passes through the semiconductor gain medium. This basic structure is referred to as a Fabry-Perot laser diode and will be discussed in detail in Section 2.3.3.

In common with all laser structures, a requirement for initiation and maintenance of laser diode oscillation is that the optical gain matches the optical losses within the cavity. The major losses within the cavity result not only from R_{abs} but also from scattering and from transmission through the mirrors (i.e., mirror losses).

Therefore, coherent light exits the cavity only if the internal gain reaches a threshold value g_{th} where optical gain and loss match each other. This threshold gain value can be obtained by imposing that the amplitude of a plane wave remains unchanged after one round trip inside the amplifying medium. It is straightforward to verify that this round-trip gain condition leads to the condition

$$R_1 R_2 \cdot \exp\{2\Gamma(g_{th} - \alpha_{int})L\} = 1 \tag{2.25}$$

where L is the length of the cavity, α_{int} is the internal loss per unit length that includes all possible loss mechanisms except those associated with the transmission through the mirrors, and Γ is the optical confinement factor that is introduced to account for the transversal spread of the optical field outside the active layer ($\Gamma \approx 0.3$–0.5).

Equation (2.25) yields the threshold gain as

$$\Gamma g_{th} = \alpha_{int} + \frac{1}{2L} \cdot \ln\left(\frac{1}{R_1 R_2}\right) \tag{2.26}$$

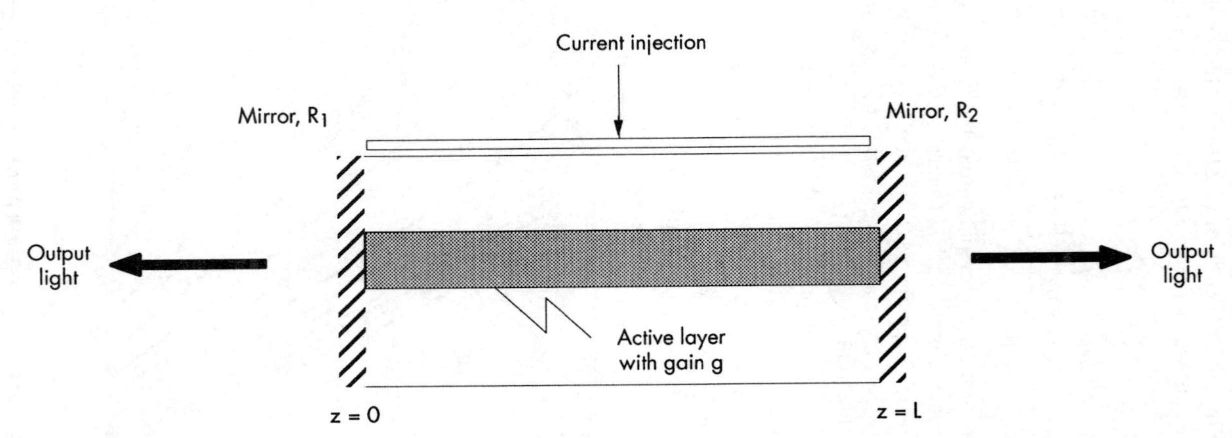

Figure 2.10 Structure of a semiconductor laser diode with partially reflecting mirrors at both end-facets. This structure is referred to as a Fabry-Perot resonant cavity.

This threshold gain is reached for a certain value of the injected carrier density N_{th} (that can be obtained by combining (2.26) with (2.24)) to which corresponds a specific value of the injection current, known as the threshold current. The light output of an LD can then be written as

$$P_{out}(I) = \eta_d \left(\frac{h\nu}{2e} \right) \cdot (I - I_{th}) \tag{2.27}$$

where η_d is the external quantum efficiency defined as the ratio of the photon escaping rate to the photon generation rate, ν is the optical frequency, I is the injection current, and I_{th} is the threshold current, which depends on the device geometry.

A typical light output versus current characteristic of a 1.3-μm InGaAsP DH laser at different temperatures is shown in Figure 2.11.

It is seen that, at typical room temperatures, the threshold current is in the range 10 mA to 30 mA. The threshold current is very temperature-dependent. A good fit with the experimental observations leads to

$$I_{th}(T) = I_0 \cdot \exp\{T/T_0\} \tag{2.28}$$

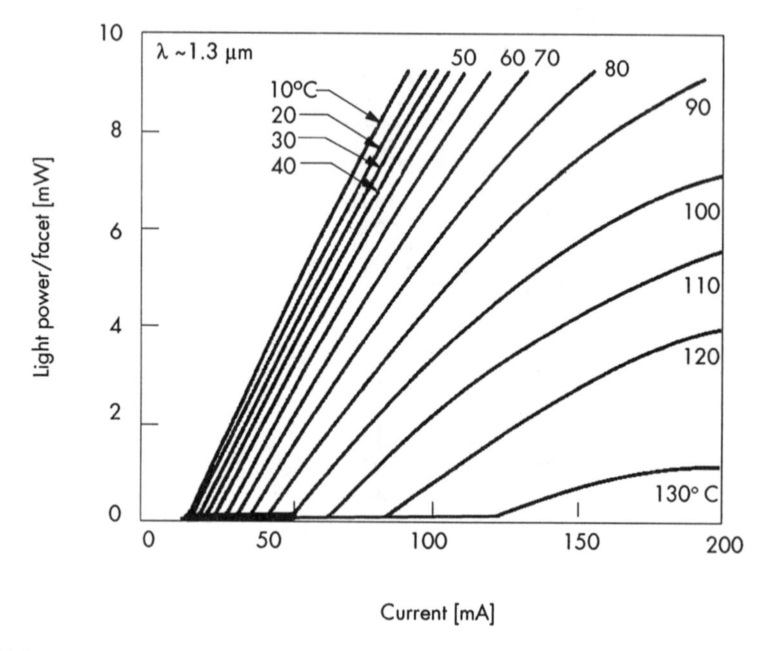

Figure 2.11 L-I curves at several absolute temperatures of a typical 1.3-μm semiconductor laser diode. [After Agrawal, 1992.]

where I_0 is a hypothetical threshold current at absolute zero temperature, and T_0 is the (threshold) *temperature coefficient,* which depends upon the material as well as the device geometry. For 1.3- or 1.5-μm InGaAsP devices, T_0 is between 50K and 70K.

For the same reason as for LEDs (i.e., heating of the active layer), the slope of the L-I curve above threshold decreases as the device temperature increases. The temperature dependencies of the threshold current and the external quantum efficiency is one of the principal reasons why LDs often have a thermoelectric cooler (Peltier) inside their package.

2.3.3 The Fabry-Perot Laser Diode

In the Fabry-Perot LD, the partial mirrors introduced in the preceding section are actually realized by cleaving the end facets of the semiconductor crystal. The partial reflectivity of the cleaved facets results from Fresnel reflection at the semiconductor air interface. Fresnel reflectivity is given as $R = \left(\frac{n-1}{n+1}\right)^2$ where n is the refractive index of the semiconductor gain medium [Jenkins, 1981]. Typically, $n \cong 3.5$ resulting in about 30% facet reflectivity.

In addition, to assure the gain for coherent light, the optical cavity also provides frequency selectivity. Indeed, since the light within the cavity is coherent, the only sustained optical frequencies are those corresponding to the standing waves of the Fabry-Perot resonator. In other words, only optical waves that experience a round-trip phase equal to some multiple of 2π will oscillate when the lasing threshold is reached. Since the round-trip phase is equal to $2kL$ where $k = 2\pi n/\lambda$ is the wave vector of the light inside the cavity, and L is the length of the cavity, this leads to the condition

$$2kL = m \cdot 2\pi \text{ or } \lambda = \lambda_m = \frac{2nL}{m} \qquad (2.29)$$

where m is an integer.

The optical waves with wavelength λ_m given by (2.29) are referred to as the longitudinal modes and are determined by the optical length of the cavity, nL.

The spacing between adjacent longitudinal modes is

$$\delta\lambda = \lambda_m - \lambda_{m+1} \cong \frac{\lambda_m^2}{2n_gL} \qquad (2.30)$$

where we replaced n in (2.29) by n_g (the group index defined in Appendix C) to take the material dispersion of the gain medium into account. Typically, $\delta\lambda$ = 0.5–1.0 nm for L = 200–400 μm.

The power of each longitudinal mode depends inversely on the difference of the overall loss and the internal gain; the closer the gain is to the loss, the higher the mode power will be. Note that the net gain (i.e., gain minus loss) differences between neighboring longitudinal modes are very small ($\approx 0.1\text{cm}^{-1}$). This is because the losses of a Fabry-Perot laser are due primarily to the mirror transmitivity, which is wavelength-independent over the wide bandwidth of the gain curve ($\approx 30 - 55$nm for bulk materials and $\approx$ 200nm for MQWs). Therefore, although the mode that is closest to the peak of the gain curve has the highest power (the main mode), many longitudinal modes can reach the lasing threshold, resulting in a multimode spectrum.

Figure 2.12 depicts the power spectrum of a Fabry-Perot LD output at different values of the injection current I. It is seen that when I is smaller than the threshold, the power spectrum looks like that of LEDs because, below threshold, the only light generating mechanism is spontaneous emission. As the current increases above threshold, the main-mode power increases faster than the side modes, and the side-mode suppression ratio (SMSR), defined as the power ratio of the main mode to that of the most dominant side mode, improves. Referring to the aspect of the power spectrum, Fabry-Perot LDs are also often called multilongitudinal mode (MLM) lasers. The FWHM of the power spectrum is typically between 2 and 5 nm.

The width of each longitudinal mode is well approximated by the following relation:

$$\Delta\nu \cdot P_0 = \frac{K}{L^2} = C^{\text{ste}} \tag{2.31}$$

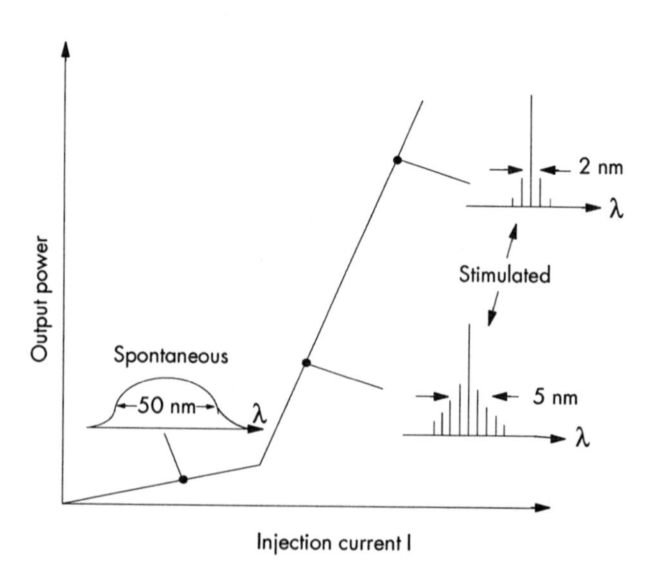

Figure 2.12 The changes in output power spectrum from a laser diode at several bias currents.

where P_0 is the output power and K is a constant (note that $|\Delta\nu| = c\,|\Delta\lambda|/\lambda^2$, as shown in Appendix B). Typical values of $\Delta\nu$ for a cavity length of 250 μm and an output power of $P_0 = 5$ mW range from 10 MHz to 100 MHz. We will return to (2.31) in Section 2.3.4 when we will deal with single-longitudinal-mode (SLM) laser diodes.

2.3.3.1 Modulation Bandwidth

Similarly to LEDs, the 3-dB bandwidth of a laser diode $f_{3\text{dB}}$ is defined as the modulation frequency at which the output light power is reduced by a factor 2 with respect to its dc value. Figure 2.13(a) shows the small-signal modulation response for a typical 1.3-μm Fabry-Perot laser diode. It is seen that the modulation response is flat for frequencies $f <<$ fr, shows a peak at $f = f_r$, and then drops sharply for $f > f_r$. This behavior indicates that the semiconductor Fabry-Perot laser acts as a filter with cutoff frequency f_r. This feature is observed for all types of semiconductor lasers. The frequency f_r is generally referred to as the relaxation oscillation frequency for reasons that will become clear shortly.

Solving the laser rate equations under the assumption of small-signal modulation above threshold, it can be shown that $f_{3\text{dB}}(\cong \sqrt{3}f_r)$ increases with an increase in the bias level as $\sqrt{P_0}$ [Saleh, 1991].

Although often given to characterize the modulation properties of LDs, the 3-dB bandwidth has little practical use. Indeed, in most optical communication systems the laser diode is typically biased close to threshold and modulated considerably above threshold to convert the input electrical data stream into optical pulses. This corresponds to the case of large-signal modulation.

The application of a current step to the LD results in a switch-on delay followed by high-frequency (of the order of several GHz) damped oscillations, known as relaxation oscillations. This transient response is illustrated in Figure 2.13(b) and can be understood qualitatively as a manifestation of the energy exchange between the electrons and photon emission. As the photon population increases, the electron population depletes and the optical gain decreases. This gain reduction results in a decrease of the photon population. As the photon population decreases, the electron population rises again, and the cycle repeats until a steady state is reached.

The switch-on delay t_d results from the fact that stimulated emission does not occur until the optical gain within the cavity has reached its threshold value g_{th}.

The value t_d is given by

$$t_d = t_0 \cdot \ln\left[\frac{I_M}{I_M + I_B - I_{\text{th}}}\right] \text{ for } I_B < I_{\text{th}} \tag{2.32}$$

where $t_0 \approx 2$ to 5 ns, and I_M, I_B and I_{th} are respectively the magnitude of the modulating, the bias, and the threshold currents.

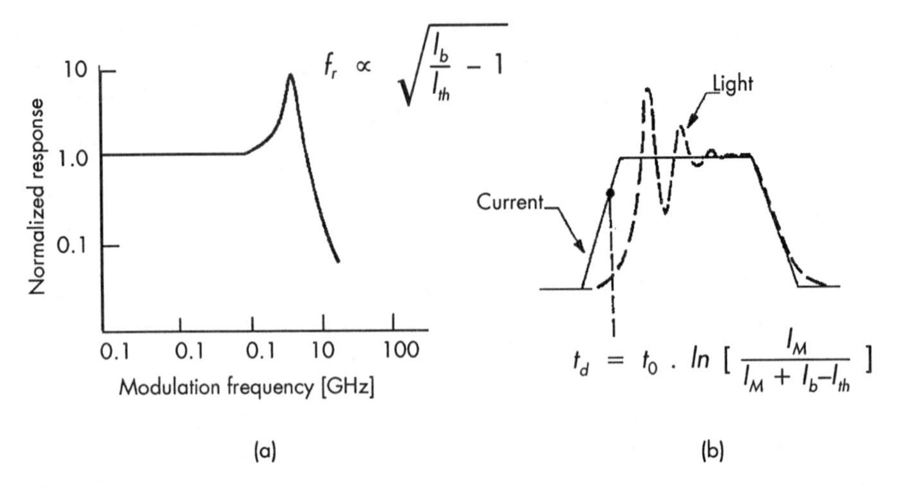

Figure 2.13 Typical modulation characteristics of a laser diode: (a) small-signal response, (b) pulse response. [After Dixon, 1991.]

The switch-on delay disappears when the LD is dc-biased at $I_B \geq I_{th}$. Since the switch-on delay can produce severe performance degradations, especially at high bit rates, the laser driving circuitry generally provides for a dc bias slightly above threshold. It is noteworthy that, since I_{th} is strongly temperature-dependent, the laser driver has to also provide some sort of feedback control. This is generally achieved with the help of a back-facet monitoring photodiode that measures the light output from one facet of the LD and uses this information to keep the mean output power of the light signal constant.

2.3.3.2 Mode Partition Noise

Transient processes as well as temperature changes may affect the position of the maximum gain of the stimulated emission. In essence, because of the small net-gain difference between neighboring longitudinal modes, the position of the dominant mode initiated by spontaneous emission changes randomly. As a consequence, the optical spectrum can vary from pulse to pulse, or even during the length of a pulse. This phenomenon is known as mode partition noise (MPN).

Even if the pulses have identical power, MPN is harmful for the fiber communication system because of fiber dispersion. Briefly, successive pulses travel down the fiber at slightly different speeds and arrive time-overlapped at the receiver. This leads to intersymbol interference and thus to system performance degradations. The effect of

MPN on system performance will be discussed in more quantitative terms in Section 3.4.1.2.

2.3.4 Single-Longitudinal-Mode Laser Diodes

The multimode spectrum together with mode partition noise of Fabry-Perot LDs leads to system penalties due to the dispersion phenomena in single-mode fibers.

In order to reduce these system penalties, design efforts have focused on high-speed SLM laser diodes operating at either 1.3- or 1.5-μm wavelengths. In addition, some structures that have been designed for single-mode operation also have the capability to be tuned in wavelength. Such wavelength tunable single-mode lasers are used as local oscillators in coherent detection receivers (Chapter 4) as well as transmitters in fast-packet switched networks (Chapter 7). This section discusses fixed-wavelength single-mode laser diodes, while wavelength tunable devices will be reviewed in Section 2.3.5.

The principle applied to obtain SLM laser operation is to increase the net gain difference between neighboring longitudinal modes such that only one mode can build up to lase while other modes having higher losses (insufficient net gain) are suppressed from oscillation.

This can be achieved by shortening the length of the Fabry-Perot LD. Indeed, from (2.30), the wavelength separation between adjacent modes, $\delta\lambda$, is inversely proportional to L. If $\delta\lambda$ is made larger than the spectral width of the net-gain curve, then only one longitudinal mode can be selected. However, as the cavity length is small, amplification gain of the stimulated emission becomes weaker and the output power decreases. As a consequence, the linewidth of the selected longitudinal mode is considerably broadened, as can be seen from (2.31). Since broad linewidths are very detrimental in applications where single-mode LDs are required, other SLM laser structures have been investigated.

This has led to the concept of distributed feedback (DFB) laser diodes, which were theoretically demonstrated in 1971 [Kogelnik, 1971] and realized as early as 1974 [Scifres, 1974], some twenty years ago.

The principle of DFB consists of incorporating a corrugated structure (a diffraction grating) along the cavity with an antireflection-coated end facet, as shown in Figure 2.14.

The periodic variation of the effective refractive index along the direction of wave propagation provides feedback of light through the phenomenon of *Bragg diffraction*. Therefore, in contrast with conventional Fabry-Perot lasers, the feedback in DFB lasers is not localized at the end facets, but is distributed all along the cavity.

Feedback occurs only for wavelength λ_B satisfying the Bragg condition

$$\lambda_B = \frac{2\bar{n}}{m} \cdot \Lambda \tag{2.33}$$

where Λ is the grating period, m is an integer that represents the order of Bragg diffraction, and $\bar{n}$ is the effective refractive index of the corrugated medium.

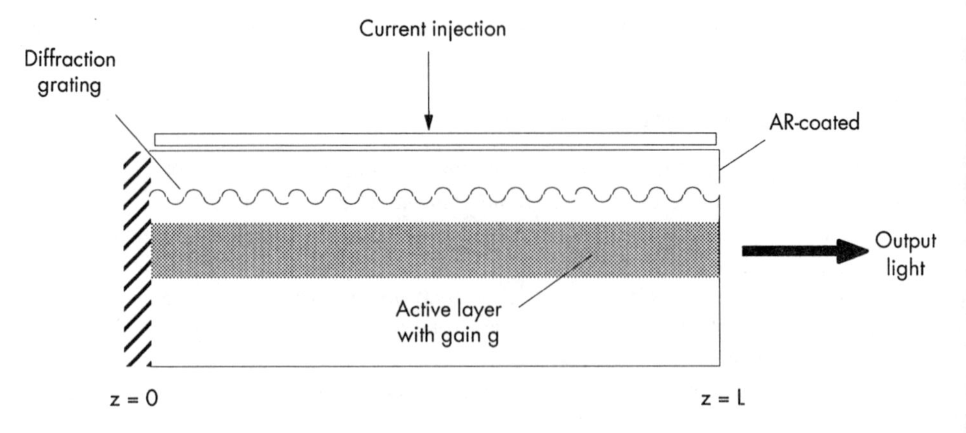

Figure 2.14 Distributed feedback laser structure. Shaded area shows the active region of the device.

The strongest feedback occurs for the first-order Bragg diffraction (m = 1). Therefore, for a DFB laser operating at 1.55 μm, (2.33) indicates that Λ must be as small as ≈ 230 nm ($\bar{n}$ = 3.4). Such gratings can be made by using either holographic techniques on photoresists or electron-beam lithography.

Variant structures of the basic one shown in Figure 2.14 have proven superior performances in terms of stability and side-mode suppression ratio (SMSR). Currently, the best structure for stable single-mode operation is the quarter-wave-shifted DFB laser, in which the grating is shifted by $\lambda_B/4$ in the middle of the cavity. These structures provide an SMSR of about 30 dB as a typical value.

Although the fabrication of DFB lasers involves advanced technology with multiple epitaxial growths, these devices are today routinely produced and available commercially. They are extensively used in applications such as single-mode fiber broadcast analog television systems and high-speed digital trunk systems operating at bit rates of 2.4 Gbps and more. The undersea transatlantic fiber system (TAT-9), which became operational in 1992, also employs DFB lasers.

2.3.4.1 Relative Intensity Noise

Even at a constant bias current with negligible current fluctuations, the output light of any semiconductor laser exhibits intensity and phase fluctuations. These fluctuations appear as noise and are particularly detrimental in applications where DFB lasers are employed [Miller, 1991]. The dominant process that is at the origin of these noise phenomena is spontaneous emission. The randomly generated spontaneous light is superimposed onto the coherent stimulated emission and thus perturbs both amplitude and phase of the laser output.

The amplitude fluctuation is characterized by the so-called relative intensity noise (RIN), defined as the power spectral density of the normalized output power fluctuations. If $\overline{P}$ is the average output power of the laser diode under constant bias and $P(t)$ is the instantaneous output power, then $\delta P(t) = (P(t) - \overline{P})/\overline{P}$ represents the normalized fluctuations of the output power. Thus, by definition, the RIN is the Fourier transform of the autocorrelation function of $\delta P(t)$ and is given by

$$\mathrm{RIN}(f) = \int_{-\infty}^{+\infty}\left(\int_{-\infty}^{+\infty} \delta P(t) \cdot \delta P(t+\tau)\, dt\right) \cdot e^{i2\pi f\tau}\, d\tau \tag{2.34}$$

The RIN can be theoretically evaluated by solving the linearized rate equations that govern the evolution of $P(t)$ and the carrier density $N(t)$ [Agrawal, 1986]. It can be shown that, at a given frequency, the RIN varies as P^{-3} at low powers and as P^{-1} at high powers. It is considerably enhanced near the relaxation oscillation frequency (f_r), but decreases rapidly for higher frequencies.

Figure 2.15(a) shows typical values of the RIN, expressed in dB/Hz, at several bias levels.

The RIN value is very sensitive to reflections that can occur at glass-air interfaces along the fiber link or at refractive index discontinuities occurring at splices. Figure 2.15(b) shows the dependency of the RIN on the reflection power level. For systems that

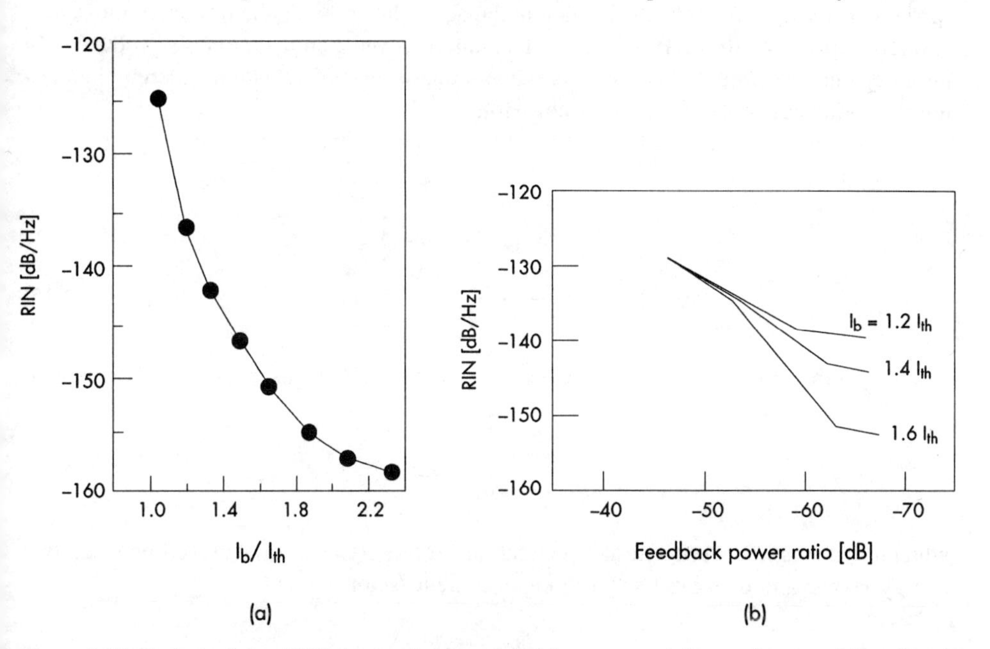

Figure 2.15 Typical values of RIN: (a) as a function of the bias current and (b) as a function of the reflected power ratio.

are particularly sensitive to phase or intensity fluctuations, such as *coherent detection* systems (Chapter 4) or SCMA networks (Chapter 8), it may be necessary to isolate the laser source from back reflections. This isolation can be achieved by means of optical isolators (Chapter 6), which can attenuate reflections by as much as 50 dB.

2.3.4.2 Laser Linewidth

The linewidth of individual Fabry-Perot longitudinal mode is not negligible but has typical values between 10 MHz and 100 MHz (Section 2.3.3). This holds also for DFB lasers.

In their first paper proposing the laser in 1958, Shawlow and Townes derived a formula for the linewidth, and showed that the line shape would be Lorentzian with a linewidth-power product $\Delta\nu \cdot P_0$ constant, as in (2.31). A few years later, the theory was refined and measurements of the linewidth of a HeNe gas laser have confirmed the theoretical predictions. Although the fit between experimental observations and theory was quite satisfactory for gas lasers, it was not at all so for semiconductor lasers. While the observed line-shape is Lorentzian as expected, the linewidth is by a factor 5 to 40 greater than that predicted by the modified Shawlow-Townes formula.

This linewidth enhancement of semiconductor lasers was explained by Henry [Henry, 1982] as being due primarily to the change in lasing frequency with gain. This can be understood qualitatively as follows. The spontaneous emission leads to fluctuations of carrier density, which produces the gain in the active layer. Since the refractive index also strongly depends on the carrier density, fluctuations of the gain will not only produce light intensity fluctuations but also a substantial amount of phase fluctuations, thereby increasing the frequency noise in the laser emission.

The linewidth formula for semiconductor lasers is given by [Henry, 1982]

$$\Delta\nu \cdot P_0 = \frac{\nu_g^2 \cdot h\nu \cdot n_{sp} \cdot \ln\left(\frac{1}{R^2}\right)}{8\pi \cdot L^2} \cdot (1 + \alpha^2) \tag{2.35}$$

where v_g is the group velocity in the cavity, n_{sp} is the spontaneous emission factor defined as

$$n_{sp} = \frac{N_2}{N_2 - N_1} \tag{2.36}$$

which expresses how complete the population inversion has been achieved between two energy levels, and α is the linewidth enhancement factor defined as

$$\alpha = \frac{dn/dN}{dg/dN} = \frac{dn/dN}{A} \tag{2.37}$$

which describes the ratio of the index of refraction change to the change in gain as the carrier density is varied. Reported values of α range from 2 to 8 [Osinski, 1987].

2.3.4.3 Frequency Chirping

An important consequence of the linewidth enhancement factor is the so-called *frequency chirp*. When a semiconductor laser is amplitude modulated (e.g., amplitude shift keying, or ASK), the carrier density changes result in both optical gain and refractive index changes. The changes in optical gain produce the ASK light pulses while the changes in refractive index produce optical phase modulation. Since a time-varying phase is equivalent to a frequency modulation around a steady-state value ν_0, the frequency chirp is defined as the instantaneous frequency shift from ν_0 and is given by [Tien Pei Lee, 1991]

$$\delta\nu(t) = \frac{\alpha}{4\pi} \cdot \left[\left(\frac{d}{dt} \ln P(t) \right) + \varkappa P(t) \right] \tag{2.38}$$

where $P(t)$ is the time variation of the optical power and $\varkappa$ is a constant.

Equation (2.38) shows that the frequency chirp comes from two parts. The first term in the right-hand side of (2.38) is referred to as the *transient* (or instantaneous) *chirp* and is a direct manifestation of relaxation oscillation. The second term, proportional to $P(t)$, is referred to as the *adiabatic chirp* and results from changes in the carrier density (at steady-state) between the high and low power levels in the optical signal. The adiabatic chirp is often useful in coherent optical systems that use frequency shift keying (FSK) modulation since it provides a simple and effective means to produce a frequency shift by changing the bias current between two distinct values (both above threshold). For ASK systems, however, the frequency chirp is very detrimental because of fiber dispersion. The consequences of the chirp on fiber system performance will be discussed in detail in Chapter 3.

Figure 2.16 shows, as an example, the laser output optical power $P(t)$ and the frequency chirp when the input is a 2-Gbps nonreturn-to-zero data stream with rectangular current pulses. The transient chirp and the adiabatic chirp offset are clearly visible. One can also see that the leading edge shifts towards higher frequencies (blue shift), while the trailing edge shifts towards lower frequencies (red shift).

2.3.5 Wavelength-Tunable Laser Diodes

For system designers, an ideal laser diode should emit a single-longitudinal mode with a very narrow linewidth and should be wavelength tunable with nanosecond (or less) tuning speed over a wavelength range of about 100 nm around 1.3 or 1.5 μm.

A number of techniques have been proven to be useful to provide tunability. How-

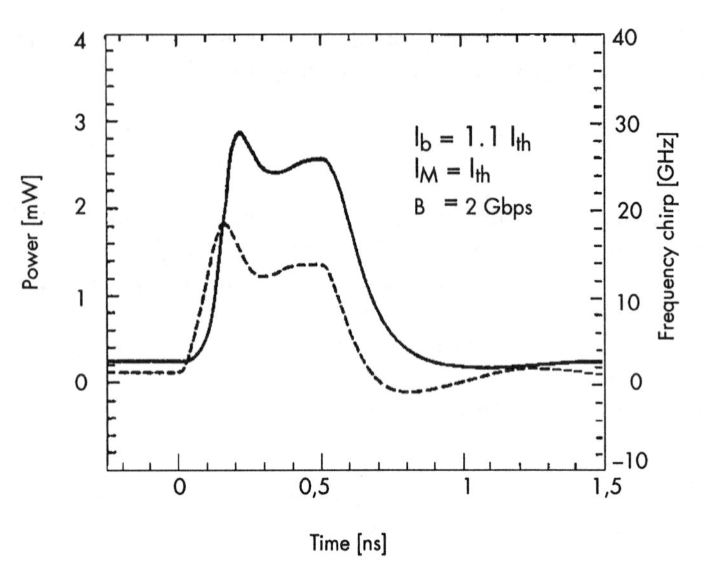

Figure 2.16 Calculated pulse modulation response: (a) optical pulse shape and (b) the frequency chirp. [After Agrawal, 1992.]

ever, up to now none of them have been able to achieve the above characteristics simultaneously. Important tradeoffs exist, especially between the speed of tuning and the wavelength range over which the laser is continuously tunable.

This section reviews several useful options for wavelength-tunable laser diodes. Except for temperature-tuned (Section 2.3.5.1) and external-cavity grating-based devices (Section 2.3.5.2), all these options are still in the research phase. Nevertheless, as technology and manufacturability improves, the described options may become important devices for future multiaccess optical-fiber networks. Of the most promising structures, those based on multielectrodes MQWs have already demonstrated satisfactory capabilities in terms of both tuning range and tuning speed.

2.3.5.1 Temperature Tuning

Since the effective refractive index of the laser diode's active layer is temperature-dependent, a simple way to make a laser tunable in wavelength is to change its temperature. The rate of wavelength tuning by temperature is about -0.1 nm/°C (+13 GHz/°C at 1.5 μm). Since the temperature change should not be much more than ± 10°C (particularly because of reliability considerations), the practical tuning range that can be achieved by this method is about 2 nm at the very most. Notice that recently a tuning range of up to

10.8 nm has been demonstrated with a MQW-DBR laser employing passive-sections heaters [Kameda, 1993].

Moreover, the tuning speed is limited by thermal impedance (a few milliseconds). Therefore, the method is very limited in terms of both tuning range and tuning speed.

2.3.5.2 External-Cavity Tunable Laser Diodes

A simple method for making a laser diode tunable over a wide range is to add an external tunable filter at one of the output facets, which is then antireflection-coated. Such lasers are referred to as external-cavity semiconductor lasers. In essence, the external filter selects one Fabry-Perot mode among the several ones sustained by the long cavity length. By tuning the filter, the wavelength of the selected mode can be fine-tuned until a jump to a new Fabry-Perot mode occurs. For a cavity length of 10 cm, the neighboring Fabry-Perot modes are separated by $\delta\nu \approx 1$ GHz (*i.e.*,$\delta\lambda \approx 0.005$ nm). Therefore, fine-tuning range is very limited and a large tuning range is obtained by jumps between Fabry-Perot modes. This behavior is observed for all tunable laser diodes.

Although many different forms of external filters have been employed [Heismann 1987, Coguin 1988], the most widely used is the diffraction grating, used as shown in Figure 2.17. One facet of the laser diode is antireflection-coated, and the light from this end facet is collimated by a lens before impinging onto the diffraction grating, which serves both as a mirror and as a narrowband filter. The lasing wavelength is tuned by moving the grating: coarse-tuning is obtained by rotating the grating while fine-tuning is obtained by longitudinal displacement of the grating.

A maximum tuning range of 55 nm at 1.55 μm has been obtained with a DH laser diode as the gain medium [Wyatt, 1983]. More recently, an extremely wide tuning range of 240 nm has been obtained using a strongly pumped MQW laser diode [Tabuchi, 1990].

A further advantage of the external-cavity structures over monolithic structures, which will be described in the following sections, is the very narrow linewidth that results from the extended overall effective cavity length, as can be seen from (2.31). Linewidths of a few kilohertz have quite frequently been reported using external grating structures.

Drawbacks of external grating laser diodes include the slow tuning speed, the relatively large physical dimension, and the difficulty to realize the mechanical stability generally required for optical transmitters.

These drawbacks have recently been overcome with a new grating-based single-output wavelength selectable semiconductor light source [Soole, 1992]. Instead of using a single laser diode with a moving grating, this new device uses a two-dimensional array of active elements integrated with a fixed diffraction grating. This device is schematically depicted in Figure 2.18 and has been called the MAGIC laser, which is an acronym for multistripe array grating-integrated cavity laser.

Each stripe can be addressed independently to cause lasing at distinct wavelengths. The grating selects and couples to the central stripe only one wavelength. The device has

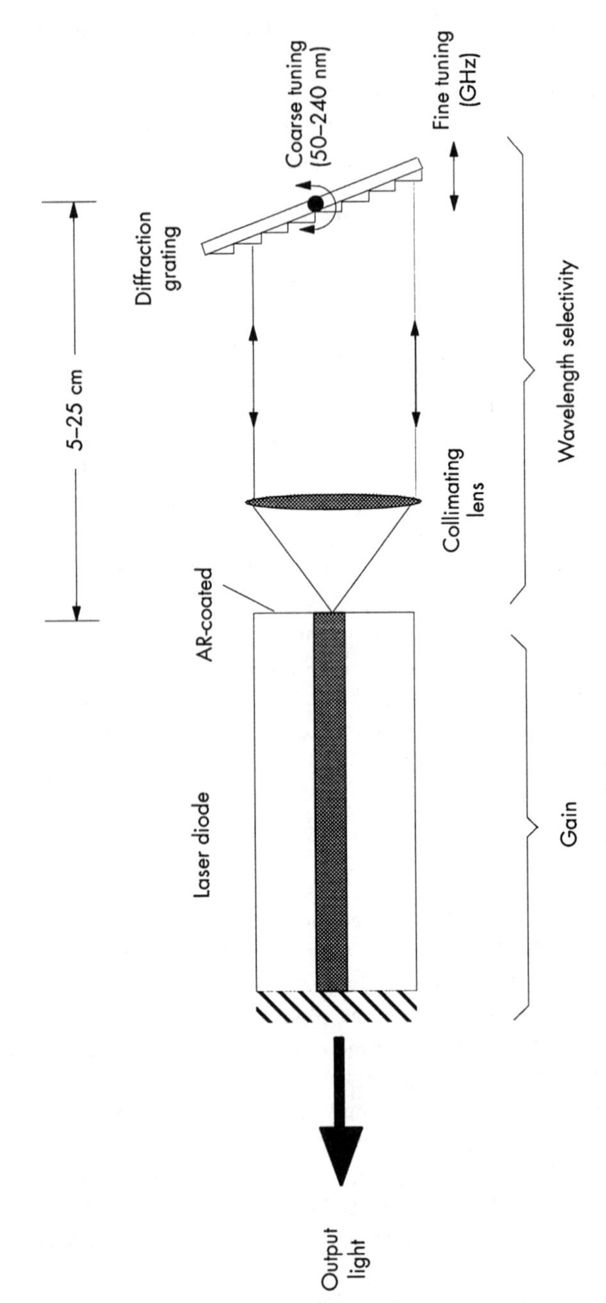

Figure 2.17 External-cavity grating-based tunable laser diode.

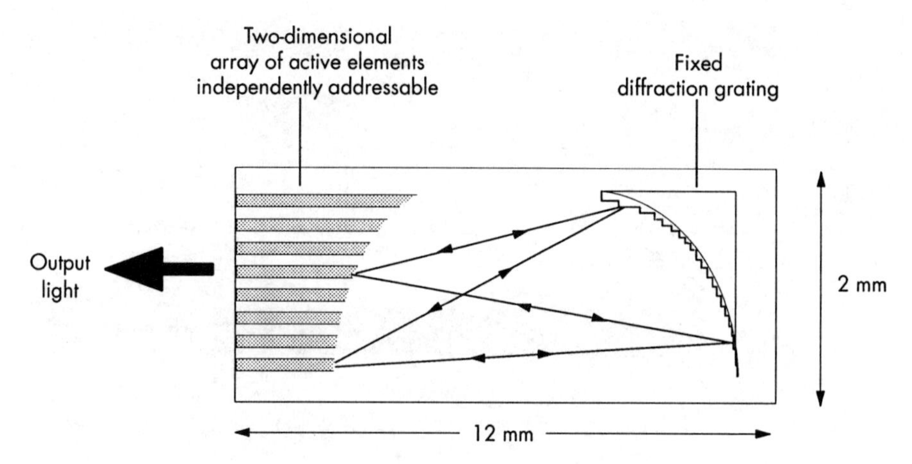

Figure 2.18 The multistripe array grating integrated cavity (MAGIC) laser.

shown the capability to address 15 discrete wavelengths separated by 1.89 nm in the 1.5-μm window. According to the inventors, the number of different wavelengths depends on the number of stripes [Jungbluth, 1993].

2.3.5.3 Two-Section DFB Laser Diodes

Rapid wavelength tuning, in the nanosecond range, can be achieved by carrier injection into the active laser medium because this reduces the effective refractive index, which in turn shifts the laser output wavelength [Delorme, 1993]. This effect is identical to the adiabatic chirp mentioned earlier in Section 2.3.4.2.

The wavelength tuning range can be estimated by $\delta\lambda/\lambda = \delta n_{eff}/n_{eff}$. Practically, the maximum relative index change is limited to about 1% because of heating problems at excessive carrier injection. Therefore, a maximum tuning range of 10 to 15 nm is expected by this method. Notice that, similarly to the external-cavity tunable laser diodes, this maximum tuning range is not achieved continuously. The cavity length of the laser diode is generally sufficiently long for there to be several Fabry-Perot cavity modes. For cavity lengths of 250–500 μm, the longitudinal modes are separated by $\delta\lambda \approx 1.5 - 3.0$nm, which corresponds to the continuous tuning range.

In order to control independently the lasing wavelength and the output power of the laser diode, at least two separate electrodes are needed: one electrode is used to vary the refractive index and thus to control the emitted wavelength, and another electrode is used to convert the electrical input signal into a modulated output light.

A schematic representation based on a DFB structure is shown in Figure 2.19. Such a laser is referred to as a two-section DFB laser diode.

The optical output power is determined by the first section biased above threshold.

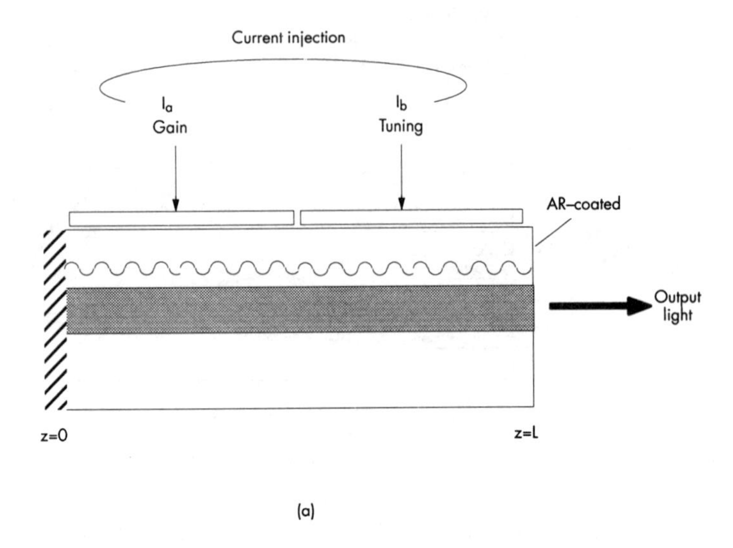

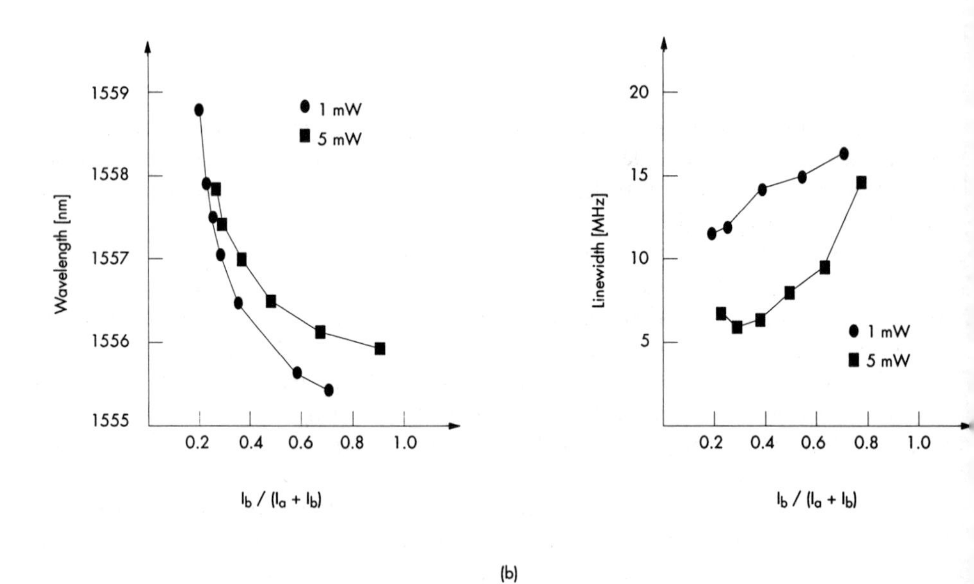

Figure 2.19 (a) Schematic of a two-section DFB tunable laser diode and (b) its tuning characteristics. [After Tien Pei Lee, 1991.]

The wavelength of the emitted light is primarily determined by the second section pumped at current densities slightly below the normalized threshold current density (i.e., under uniform pumping of the whole structure). In effect, this section where the optical power is high (near the output facet) serves as an efficient tunable Bragg reflector because at low pumping level the injected carriers do not contribute significantly to photon generation, thereby resulting in a large change of the refractive index.

A continuous tuning range of 3.3 nm with 15-MHz linewidth at 1-mW output power has been demonstrated [Okai, 1989]. The tuning range is essentially limited by the maximum allowed amount of carrier injection into the wavelength-control section.

2.3.5.4 Two-Section and Three-Section DBR Laser Diodes

Improvement of the wavelength tuning range has been achieved by separating the wavelength-selective Bragg region from the gain region inside the laser cavity. The Bragg region is made with a larger bandgap compound than the gain region. Therefore, the Bragg region can be strongly pumped without contributing to photon generation, resulting in a wider tuning range [Eddolls, 1992]. This structure is referred to as a two-section distributed Bragg reflector (DBR).

A further improvement is obtained by adding a third section, intended to control the phase of the lightwave inside the laser cavity. A schematic representation of such an arrangement as well as its tuning characteristics are shown in Figure 2.20.

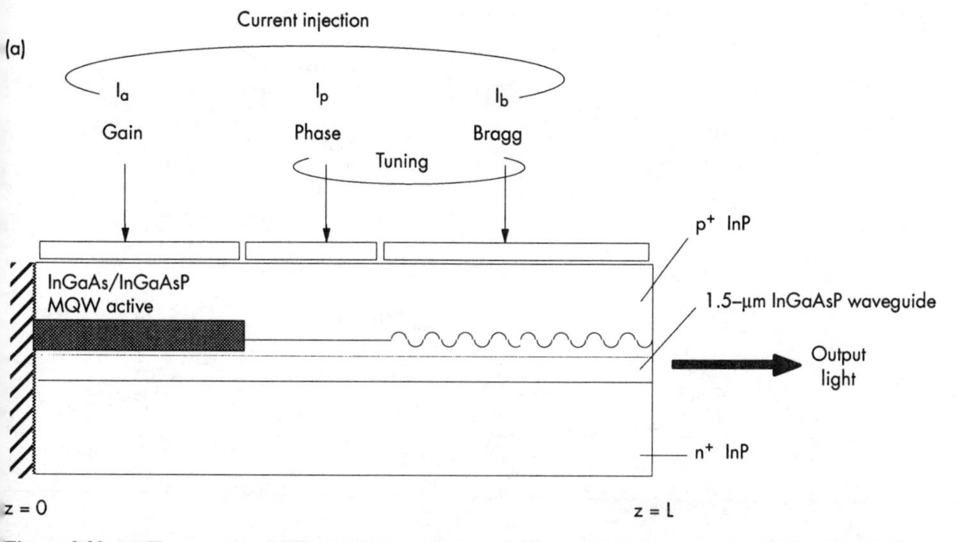

Figure 2.20 (a) Three-section DBR tunable laser diode and (b) its tuning characteristics. [After Tien Pei Lee, 1989.]

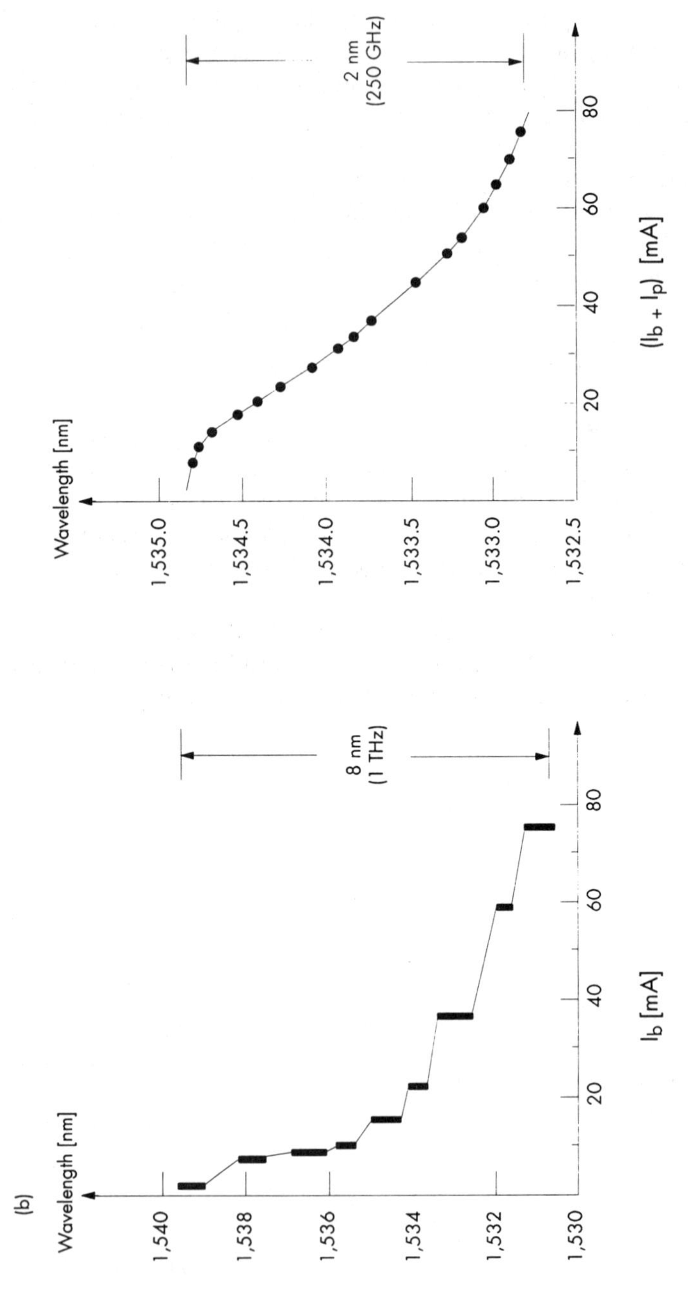

Figure 2.20 (continued)

The operating principle of wavelength tuning in the so-called three-section DBR can be qualitatively understood as follows. The DBR section exhibits a high reflection within a limited bandwidth, typically about 3 nm. The mode that is nearest to the maximum reflection of the DBR reflector will lase if its round-trip phase is a multiple of 2π, as explained in Section 2.3.3. The phase-shift section provides for the control of the round-trip phase such that the lasing wavelength can be tuned around each Bragg reflection bandwidth.

With independent adjustment of the three currents in the active, Bragg, and phase sections, quasi-continuous tuning ranges from 8 to 10 nm have been achieved [Kotaki, 1988].

Very recently, wider wavelength tuning of more than 10 nm has been reported. In [Kim, 1993], a tuning range of 48 nm has been achieved with a tuning current of 40 mA in a 1.55-μm InGaAsP/InP MQW laser based on a grating-assisted vertical coupler intracavity filter.

With the "superstructure grating" based on linear chirping in the grating pitches, 83-nm tuning has been observed in single-mode operation of a three-section DBR laser [Tohmori, 1993].

2.3.5.5 The Y-Laser

A new laser structure having not only high-speed broadband wavelength tuning capability but also, among others, space switching and wavelength conversion capabilities, has been demonstrated in [Hildebrand 1991, Schilling 1991, Wunstel 1992, Hildebrand 1993]. This photonic device is built up of a *Y*-shaped waveguide and controlled with four electrodes.

The principle of operation as a laser has been explained using different approaches [Hildebrand 1991, Mwarania 1992]. In a simplified view, the device can be understood as two standard Fabry-Perot (FP) laser diodes, each with three electrodes, put together to form a *Y*-laser. Due to interferometric effects between the longitudinal modes of the two FP resonators, the emission spectrum of the *Y*-laser is single-longitudinal mode.

Wavelength tuning is achieved by controlling the currents injected into the distinct segments. Continuous tuning (≤ 1 nm) is achieved by fine-tuning of the selected FP modes, while step-tuning is accomplished by jumping to another FP mode. An overall tuning range of up to 38 nm from 1,532 to 1,570 nm with a dc linewidth of 8 to 30 MHz has been achieved. The tuning speed is of the order of 5 to 10 ns for a wavelength separation of less than about 10 nm.

2.3.5.6 Vertical-Cavity Surface-Emitting Laser Diodes

All the laser devices that have been discussed so far are edge-emitters. Advances in epitaxial growth techniques have made possible the fabrication of vertical-cavity surface-emitting lasers (VCSELs) [Chang-Hasnain, 1991].

The structure of VCSELs typically consists of a quantum-well active region sandwiched between two Bragg sections, which serve as reflectors. VCSELs offer numerous advantages, two of which are particularly important in the present context. First, the circular emission pattern of each VCSEL leads to excellent coupling with a single-mode optical fiber. Second, since the cavity length is as short as a few tens of microns (compared with several hundred microns for edge emitter lasers), VCSELs are SLM lasers. The laser emission wavelength is determined by the cavity length and can therefore be tailored during fabrication.

Large VCSEL arrays with distinct and uniformly spaced wavelengths have been achieved by growing part of the Bragg sections with gradual thickness. As many as 140 wavelengths with spacings from 0.1 nm to tens of nm around 0.9 μm have been obtained with a 7×20 VCSEL array. According to [Chang-Hasnain, 1992], the fabrication technique can be readily extended to longer wavelength VCSELs.

Such devices can be used as tunable laser sources by addressing one laser at a time. An independently addressable 32×32 array has been demonstrated with a tuning speed essentially determined by the speed of electrical switching from one laser to another.

A problem with VCSEL arrays that remains to be solved before they become practical within multiaccess optical-fiber networks is the coupling of the light from all possible active lasers into a single output fiber.

To conclude this section, Table 2.1 summarizes the relevant characteristics (only orders of magnitude are indicated) of wavelength tunable semiconductor lasers that have been reported so far.

2.4 SUMMARY

We have seen that by suitable design, semiconductor compounds can be made to emit light through current injection. The characteristics of the emitted light depend upon the device

Table 2.1
Wavelength Tunable Semiconductor Lasers

Configuration	*Tuning Method*	*Tuning Range (nm)*	*Tuning Speed*	*Linewidth*
External cavity:				
a) Grating-based	Mechanical	55 (DH)	0.1 sec	10kHz
		240 (MQW)	0.1 sec	
	Electronic	25 (MAGIC)	nsec	
b) Electro-optic	Electronic	7	μsec	60 kHz
c) Acousto-optic	Electronic	80	μsec	10 kHz
1-section DFB	Temperature	2	msec	few MHz
2-section DFB	Electronic	3	nsec	few MHz
3-section DBR	Electronic	10–100	nsec	few MHz
Y-laser	Electronic	40	nsec	few MHz
VCSEL	Electronic	10–200	nsec	

structure. On the one hand, there are low-cost LEDs that emit incoherent light. Their use, however, results in poor system performance, especially for single-mode fiber-based systems.

On the other hand, there are more expensive LDs that emit coherent light through the stimulated emission process. This coherent-light output is obtained by adding some sort of optical feedback to the basic structure of LEDs.

When a highly reflective facet is provided at each end of the device, the output of the so formed Fabry-Perot LD exhibits a multimode spectrum with mode partition noise.

Single-mode operation can be achieved by selecting out one of the Fabry-Perot cavity resonances. DFB, DBR, and VCSEL are a few examples of single-mode lasers essential for high-capacity networks. However, the presence of some level of undesired spontaneous emission causes the spectral line of the output light to have a nonzero width, even under constant bias current. When the laser is on-off keyed, it provides undesirable frequency-chirp effects whose impact on fiber system performance will be discussed in the next chapter. In addition, again because of spontaneous emission, the laser output exhibits intensity noise, which is particularly detrimental in *coherent detection* systems and in SCMA networks.

Finally, wavelength tunability over useful ranges has been proven feasible. With MQWs, tuning of over 200 nm can be covered within a few nanoseconds. MQWs undoubtedly will be the key material in several multiaccess fiber network designs because of their superior performance over conventional bulk laser diodes. This is not only of considerable importance for laser sources, but for photonic amplification as well.

LDs are basically just like any other types of devices in the semiconductor components family. With the present trends in system requirements and improvements in device fabrication, LDs have considerable potential for cost reduction just like with other semiconductor devices, such as memory and logic, which are now universally integrated in computing and communication systems.

REFERENCES

[Hall, 1962] Hall, R. N. et al. "Coherent light emission from *GaAs* junctions," *Phys. Rev. Letters*, Vol. 9, 1962, pp. 336–368.

[Nathan, 1962] Nathan, M. I., et al. "Stimulated emission of radiation from *GaAs p-n* junctions," *Applied Phys. Letters*, Vol. 1, 1962, pp. 62–64.

[Kogelnick, 1971] Kogelnik, H., et al. "Stimulated emission via periodic structure," *Applied Phys. Letters*, Vol. 18, 1971, pp. 152–154.

[Scifres, 1974] Scifres, D. R., et al. "Distributed feedback single heterostructure laser," *Applied Phys. Letters*, Vol. 25, 1974, p.25.

[Kressel, 1977] Kressel, H., and J. K. Butler. *Semiconductor Lasers and Heterojunction LEDs*, London: Academic Press, 1977.

[Jenkins, 1981] Jenkins, F. A. and H. E. White. *Fundamentals of Optics*, New York, NY: McGraw-Hill, Fourth edition, 1981.

[Henry, 1982] Henry, C. H. "Theory of the linewidth of semiconductor lasers", *IEEE J. Quant. Electron.*, Vol. QE-18, No. 2, Feb. 1982, pp. 259–264.

[Wyatt, 1983] Wyatt, R., et al. "10*kHz* linewidth 1.5μ*m* *InGaAsP* external cavity laser with 55*nm* tuning range," *Electron. Letters*, Vol. 19, 1983, pp. 110–112.

[Senior, 1985] Senior, J. M. *Optical Fiber Communications*, New York, NY: Prentice Hall, 1985.

[Agrawal, 1986] Agrawal, G. P., and N. K Dutta. *Long-Wavelength Semiconductor Lasers*, New York, NY: Van Nostrand Reinhold, 1986.

[Arakawa, 1986] Arakawa, Y., et al. "Quantum well lasers—Gain, Spectra, Dynamics," *IEEE J. Quant. Electron.*, Vol. QE-22, No. 9, Sept. 1986, pp. 1887–1899.

[Osinski, 1987] Osinski, M., and J. Buus. "Linewidth broadening factor in semiconductor lasers," *IEEE J. Quantum Electron.*, Vol. QE-23, No. 1, Jan. 1987.

[Heismann, 1987] Heismann, F., et al. "Narrow-linewidth electro-optically tunable $InGaAsP\text{-}Ti{:}LiNbO_3$ extended cavity laser," *Applied Phys. Letters*, Vol. 51, 1987, pp. 164–165.

[Kobayashi, 1988] Kobayashi, K., et al. "Single frequency and tunable laser diodes" *IEEE J. Light. Technology*, Vol. 6, No. 11, 1988, pp. 1623–1633.

[Tien Pei Lee, 1988] Tien Pei Lee, et al. in *Optical Telecommunications II*, Chapter 12: "Light Emitting Diodes for Telecommunication," Edited by S. E. Miller and I. P Kaminow, London: Academic Press, 1988.

[Kotaki, 1988] Kotaki, Y., et al. "Tunable *DBR* laser with wide tuning range," *Electron. Letters*, Vol. 24, April 1988, No. 8, pp. 503–505.

[Coguin, 1988] Coguin, G., et al. "Single- and multiple-wavelength operation of acousto-optically tuned semiconductor lasers at 1.3 microns," *Proc. 11th IEEE Intl. Semiconductor Laser Conf.*, Boston, MA, 1988, pp. 130–131.

[Okai, 1989] Okai, M., et al. "Wide-range continuous tunable double-sectioned distributed feedback lasers," *Proc. 15th European Conf. on Opt. Communication, ECOC89*, Gothenborg, Sweden, 1989.

[Tien Pei Lee, 1989] Tien Pei Lee, et al. "Wavelength tunable and single-frequency semiconductor lasers for photonic communication networks," *IEEE Communication Magazine*, Oct. 1989, pp. 42–52.

[Koch, 1990] Koch, T. L., et al. "Semiconductor lasers for coherent optical fiber communications," *IEEE J. Light. Technology*, Vol. 8, No. 3, March 1990, pp. 274–293.

[Tabuchi, 1990] Tabuchi, H., et al. "External grating tunable *MQW* laser with wide tuning range of 240*nm*," *Electron. Letters*, Vol. 26, No. 11, May 1990, pp. 742–743.

[Saleh, 1991] Saleh, B. A., and M. Teich. *Fundamentals of Photonics*, New York, NY: Wiley & Sons, 1991.

[Dixon, 1991] Dixon, M., and J. L. Hokanson. "Optical transmitter design," in *Topics in Lightwave Transmission Systems*, Edited by T. Li, London: Academic Press, 1991, pp. 1–77.

[Wiesbuch, 1991] Wiesbuch, C., and B. Vinter. *Quantum Semiconductor Structures: Fundamentals and Applications*, London: Academic Press, 1991.

[Tien Pei Lee, 1991] Tien Pei Lee. "Recent advances in long-wavelength semiconductor lasers for optical fiber communication," *IRE Proc.*, Vol. 19, No. 3, March 1991, pp. 253–276.

[Miller, 1991] Miller, C. M. "Intensity modulation and noise characterization of high-speed semiconductor lasers," *IEEE Light. Trans. Systems*, May 1991, pp. 44–53.

[Hildebrand, 1991] Hildebrand, O., et al. "Integrated interferometric injection laser (*Y*-laser): one device concept for numerous system applications," *17th European Conf. on Opt. Communication, ECOC/IOOC91*, Paris, Tu.A.5.1, Sept. 1991.

[Schilling, 1991] Schilling, M., et al. "Multifunctional photonic switching operation of 1500*nm* *Y*-coupled cavity laser (*YCCL*) with 28*nm* tuning capability," *IEEE Photon. Technology Letters*, Vol. 3, 1991, pp. 1054–1057.

[Chang-Hasnain, 1991] Chang-Hasnain, C. "Monolithic multiple wavelength surface emitting laser arrays," *IEEE J. Light. Technology*, Vol. LT-9, No. 12, Dec. 1991, pp. 1665–1673.

[Chang-Hasnain, 1992] Chang-Hasnain, C. "Vertical-cavity surface-emitting lasers: *2D* arrays," *Proc. Opt. Fiber Communication Conf., OFC'92*, WB1, Feb. 1992, p.100.

[Wunstel, 1992] Wunstel, K., et al. "*Y*-shaped semiconductor device as a basis for various photonic switching applications," *Proc. Opt. Fiber Communication Conf., OFC'92*, WG6, Feb. 1992, p.125.

[Eddolls, 1992] Eddolls, D. V., et al. "Two-segment multiquantum well lasers with 7*nm* tuning range and narrow linewidth," *Electron. Letters*, Vol. 28, No. 11, May 1992, pp. 1057–1058.

[Soole, 1992] Soole, J.B.D., et al. ''Multistripe array grating integrated cavity (*MAGIC*) laser: a new semiconductor laser for *WDM* applications,'' *Electron. Letters*, Vol. 28, No. 19, Sept. 1992, pp. 1805–1807.

[Mwarania, 1992] Mwarania, E. K., et al. ''Modeling of *Y*-junction waveguide resonators,'' *IEEE J. Light. Technology*, Vol. 10, No. 11, Nov. 1992, pp. 1700–1707.

[Morton, 1992] Morton, P. A., et al. ''25*GHz* bandwidth 1.55μ*m GaInAsP p*-doped strained multiquantum-well lasers,'' *Electron. Letters*, Vol. 28, No. 23, Nov. 1992, pp. 2156–2157.

[Agrawal, 1992] Agrawal, G. *Fiber-Optic Communication Systems*, New York, NY: Wiley & Sons, 1992.

[Green, 1993] Green, P. *Fiber Optic Networks*, New Jersey: Prentice Hall, 1993.

[Delorme, 1993] Delorme, F., et al. ''Fast tunable 1.5μ*m* distributed Bragg reflector laser for optical switching applications,'' *Electron. Letters*, Vol. 29, No. 1, Jan. 1993, pp. 41–43.

[Tohmori, 1993] Tohmori, Y., et al. ''Over 100*nm* wavelength tuning in superstructure grating (*SSG*) *DBR* lasers,'' *Electron. Letters*, Vol. 29, No. 4, Feb. 1993, pp. 352–354.

[Kim, 1993] Kim, I., et al. ''Broadly tunable *InGaAsP/InP* vertical-coupler filtered laser with low tuning current,'' *Electron. Letters*, Vol. 29, No. 8, April 1993, pp. 664–666.

[Jungbluth, 1993] Jungbluth, E. ''*MAGIC* laser emits at 15 wavelengths,'' *Laser Focus World*, May 1993, pp. 20–23.

[Kameda, 1993] Kameda, T., et al. ''A *DBR* laser employing passive-section heaters, with 10.8*nm* tuning range and 1.6*MHz* linewidth,'' *IEEE Photon. Technology Letters*, Vol. 5, No. 6, June 1993, pp. 608–610.

[Hildebrand, 1993] Hildebrand, O., et al. ''The *Y*-laser: a multifunctional device for optical communication systems and switching networks,'' *IEEE J. Light. Technology*, Vol. LT-11, No. 2, Dec. 1993, pp. 2066–2075.

Chapter 3
Single-Mode Optical Fibers

This chapter is intended to provide an overview of the single-mode fiber characteristics that are important for understanding system performance limitations. Section 3.1 presents the geometrical aspect of a single-mode optical fiber. Section 3.2 introduces Maxwell's equations and important concepts such as fiber modes and the single-mode condition. In Sections 3.3 and 3.4, particular attention will be paid to fiber losses and dispersion characteristics because of their importance in optical-fiber systems. The effect of chromatic dispersion will be discussed with respect to two distinct situations: (1) fiber transmission systems with single-longitudinal mode (SLM) laser sources (chirp effect), and (2) fiber transmission systems with multilongitudinal mode (MLM) laser sources (mode partition noise).

If the optical power launched into a single-mode fiber increases, a point is reached where nonlinear effects can no longer be neglected. A discussion of these nonlinear effects is presented in Section 3.5. Their implications on the performance of multiaccess optical-fiber communication systems are reported in Chapter 7.

3.1 THE SINGLE-MODE FIBER

Figure 3.1 shows the structure of a typical step-index single-mode optical fiber. It consists of a central cylindrical core with refractive index n_0 in which the light is confined, surrounded by a concentric cladding layer with refractive index n_1 slightly lower than n_0.

In principle, the cladding layer is not essential for light guidance in optical fibers provided that n_0 is higher than the refractive index of the surrounding medium. However, as will become apparent in the following section, the presence of the cladding layer with index close to core index allows single-mode operation at a wavelength of 1.3 μm or above within fibers whose core diameter is of the order of several microns. Without the cladding layer, the core diameter would have to be of the order of 1 μm or less. In addition

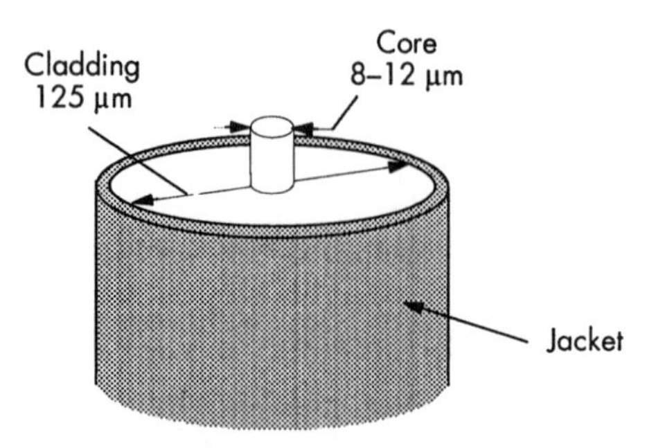

Figure 3.1 Structure of a step-index single-mode optical fiber.

to simplicity of fabrication, a larger core diameter allows more light to be coupled into the fiber, resulting in improvement of the system power budget. Notice finally that unclad fibers would be impractical due to fragility and lack of protection of core from surface contamination.

In order to have a complete picture of the propagation mechanism in single-mode fibers with transverse dimensions approaching the wavelength of the propagating fields, electromagnetic mode theory must be used. This is the subject of the next section.

3.2 GUIDED MODES

Like all electromagnetic phenomena, the propagation of optical fields in fibers is governed by Maxwell's equations [Clemmow, 1973]:

$$\Delta \times \boldsymbol{E} = -\frac{\partial \boldsymbol{B}}{\partial t} \tag{3.1}$$

$$\Delta \times \boldsymbol{H} = \boldsymbol{J}_c + \frac{\partial \boldsymbol{D}}{\partial t} \tag{3.2}$$

$$\Delta \cdot \boldsymbol{D} = \varrho_c \tag{3.3}$$

$$\Delta \cdot \boldsymbol{B} = 0 \tag{3.4}$$

where $\boldsymbol{E}$ and $\boldsymbol{H}$ are the electric and magnetic field vectors, and $\boldsymbol{D}$ and $\boldsymbol{B}$ the corresponding electric displacement and magnetic flux densities, respectively. $\boldsymbol{J}_c$ and ϱ_c represent the current density vector and the charge density, respectively. They vanish in optical fibers because of the absence of free charges in the constituent medium (SiO_2). ***D*** and ***B*** are related to ***E*** and ***H*** by the constitutive relations

$$\boldsymbol{D} = \varepsilon_0 \boldsymbol{E} + \boldsymbol{P} \tag{3.5}$$

$$\boldsymbol{B} = \mu_0 \boldsymbol{H} + \boldsymbol{M} \tag{3.6}$$

where ε_0 is the vacuum permittivity, μ_0 is the vacuum permeability, and ***P*** and ***M*** are the induced electric and magnetic polarizations. Optical fibers are nonmagnetic, therefore $\boldsymbol{M} = \boldsymbol{0}$.

Taking the curl of (3.1) and using (3.2), (3.5), and (3.6), we can eliminate ***B*** and ***D*** in favor of ***E*** and ***P*** to obtain

$$\nabla \times \nabla \times \boldsymbol{E} = -\frac{1}{c^2}\frac{\partial^2 \boldsymbol{E}}{\partial t^2} - \mu_0 \frac{\partial^2 \boldsymbol{P}}{\partial t^2} \tag{3.7}$$

where we used the relation $\mu_0\varepsilon_0 = 1/c^2$, c being the light velocity in vacuum.

It can be shown that, far from medium resonance (this is the case for optical fibers in the wavelength range 0.5–2 μm) and when nonlinear effects can be neglected, ***P*** and ***E*** are related through the phenomenological relation [Butcher, 1991]

$$\boldsymbol{P}(\boldsymbol{r}, t) = \varepsilon_0 \int_{-\infty}^{+\infty} \chi(t - t')\, \boldsymbol{E}(\boldsymbol{r}, t')dt' \tag{3.8}$$

where $\chi(t)$ is the (linear) dielectric susceptibility.

Inserting (3.8) into (3.7) and taking the Fourier transform, we obtain

$$\nabla \times \nabla \times \tilde{\boldsymbol{E}}(\boldsymbol{r}, \omega) + \varepsilon_r(\omega)\frac{\omega^2}{c^2}\tilde{\boldsymbol{E}}(\boldsymbol{r}, \omega) = \boldsymbol{0} \tag{3.9}$$

where $\tilde{\boldsymbol{E}}(\boldsymbol{r}, \omega)$ is the Fourier transform of $\boldsymbol{E}(\boldsymbol{r}, t)$ defined by

$$\tilde{\boldsymbol{E}}(\boldsymbol{r}, \omega) = \int_{-\infty}^{+\infty} \boldsymbol{E}(\boldsymbol{r}, t)\exp(j\omega t)dt \tag{3.10}$$

The frequency-dependent dielectric constant $\varepsilon_r(\omega)$ in (3.9) is defined by

$$\varepsilon_r(\omega) = 1 + \tilde{\chi}(\omega) \tag{3.11}$$

where $\tilde{\chi}(\omega)$ is the Fourier transform of $\chi(t)$.

In general, $\varepsilon_r(\omega)$ is complex and its real and imaginary parts are related to the refractive index $n(\omega)$ and the absorption coefficient $\alpha(\omega)$ by using the definition [Agrawal 1989, Butcher 1991]

$$\varepsilon_r(\omega) = \left\{ n(\omega) + j\frac{\alpha(\omega)c}{2\omega} \right\}^2 \tag{3.12}$$

The frequency dependence of α and n in optical fibers will be discussed in Sections 3.3 and 3.4, respectively.

For step-index fibers, several simplifications can be made before solving (3.9). First, optical fibers have very low losses in the wavelength range 0.5–2.0 μm and the imaginary part of $\tilde{\boldsymbol{E}}(\omega)$ is negligible in comparison with the real part. Therefore, losses are neglected here and (3.12) reduces to $\varepsilon_r(\omega) = n^2(\omega)$. The influence of losses will be discussed in Section 3.3. Second, since $n(\omega)$ is independent of the spatial coordinates in both the core and the cladding, we have

$$\nabla \cdot \tilde{\boldsymbol{D}} = \varepsilon_r(\omega) \nabla \cdot \tilde{\boldsymbol{E}} = 0 \tag{3.13}$$

Using the mathematical identity

$$\nabla \times \nabla \times \tilde{\boldsymbol{E}} = \nabla(\nabla \cdot \tilde{\boldsymbol{E}}) - \nabla^2 \tilde{\boldsymbol{E}} = -\nabla^2 \tilde{\boldsymbol{E}} \tag{3.14}$$

together with the above simplifications, (3.9) becomes

$$\nabla^2 \tilde{\boldsymbol{E}} + n^2(\omega)\frac{\omega^2}{c^2}\tilde{\boldsymbol{E}} = \boldsymbol{0} \tag{3.15}$$

Similarly, following the same procedure for the magnetic field $\boldsymbol{H}$, we obtain

$$\nabla^2 \tilde{\boldsymbol{H}} + n^2(\omega)\frac{\omega^2}{c^2}\tilde{\boldsymbol{H}} = \boldsymbol{0} \tag{3.16}$$

where $\tilde{\boldsymbol{H}}(\boldsymbol{r}, \omega)$ is the Fourier transform of $\boldsymbol{H}(\boldsymbol{r}, t)$.

Because of the cylindrical symmetry of optical fibers, it is convenient to express the wave equations (3.15) and (3.16) in cylindrical coordinates ϱ, θ and z

$$\frac{\partial^2 \psi}{\partial \varrho^2} + \frac{1}{\varrho}\frac{\partial \psi}{\partial \varrho} + \frac{1}{\varrho^2}\frac{\partial^2 \psi}{\partial \theta^2} + \frac{\partial^2 \psi}{\partial z^2} + n^2(\omega)\frac{\omega^2}{c^2}\psi = 0 \tag{3.17}$$

Since $\boldsymbol{E}$ and $\boldsymbol{H}$ satisfy Maxwell's equations (3.1–3.4), only two components out of six are independent. It is customary to choose ψ_z ($\equiv \tilde{E}_z$ or $\tilde{H}_z$) as the independent components and express $\tilde{E}_\varrho$, $\tilde{E}_\theta$, $\tilde{H}_\varrho$, $\tilde{H}_\theta$ in terms of $\tilde{E}_z$ and $\tilde{H}_z$.

The wave equation for $\tilde{E}_z$ is solved by making the substitution (separation of the variables)

$$\tilde{\boldsymbol{E}}_z = \psi_1(\varrho)\psi_2(\theta)e^{j(\beta(\omega)z-\omega t)} \tag{3.18}$$

where β is the propagation constant.

Substituting (3.18) into (3.17), we find

$$\psi_2(\theta) = e^{jm\theta} \tag{3.19}$$

$$\frac{\partial^2\psi_1}{\partial\varrho^2} + \frac{1}{\varrho}\frac{\partial\psi_1}{\partial\varrho} + \left[n^2k_0^2 - \beta^2 - \frac{m^2}{\varrho^2}\right]\psi_1 = 0 \tag{3.20}$$

where $k_0 = \dfrac{\omega_0}{c} = \dfrac{2\pi}{\lambda_0}$.

Equation (3.20) is the well-known Bessel equation of order m [*CRC*, 1978]. Replacing ψ_1 by $\tilde{E}_z$ and $\tilde{H}_z$, we obtain the following solutions:

$$\varrho \le a \quad \begin{cases} \tilde{E}_z = a_m J_m(\gamma\varrho)e^{jm\theta}e^{j(\beta z-\omega t)} \\ \\ \tilde{H}_z = b_m J_m(\gamma\varrho)e^{jm\theta}e^{j(\beta z-\omega t)} \end{cases} \tag{3.21}$$

$$\varrho \ge a \quad \begin{cases} \tilde{E}_z = a'_m K_m(\sigma\varrho)e^{jm\theta}e^{j(\beta z-\omega t)} \\ \\ \tilde{H}_z = b'_m K_m(\sigma\varrho)e^{jm\theta}e^{j(\beta z-\omega t)} \end{cases} \tag{3.22}$$

where a is the core radius, J_m is the Bessel function of order m, and K_m is the modified Bessel function for which a simplified expression will be given later on. The parameters γ and σ are defined by

$$\gamma = \sqrt{n_0^2k_0^2 - \beta^2}, \qquad \sigma = \sqrt{\beta^2 - n_1^2k_0^2} \tag{3.23}$$

The condition that the wave will be guided requires that both γ and σ take real values. At this point, it is useful to define the parameter V, called the normalized frequency, of which the physical significance will become apparent later.

$$V = k_0 a\sqrt{n_0^2 - n_1^2} = a\sqrt{\gamma^2 + \sigma^2} \tag{3.24}$$

The second equality in (3.24) is obtained using (3.23).

It can be shown that the $J_m(\gamma\varrho)$ functions somewhat look like decaying sinusoids as ϱ increases, while the $K_m(\sigma\varrho)$ functions look like decaying exponentials. The asymptotic approximation of $K_m(\sigma\varrho)$ for large $\sigma\varrho$ is proportional to $\exp(-\sigma\varrho)$. Thus, for large σ, the field is tightly confined inside the core, while for decreasing values of σ the field is spread more and more into the cladding.

In expressions (3.22) and (3.21), a_m, b_m, a'_m, and b'_m are constants. The solutions given by (3.22) and (3.21) are guided modes of the fibers if they fulfill the continuity conditions at the core boundary $\varrho = a$: the tangential components must be the same when $\varrho = a$ is approached from inside or outside the core. This condition leads to the so-called "characteristic equation" whose solutions determine the propagation constant β of the fiber modes. The procedure is well-known and is presented in [Marcuse, 1982].

A simplified expression can be obtained if the relative refractive index difference is small; that is, if

$$\delta = \frac{n_0 - n_1}{n_0} \ll 1 \tag{3.25}$$

Typically, $\delta \leq 0.01$.

Then the characteristic equation becomes

$$\frac{J_{m\mp1}(\gamma a)}{\gamma a J_m(\gamma a)} = \pm \frac{K_{m\mp1}(\sigma a)}{\sigma a K_m(\sigma a)} \tag{3.26}$$

For the up-signs, solutions of this equation correspond to the so-called HE modes, and the down-signs correspond to the so-called EH modes. By convention, EH is used if the E_z component has the larger value, and HE is used if the H_z component dominates.

In general, the characteristic equation (3.26) may have several solutions for each integer value of m. It is customary to denote these solutions by β_{mn} (and the corresponding mode fields by HE_{mn} and EH_{mn}), where both m and n are integers.

3.2.1 The Single-Mode Condition

An important characteristic for each mode is its cutoff frequency. A mode is said to be cutoff when its field no longer decays in the cladding for increasing ϱ. Since, as we mentioned before, the decay rate of the field is determined by the value of σ, the cutoff frequency is obtained by setting the condition $\sigma = 0$. The cutoff frequency of the first guided mode (i.e., the single-mode condition) will now be determined using an approximate expression for the modified Bessel functions when σa is small (i.e., $\sigma \to 0$) [*CRC*, 1978]:

$$K_m(\sigma a) \cong \frac{2^{m-1}}{\sigma a^m}(m-1)! \tag{3.27}$$

Inserting (3.27) into (3.26), we obtain for HE_{mn} modes with $m > 1$

$$2(m-1)J_{m-1}(\gamma a) = \gamma a J_m(\gamma a) \tag{3.28}$$

Solutions of (3.28) lead to the cutoff conditions for the HE_{mn} modes. Since for Bessel functions the following identity exists [CRC, 1978]:

$$J_{m-1}(\gamma a) = \frac{\gamma a}{2(m-1)}[J_{m-2}(\gamma a) + J_m(\gamma a)] \tag{3.29}$$

we can replace $J_{m-1}(\gamma a)$ in (3.28) and obtain

$$J_{m-2}(\gamma a) = 0 \tag{3.30}$$

The lowest order mode is the HE_{11} mode, also known as the *fundamental mode*. To allow the fiber to support only this mode, all higher order modes must be beyond cutoff.

This leads to the condition: $J_0(\gamma a) = 0$ or $\gamma a = 2.405$ (3.31)

Returning to (3.24), and using (3.31), we obtain the single-mode condition in terms of the normalized frequency

$$V < V_C = 2.405 \tag{3.32}$$

Figure 3.2 shows the propagation constants β_{mn} as a function of V for the first few orders of fiber modes.

Also indicated in Figure 3.2 are the transverse-electric (TE) and transverse-magnetic (TM) fields that correspond to the HE_{0n} and EH_{0n} modes, respectively.

By inspection of (3.24) and (3.32), we conclude that to make the fiber single-mode at an operating wavelength λ, we can reduce either the core diameter or the relative refractive index (3.25), or both. Reducing the relative refractive index allows the core diameter to be increased while keeping single-mode operation. This explains one of the major reasons why the cladding layer is necessary in practice.

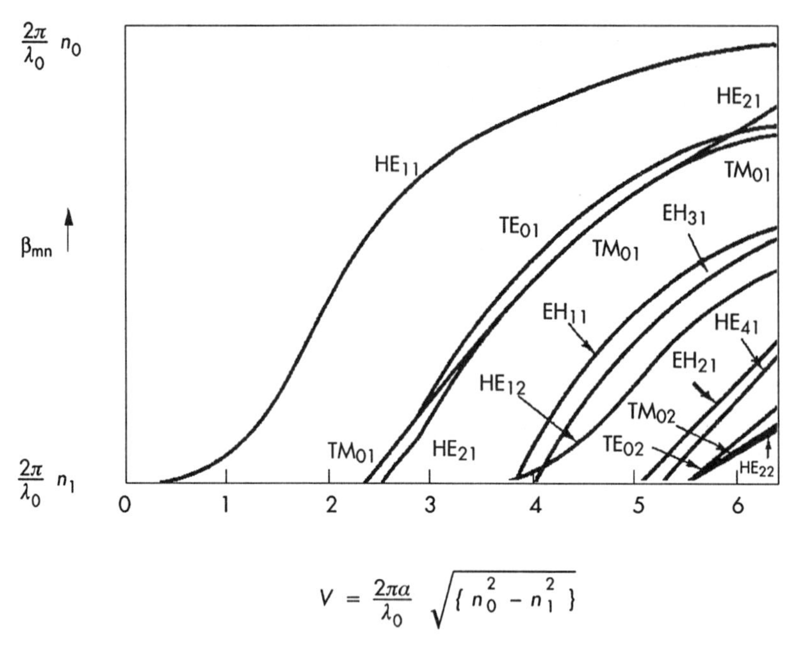

Figure 3.2 Mode propagation constants β_{mn} versus the normalized frequency V for the first few orders of fiber modes.

Notice that a fiber may be single-mode at a given wavelength but not at shorter wavelengths for which the condition (3.32) may no longer be valid. Thus, the single-mode character is not only an intrinsic property of the fiber itself, but also of the operating wavelength.

3.2.2 Birefringence

The fundamental fiber mode HE_{11} is linearly polarized in either x or y direction (the z component is about 100 times smaller). In fact, even a single-mode fiber is not truly single-mode since it can support two degenerate modes of orthogonal polarizations. Under ideal conditions (perfect cylindrical symmetry and isotropic material), a mode excited with its polarization in the x direction would not couple to the mode with the orthogonal y-polarization state. However, in practice, irregularities such as random variations in the core diameter along the fiber length, small fluctuations in material anisotropy, or stress-induced anisotropy result in a mixing of the two polarization states. The two polarization components mix randomly, and scramble the polarization of the incident light as it propagates down the fiber. Consequently, light launched into the fiber with linear polarization quickly reaches a state of arbitrary polarization.

It is interesting to note that if two light signals having orthogonal states of polarization (SOP) are launched into a single-mode fiber, they will exit the fiber with orthogonal SOPs. This property is used in the so-called polarization shift keying (POLSK) modulation where binary "0" and "1′" are represented by two orthogonal SOPs [Benedetto 1991, Hill 1992].

The degree of modal birefringence B is defined by [Kaminow, 1981]

$$B = \frac{|\beta_x - \beta_y|}{k_0} = |n_x - n_y| \tag{3.33}$$

where β_x and β_y are the mode propagation constants for the modes polarized in the x and y directions, respectively, and n_x and n_y are the effective mode indices in the two orthogonal polarization states.

It can be shown that for a given value of B, the power between the two modes is exchanged periodically with a period L_B, referred to as the beat length [Kaminow, 1981]

$$L_B = \frac{2\pi}{|\beta_x - \beta_y|} = \frac{\lambda}{B} \tag{3.34}$$

Typical values for L_B in standard single-mode fibers range from one to several meters.

For some applications, like coherent optical communication (Chapter 4), it is desirable that fibers transmit light without changing its state of polarization. Such fibers, called polarization-maintaining or polarization-preserving fibers, are designed by intentionally introducing a large amount of birefringence (high value of B) so that small and random fluctuations will not significantly affect the light polarization as it propagates down the fiber. A light signal whose polarization is linear and oriented along one of the principal axes remains linearly polarized during propagation down the polarization-maintaining fiber. A value of $B = 2.10^{-4}$ has been achieved with the so-called "PANDA" fibers, which corresponds to a beat length of the order of 1 cm.

3.3 ATTENUATION

Up to now, losses have been neglected in the discussion of wave propagation down the fiber. Although fiber losses are very weak compared to other transmission media (e.g., twisted pairs or coaxial cables), they must be taken into account for the design of communication systems.

Optical-fiber loss has two principal causes: absorption and scattering. The physical mechanisms of these contributions, which lead to signal attenuation, are described in [Senior 1985, Geckeler 1987].

The resulting effect on signal power during propagation along the fiber is characterized by a single coefficient α (m^{-1}). If P_{in} is the power launched at the input of the fiber, the power $P(L)$ at a distance L inside the fiber is given by

$$P(L) = P_{\text{in}} \exp(-\alpha_{[m^{-1}]} L) \tag{3.35}$$

To reflect this exponential law of power attenuation, it is customary to express α in units of dB/km by using the relation

$$\alpha[dB/km] = -\frac{10}{L} \log_{10}\left[\frac{P(L)}{P_{\text{in}}}\right] = 4{,}343\alpha_{[m^{-1}]} \tag{3.36}$$

The coefficient α(dB/km) depends upon the wavelength as can be seen from Figure 3.3, which shows the loss spectrum of a standard single-mode fiber in the wavelength range 0.7–1.8 µm.

We can distinguish two regions of minimum attenuation around 1.3 and 1.55 µm, separated by a peak centered at 1.37 µm (absorption due to the residual OH^-- ions).

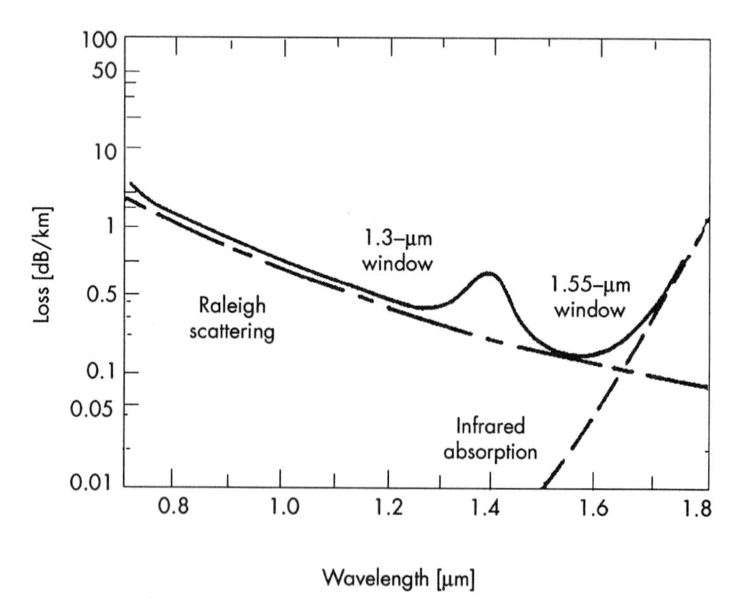

Figure 3.3 Loss spectra of low-loss silica glass fiber.

The minimum loss in the 1.3-μm-window is 0.35 dB/km, while in the 1.55-μm window it is 0.2 dB/km.

The widths of the 1.3- and 1.55-μm windows, as shown in Figure 3.3, are 100 and 150 nm, respectively (see Appendix B). These window widths are chosen such that the losses do not exceed some predetermined values. These widths will thus, in general, depend on system design constraints.

Fiber losses are not the only cause of signal power attenuation in fiber-optic communication system. In common with any cabled communication systems, optical-fiber links or networks have requirements for jointing and termination of the transmission medium. Presently, there exist two major categories of fiber joint used in practice:

1. Fiber splices, which are permanent or semipermanent joints formed between two individual optical fibers. Depending on the technique utilized, we can distinguish fusion splices (permanent) and mechanical splices (semipermanent). Typical mean loss of single-mode fusion splices lies in the range from 0.05 to 0.1 dB with a standard deviation of 0.05 dB. Mechanical splices have slightly higher losses in the range from 0.1 to 0.15 dB.
2. Connectors, which are removable joints that allow many repeated connections and disconnections with negligible degradation in the performance of the transmission line at the joint. Typical mean loss of a single-mode connector is 0.3 dB and the standard deviation is typically of the order of 0.1 dB. For some applications, the connector return loss; that is, the fraction of the power which is reflected back into the fiber, is an important specification (e.g., for full-duplex transmission at the same wavelength or for analog transmission of video signals). Special connector designs with angled-fiber end faces and physical contact between the fiber ends that form the joint allow the return loss to be reduced to below −60 dB.

The number of splices and connectors depends upon the link length and/or network topology. The choice between splices (fusion or mechanical) and connectors will mainly be dictated by the required connection flexibility.

In any case, joints add losses in the optical path and these losses must be carefully taken into account when calculating the power budget of the system.

In multiaccess optical-fiber networks, an important additional contribution to the total system loss is due to the passive components, such as splitters, taps, and wavelength multiplexers/demultiplexers. These components and their corresponding losses will be discussed in Chapter 6.

3.4 DISPERSION IN SINGLE-MODE FIBERS

As already mentioned in Section 3.2, dispersion effects in single-mode fibers result from the frequency dependence of the propagation constant $\beta(\omega)$ [Gloge, 1971]. As can be seen from (3.23), this frequency dependence results not only from the ω-dependence of n_0 and

n_1 but also from the ω-dependence of σ and γ. The former is referred to as material dispersion while the latter is called waveguide dispersion.

The material dispersion, or ω-dependence of the refractive index, is related to the characteristic resonance frequencies ω_j at which the medium absorbs the electromagnetic radiation through oscillations of bound electrons. The refractive index is well-approximated by the Sellmeier equation [Jenkins, 1981]

$$n_0^2(\omega) = 1 + \sum_{j=1}^{m} \frac{B_j \omega_j^2}{\omega_j^2 - \omega^2} \tag{3.37}$$

where ω_j (which coincides with peaks of material absorption) and B_j are constants that can be found by fitting the experimentally obtained dispersion curves with the theoretical one.

As will be discussed in Section 3.4.1 and Appendix C, the effect of material dispersion is characterized, to a first order, by the second derivative of n with respect to ω, or equivalently by the dispersion coefficient defined by

$$D(\lambda) = -\frac{\lambda}{c} \frac{d^2 n}{d\lambda^2} \tag{3.38}$$

$D(\lambda)$ is generally expressed in units of (ps/km · nm). For example, $D(\lambda) = 20$ ps/km nm signifies that an optical pulse having a spectral width of 1 nm will broaden by 20 ps for each kilometer of propagation along the fiber.

For fused silica, $D(\lambda)$ vanishes at a wavelength of $\lambda = 1.27$ μm. This wavelength is often referred to as the zero-dispersion wavelength λ_D. As the fiber core may have small amounts of dopants such as GeO_2 and P_2O_5, the actual value of λ_D for glass fiber deviates from 1.27 μm towards longer wavelengths.

Thus, in principle, it should be possible by proper doping of step-index fibers to shift the zero-dispersion wavelength in the range from 1.3 to 1.7 μm. However, the high refractive index difference between the core and the cladding, resulting from high dopant concentrations, leads to a prohibitive attenuation caused by absorption and diffusion processes due to the dopants.

Waveguide dispersion, which is directly related to the guided nature of the mode propagating down the fiber, can help to shift λ_D towards longer wavelengths. One objective would be to have $\lambda_D = 1.55$ μm, which coincides with the minimum of the fiber attenuation.

The waveguide dispersion is dependent on the refractive-index profile of the optical fiber. The frequency dependence of the waveguide-dispersion has two distinct origins. First, the refractive indices of the various types of glass used in the formation of the optical fiber, with varying refractive-index profile, do not depend on ω in precisely the same way.

Second, from (3.21) and (3.22) it can be shown that the radial variation of the fiber-mode amplitude is also wavelength-dependent: for single-mode fibers, the farther the mode is from its cutoff, the more tightly its energy is concentrated inside the core. These features are extensively used to shift the zero-dispersion wavelength in the vicinity of 1.55 μm while minimizing the excess attenuation due to the dopants, which would have been obtained by step-index fibers with high core-cladding refractive-index differences.

Using the same principle, it is also possible to design ''dispersion-flattened'' optical fibers having low dispersion over a relatively large wavelength range. This is achieved by the use of multiple cladding layers [Miller, 1988]. Figure 3.4 shows typical dispersion coefficient $D(\lambda)$ curves for four kinds of fibers. For comparison, the dispersion-flattened fibers have low dispersion over a wide wavelength range extending from 1.25 to 1.65 μm.

3.4.1 Dispersion-Induced Pulse Broadening

In the preceding section we discussed the physical origin of dispersion in single-mode fibers. We will now see how the dispersion affects the pulse shape of the information bearing-signal. To keep the analysis simple, we will first proceed with pulses having Gaussian-shaped envelopes originating from an SLM laser. This will help us to draw

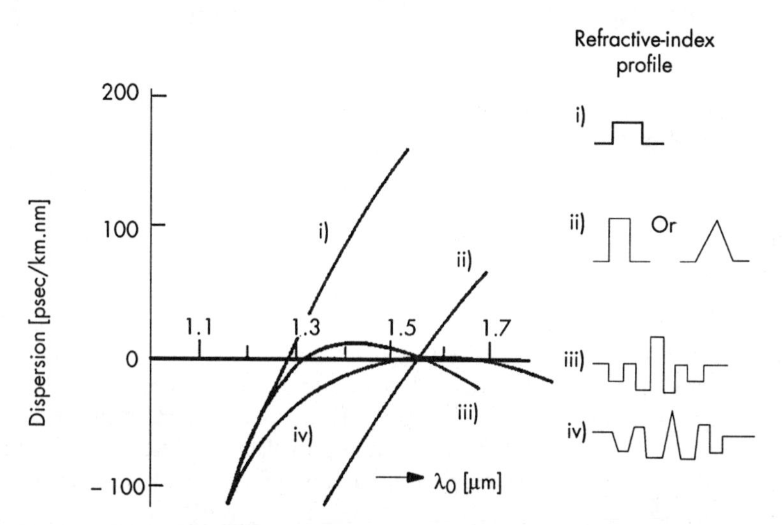

Figure 3.4 Dispersion coefficient $D(\lambda)$ as a function of the wavelength for four kinds of fibers: (i) standard step-index fiber, (ii) dispersion-shifted fiber (step index or triangular profile), and (iii and iv) dispersion-flattened fiber or broadband low-dispersion fiber. [After Lin, 1991.]

general qualitative conclusions on dispersion-induced pulse distortion. Thereafter, these results will be used to discuss the dispersion effect for more realistic pulse shapes with steeper leading and trailing edges such as those emitted by directly modulated SLM semiconductor lasers.

Finally, this section ends with a discussion of dispersion effect when multilongitudinal mode lasers are used (mode partition noise).

3.4.1.1 SLM Lasers

Consider a monochromatic wave of pulsation ω_0 and whose amplitude is Gaussian-modulated. At the input of the fiber ($z = 0$), the optical field can be expressed as

$$E(z = 0,t) = A(0,t)\exp\{-j\omega_0 t\} = A_0 \exp\left\{-\frac{t^2}{\tau_0^2}\right\}\exp\{-j\omega_0 t\} \tag{3.39}$$

where A_0 is the peak amplitude, $\tau_0 = T_0/\sqrt{2\ln 2}$, T_0 being the intensity FWHM of the Gaussian-pulse.

The pulse spectrum is determined by Fourier transform and is given by

$$\tilde{E}(0, \omega) = A_0 \int_{-\infty}^{+\infty} \exp\left\{-\frac{t^2}{\tau_0^2}\right\} e^{j(\omega-\omega_0)t}\, dt = A_0 \tau_0 \sqrt{\pi}\exp\left\{-\frac{\tau_0^2}{4}(\omega - \omega_0)^2\right\} \tag{3.40}$$

The spectrum thus also has a Gaussian shape, which is centered at ω_0, and whose intensity FWHM is given by

$$\Delta\Omega = \Delta(\omega - \omega_0) = \frac{4\ln 2}{T_0} \tag{3.41}$$

Equation (3.41) indicates that the shorter the pulse is, the broader its spectrum is. Gaussian-shaped pulses that verify (3.41) are referred to as *transform-limited* pulses.

As already mentioned, the effect of dispersion is fully determined by the ω-dependence of the mode propagation constant $\beta(\omega)$. Each spectral component of the launched pulse will experience a phase shift proportional to $\beta(\omega)z$, where z is the propagation distance along the fiber. The pulse spectrum at any point z along the fiber is thus given by

$$\tilde{E}(z, \omega) = \tilde{E}(0, \omega)\exp\{j\beta(\omega)z\} \tag{3.42}$$

The corresponding pulse in the time domain is obtained by inverse Fourier transform of (3.42):

$$E(z, t) = \frac{1}{2\pi} \int_{-\infty}^{+\infty} \tilde{E}(0, \omega) \exp \{j\beta(\omega)z\} e^{-j\omega t} d\omega \tag{3.43}$$

In principle, if $\beta(\omega)$ is known, the integration in (3.43) can be carried out either analytically or numerically. However, in most cases $\beta(\omega)$ is unknown and the integration in (3.43) can only be approximated under certain assumptions.

With bit rates of current fiber-optic communication systems, the following assumption can be made:

$$\frac{\Delta\Omega}{\omega_0} \ll 1 \tag{3.44}$$

This can be justified as long as $T_0 \geq 0.1$ps (or bit rates < 10,000 Gbps), and is therefore a very conservative assumption.

When (3.44) is verified, $\beta(\omega)$ can be expanded in Taylor's series about ω_0 and limited to the second-order term

$$\begin{aligned} \beta(\omega) &= \beta(\omega_0) + \left(\frac{\partial\beta}{\partial\omega}\right)_{\omega_0} (\omega - \omega_0) + \frac{1}{2}\left(\frac{\partial^2\beta}{\partial\omega^2}\right)_{\omega_0} (\omega - \omega_0)^2 \\ &= \beta_0 + \beta_0'(\omega - \omega_0) + \frac{1}{2}\beta_0'' (\omega - \omega_0)^2 \end{aligned} \tag{3.45}$$

Inserting (3.45) into (3.43), and using (3.40), we obtain

$$\begin{aligned} E(z, t) &= A_0 \frac{\tau_0}{\sqrt{\tau_0^2 - 2j\beta_0''z}} \exp \left\{ \frac{-(t - \beta_0'z)^2}{\tau_0^2 - 2j\beta_0''z} \right\} \exp \{j(\beta_0 z - \omega_0 t)\} \\ &= A(z, t) \exp \{j(\beta_0 z - \omega_0 t)\} \end{aligned} \tag{3.46}$$

where $A(z, t)$ is the complex pulse envelope at a distance z and time t along the fiber.

Equation (3.46) shows that the pulse envelope propagates with a group velocity $V_g = 1/\beta_0'$.

Defining $\tau = t - \beta_0' z$, we finally obtain

$$A(z, \tau) = |A(z, \tau)| e^{j\Phi(z,\tau)} \tag{3.47}$$

where

$$|A(z, \tau)| = A_0 \frac{\tau_0}{(\tau_0^4 + 4\beta_0''^2 z^2)^{1/4}} \exp\left\{ -\frac{\tau^2}{\tau_0^2 + 4\left(\frac{\beta_0''^2}{\tau_0^2}\right) z^2} \right\} \tag{3.48}$$

and

$$\Phi(z, \tau) = -\frac{2\beta_0'' z}{\tau_0^4 + 4\beta_0'' z^2} \tau^2 + \frac{1}{2} \tan^{-1}\left[\frac{2\beta_0'' z}{\tau_0^2}\right] \tag{3.49}$$

From (3.48) we see that a Gaussian pulse maintains its shape but its width increases and becomes

$$T(z) = T_0 \sqrt{1 + \left(4(\ln 2)\frac{\beta_0'' z}{T_0^2}\right)^2} \tag{3.50}$$

In addition to pulse broadening, dispersion induces a time-dependent phase given by (3.49). This time-dependent phase, $\Phi(z, \tau)$, implies an instantaneous frequency variation around $\nu_0 = \omega_0 / 2\pi$ given by (minus sign is due to the choice $\exp\{-j\omega_0 t\}$ in (3.39))

$$\delta\nu(z, \tau) = -\frac{1}{2\pi}\frac{\partial\Phi(z, \tau)}{\partial\tau} = \frac{2}{\pi}\frac{\beta_0'' z}{\tau_0^4 + 4\beta_0''^2 z^2}\tau \tag{3.51}$$

Thus, the instantaneous frequency, $\nu_0 + \delta\nu(z, \tau)$, changes linearly across the pulse ((3.51) is linear in τ). This is referred to as a *linear chirp*. The effect of dispersion on transform-limited Gaussian pulses is shown in Figure 3.5. The linear chirp depends on the sign of β_0''. When $\beta_0'' > 0$ (this is referred to as the *normal dispersion regime*), $\delta\nu$ is an increasing function of τ. The instantaneous frequencies of the leading edge of the pulse

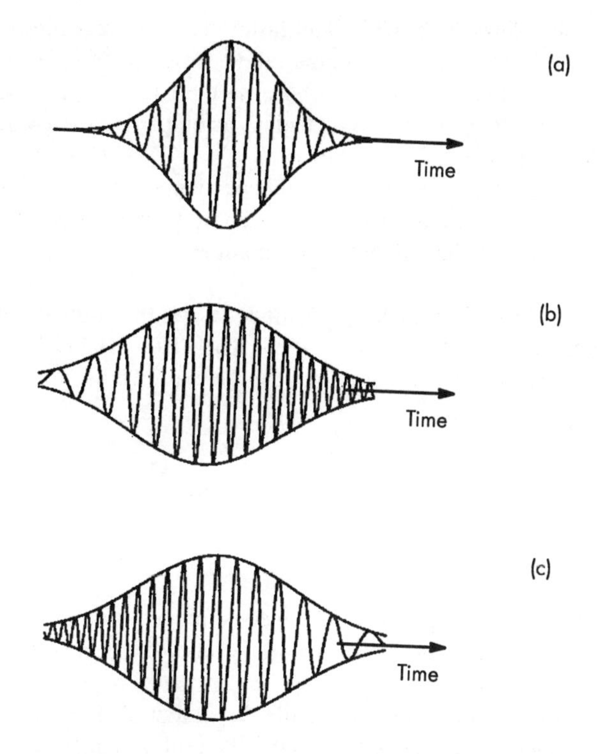

Figure 3.5 Dispersion effect on a transform-limited Gaussian pulse. (a) Input pulse, (b) effect of normal dispersion, and (c) effect of anomalous dispersion.

($\tau < 0$) are lower than ν_0, while the falling edge contains higher instantaneous frequencies $\nu > \nu_0$. The opposite occurs when $\beta_0'' < 0$ (this is referred to as the *anomalous dispersion regime*).

The pulse broadening and the dispersion-induced chirp can be understood intuitively by recalling that the group velocity of the pulse is related to β_0 by $V_g = 1/\beta_0'$. Therefore,

$$\beta_0'' = \left[\frac{\partial}{\partial\omega}\left(\frac{\partial\beta}{\partial\omega}\right)\right]_{\omega_0} = \left[\frac{\partial}{\partial\omega}\left(\frac{1}{V_g}\right)\right]_{\omega_0} = -\frac{1}{V_g^2}\left[\frac{\partial V_g}{\partial\omega}\right]_{\omega_0} \tag{3.52}$$

β_0'' thus represents the variation of the group velocity with respect to ω. For this reason, it is also called the group velocity dispersion (GVD). The relation between β_0'' and $D(\lambda)$ defined in (3.38) is reported in Appendix C.

For $\beta_0'' < 0$, the group velocity increases for increasing ω. The different frequency components of the input pulse will travel down the fiber at different speeds because of

GVD: high-frequency components will travel faster than low-frequency components. The opposite occurs for $\beta_0'' > 0$. At the fiber output, these different components will no longer be in-phase (they are in-phase at the input because the launched pulse is assumed to be transform-limited), resulting in pulse broadening accompanied by a variation of the instantaneous carrier frequencies (i.e., a chirp).

The pulse broadening factor $T(z)/T_0$ is independent of the sign of β_0'' as can be seen from (3.50). However, the chirp depends on the sign of β_0'' and this will become important when considering initially chirped pulses or nonlinear effects (e.g., the formation of soliton pulses).

For directly modulated semiconductor diode lasers, the emitted light pulses generally exhibit a frequency chirp [Heidemann 1988, Koch 1988, Hakki 1992]. The complex envelope of linearly chirped Gaussian pulses can be expressed as

$$A(0, \tau) = A_0 \exp\left\{ -(1 + jC) \frac{\tau^2}{\tau_0^2} \right\} \tag{3.53}$$

where C is the chirp parameter.

When $C > 0$ (positive chirp), the instantaneous frequency increases linearly from the leading to the trailing edge of the pulse, while the opposite occurs for $C < 0$ (negative chirp).

The evolution of $A(0, \tau)$ (3.53) along the dispersive fiber can be evaluated by the same procedure as that outlined above for $C = 0$. The result is that even a chirped Gaussian pulse maintains its Gaussian shape on propagation, but its width is now given by

$$T(z) = T_0 \sqrt{\left[1 + 4(\ln 2)\frac{C\beta_0'' z}{T_0^2}\right]^2 + \left[4(\ln 2)\frac{\beta_0'' z}{T_0^2}\right]^2} \tag{3.54}$$

Thus, in contrast to the case $C = 0$, (3.54) shows that the broadening factor $T(z)/T_0$, depends on the sign of the product $C\beta_0''$.

By examining (3.54), we see that when $C\beta_0'' < 0$, the pulse experiences an initial narrowing stage, and thereafter a monotonical broadening. The initial narrowing can be understood intuitively by remembering the dispersion-induced chirp effect experienced by initially unchirped pulses. When $C\beta_0'' < 0$, the initial chirp is in the opposite direction to that of the induced-dispersion chirp. For example, with $C > 0$ and $\beta_0'' < 0$, the low frequencies of the leading edge of the initial pulse travel slower than the high frequencies of the trailing edge. The net result is pulse narrowing until the two chirps cancel each other. At that point, where minimum pulse width is reached, the pulse is transform-limited and the initial chirp is completely canceled out. For longer propagation distances, the pulse broadens monotonically. The broadening factor as a function of the normalized propagation length, z/L_D, where $L_D = T_0^2 / [4(\ln 2)|\beta_0''|]$, is depicted in Figure 3.6 for two opposite signs of $C\beta_0''$ and is compared with the case of transform-limited pulses (i.e., $C = 0$).

A useful measure of transmission system capacity is the bit rate distance product BL. Assuming 20% tolerable pulse broadening, and using $B = 1/(2T_0)$, (3.54) gives (for sufficiently large values of $|C|$)

$$B \cdot L \cong 0.18 \frac{[1.2\mathrm{sign}(C\beta_0'') - 1]}{C} \frac{T_0}{\beta_0''} \tag{3.55}$$

Figure 3.7 shows the product BL calculated with (3.55) as a function of the GVD for different negative values of the chirp parameter (in general, semiconductor lasers have negative C). The curves are obtained for $T_0 = 125$ ps, which corresponds to a bit rate of 4 Gbps in the return-to-zero format.

The most noteworthy feature of Figure 3.7 is its asymmetry due to pulse narrowing when $C\beta_0''$ is negative and enhanced pulse broadening when $C\beta_0''$ is positive.

The figure also clearly shows that the BL product increases as β_0'' approaches zero [ps^2/km]. It must be noticed that when $\beta_0'' = 0$, (3.55) no longer holds. Indeed, in this case, (3.45) must include the higher order term proportional to $\beta_0''' = \left(\frac{\partial^3\beta}{\partial\omega^3}\right)_{\omega_0}$.

Nevertheless, in practice, the effect produced by the β_0'''-term can generally be

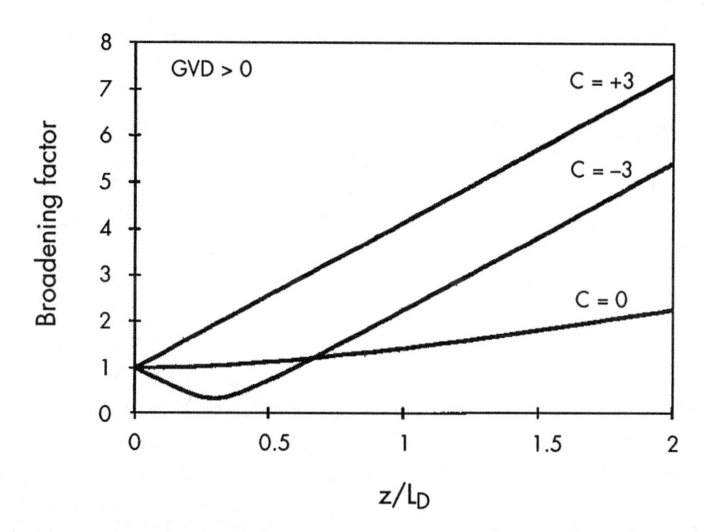

Figure 3.6 Broadening factor for Gaussian-pulses: $C = 0$ corresponds to transform limited pulses, $C = +3$ corresponds to positively chirped pulses, and $C = -3$ corresponds to negatively chirped pulses. The GVD is assumed to be positive.

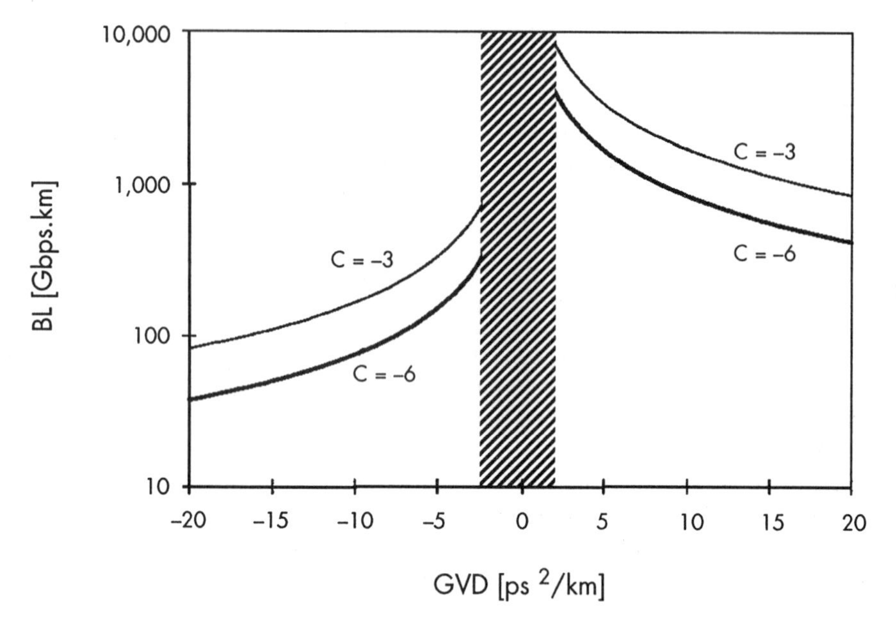

Figure 3.7 Dispersion-limited bit rate distance product for chirped Gaussian pulses as a function of the group velocity dispersion parameter for negatively chirped Gaussian pulses.

neglected in comparison with the β_0'' because impairments such as impurities, refractive index fluctuations, and core diameter variations in long-length optical fibers prevent the average value of $|\beta_0''|$ from falling to below 2 ps^2/km or so.

Thus, from Figure 3.7, we see that the best performance is achieved by the use of signal wavelength for which β_0'' is slightly positive. For $\beta_0'' = +2$ ps^2/km (1.3 μm), the *BL* product for standard single-mode fiber is as large as about 3,000 Gbps · km when $C = -6$. However, for more realistic pulse shapes with steeper leading and trailing edges this value must be reduced by a factor roughly equal to 3, resulting in a *BL* product of about 1,000 Gbps · km. For $L = 100$ km, such systems can operate at bit rates B ≈ 10 Gbps. Unfortunately, such performance is difficult to achieve without the use of optical amplifiers because of power budget limitations (Chapter 4).

In the 1.55-μm window, similar performances can be achieved provided that dispersion-shifted fibers or dispersion-flattened fibers are used. With standard single-mode fibers, however, $\beta_0'' \cong -20$ ps^2/km at 1.55 μm and the *BL* product is restricted to values ≤ 100 Gbps · km for $C = -3$. Again, for more realistic pulse shapes with steeper leading and trailing edges, this value reduces to about 60 Gbps · km. For $L = 50$ km, such systems are dispersion-limited to bit rates ≤ 1.2 Gbps. Improvements in the bit rate are possible by the use of an external modulator, which allows the laser to operate in a continuous mode resulting in a significant reduction of the chirp.

Another way to improve the bit rate distance product consists of implementing electronic equalization at the receiver circuit or pre-equalization at the transmitter circuit. The basic idea is to pass the distorted received signal through a filter (usually adaptive) whose transfer function is the inverse of the transfer function of the fiber [Winters 1990, Gitlin 1992]. Fiber dispersion compensation can also be performed in the optical domain prior to optical-to-electronic conversion [Gnauck 1990, Cimini 1990 Gnauck 1991].

The above discussion essentially applies to direct detection lightwave systems (IM/DD, see Chapter 4). Fiber dispersion also degrades the performance of systems that use coherent detection. However, since coherent detection systems necessitate SLM lasers with very narrow linewidth, dispersion effects are less severe than for intensity modulation/direct detection (IM/DD). Moreover, for some modulation formats an external modulator must be used, thereby avoiding frequency chirping.

The analysis of coherent system degradation for various modulation formats is explained in [Elrefaie, 1988]. It is shown that the system performance depends on the product B^2L rather than BL. For conventional single-mode fiber, the product B^2L is limited to 5,000 $(\text{Gbps})^2 \cdot$ km at an operating wavelength of 1.55 μm. For B = 10 Gbps, L is thus limited to 50 km. By operating the system near the zero-dispersion wavelength (i.e., 1.3 μm), the B^2L product can be improved by an order of magnitude.

3.4.1.2 Multilongitudinal Mode Lasers

With ordinary Fabry-Perot lasers, the power distribution among the numerous longitudinal modes fluctuates more or less rapidly. In other words, successive pulses of an optical bit stream exhibit a different power spectrum. When such a bit stream is transmitted down a single-mode fiber having nonzero dispersion, each pulse will be distorted differently according to its own spectral distribution. This phenomenon is referred to as *mode partition noise*.

The net result at the output of the fiber is essentially a pulse-delay fluctuation that causes a reduction of the signal-to-noise ratio (SNR) at the receiver input, which cannot be overcome by increasing the signal power. As a consequence, the maximum achievable bit rate distance product for a specified bit error rate (BER) cannot be improved by increasing the signal power.

Mode partition noise in single-mode fibers has been extensively studied [Ogawa, 1982, Agrawal 1988, Kikushima 1990, Wentworth 1992]. The results presented here are deduced from the analysis presented in [Ogawa, 1982].

In this reference, a simple formula is derived for the maximum bit rate distance product using the following assumptions:

1. The optical power of each pulse is constant.
2. The spectrum averaged over many pulses is Gaussian, centered about λ_0 with an rms-full-width represented by σ (typically between 2 to 5 nm).

3. Each pulse contains only a single dominant longitudinal mode, which varies randomly from pulse to pulse.
4. The received pulse waveform is described by a $\cos(\pi Bt)$ function where B is the bit rate, and the decision circuit samples the received signal at regular times of n/B ($n = 0, 1, \ldots$).

Assumption (3) corresponds to the worst case of mode partition noise because it leads to a maximum differential delay of pulse propagation due to fiber dispersion (the analysis can easily be extended to include a certain degree of mode fluctuations, as also described in [Ogawa, 1982]).

With these assumptions, the maximum achievable bit rate distance product is given by

$$B \cdot L[\text{Gb/s} \cdot \text{km}] = \frac{307}{D(\lambda) \cdot \sigma} \tag{3.56}$$

where $D(\lambda)$ is expressed in [ps/km · nm] at the wavelength λ_0 and σ is expressed in [nm].

As an example, with $D(\lambda)$ = 2 ps/km · nm at 1.3-μm signal wavelength (conventional single-mode fiber), and σ = 2 nm, (3.56) gives $B \cdot L \cong 77$ Gbps· km. Thus, mode partition noise restricts the bit rate distance product much more severely than do systems using SLM lasers whose center wavelength matches the minimum dispersion wavelength of single-mode fibers.

Terrestrial links operating up to 1.7 Gbps over 40 km (close to the limit imposed by mode partition noise) and using multilongitudinal mode lasers have been developed and are now currently used in the field.

3.5 NONLINEAR EFFECTS

The linear law expressed by (3.8) is only valid when the optical power levels inside the fiber are sufficiently low (typically below about 1 mW) so that the induced polarization $\boldsymbol{P}$ is proportional to the electric fields present inside the medium.

For higher power levels the response of the dielectric medium that constitutes the optical fiber becomes nonlinear and the induced polarization $\boldsymbol{P}$ satisfies the more general relation [Butcher, 1991]

$$\boldsymbol{P} = \varepsilon_0 \{\tilde{\chi}^{(1)} \cdot \boldsymbol{E} + \tilde{\chi}^{(3)}\!:\!\boldsymbol{E} \cdot \boldsymbol{E} \cdot \boldsymbol{E}\} \tag{3.57}$$

where $\tilde{\chi}^{(j)}$(j = 1, 2, 3 . . .) is the jth order susceptibility, which is dependent on the frequency of the fields present in the medium.

The first term in (3.57), $\varepsilon_0\tilde{\chi}^{(1)} \cdot \boldsymbol{E}$, corresponds to the linear regime discussed in the previous sections and represents the dominant contribution to $\boldsymbol{P}$.

The second term, $\varepsilon_0\tilde{\chi}^{(3)}\colon \boldsymbol{E} \cdot \boldsymbol{E} \cdot \boldsymbol{E}$, is responsible for the lowest order nonlinear effects in optical fibers (the term proportional to $\tilde{\chi}^{(2)}$ vanishes in silica glasses because of symmetry of the amorphous constituent medium). The third-order susceptibility $\tilde{\chi}^{(3)}$ is at the origin of such phenomena as *nonlinear refraction* and *four-photon mixing* (including third harmonic generation).

Although the absolute value of $\tilde{\chi}^{(3)}$ in silica fibers is at least two orders of magnitude smaller than that of most nonlinear optical media, nonlinear effects can be observed at relatively low power levels because the small core diameter of single-mode fibers allows high-power densities, which can be maintained over long distances owing to the very low attenuation of single-mode fibers.

Two other important phenomena in optical fibers fall into a second class of nonlinear effects. They are known as *stimulated Raman scattering* and *stimulated Brillouin scattering;* both of them are related to the vibrational excitation modes of silica.

It is important to note that, since nonlinear effects involve the power density of optical fields, the attenuation of the fiber (although very small) cannot, in general, be neglected. As the waves propagate along the fiber their powers decrease, reducing the amount of the nonlinear effects accordingly. In order to account for this exponential decay of lightwave power with length along lossy fibers (3.35), an effective length is defined so that [Stolen, 1979]

$$P(0) \cdot L_{\text{eff}} = \int_0^L P(z)dz \tag{3.58}$$

Using (3.35), we obtain after integration of (3.58)

$$L_{\text{eff}} = \frac{1 - e^{-\alpha L}}{\alpha} \tag{3.59}$$

For $\alpha L << 1$ (negligible attenuation): $L_{\text{eff}} \cong L$, while for $\alpha L >> 1$: $L_{\text{eff}} \cong 1/\alpha$.

The following description of nonlinear effects will assume lossless fibers, i.e. $L_{\text{eff}} = L$. It is straightforward to adapt the obtained results to lossy fibers, and this is left to the reader as an exercise.

Nonlinear effects can limit lightwave system performance, or be used to their advantage as, for example, in full-optical amplification, pulse compression, or soliton transmission.

These nonlinear effects will now be briefly discussed and the corresponding optical powers at which they become significant will be evaluated. We restrict the discussion to the nonlinear refraction, four-photon mixing, and stimulated Raman and Brillouin scatterings. Performance degradations caused by each nonlinear effect in multiaccess optical-fiber networks will be discussed in Chapter 7.

3.5.1 Nonlinear Refraction

Nonlinear refraction is characterized by an intensity-dependent refractive index following [Stolen, 1979]

$$n(\omega, |E|^2) = n_0(\omega) + \Delta n_{NL} = n_0(\omega) + n_2 |E(z, t)|^2 \tag{3.60}$$

where $|E(z, t)|^2$ is the square modulus of the electric field associated with the lightwave propagating inside the fiber, $n_0(\omega)$ is the refractive index in the linear regime (3.38), and n_2 is the nonlinear index coefficient, which takes the value $n_2 = 1.22\ 10^{-22}\ (\text{V/m})^{-2}$ in silica fibers. It can be shown that n_2 is related to $\tilde{\chi}^{(3)}$ in (3.57) by the relation [Butcher, 1991]

$$n_2 = \frac{3}{8n_0} \tilde{\chi}^{(3)}_{xxxx} \tag{3.61}$$

Nonlinear refraction leads to a large number of interesting nonlinear effects such as self-phase modulation (SPM), which permits the existence of optical solitons, and cross-phase modulation (XPM), which is detrimental in multiaccess optical-fiber networks.

3.5.1.1 Self-Phase Modulation

SPM refers to the self-induced phase shift experienced by an optical field during its propagation in optical fibers [Stolen, 1978].

Assuming a lossless fiber ($\alpha = 0$), this self-induced phase shift after a propagation over a distance L is readily obtained as

$$\Phi_{NL} = \frac{2\pi}{\lambda_0} n_2 L |E|^2 \tag{3.62}$$

For constant optical field strength, SPM just adds a rotation of the phase with L and thus has little consequence on system performance [Tajima, 1986].

For varying optical field strengths, such as those pulses used in digital communication systems, however, the nonlinear phase Φ_{NL} will depend on time. This temporally varying phase implies that the instantaneous optical frequency differs across the pulse from its central value ν_0 by an amount $\delta\nu_{NL}$ given by

$$\delta\nu_{NL} = -\frac{1}{2\pi} \frac{\partial \Phi_{NL}}{\partial t} \tag{3.63}$$

Since Φ_{NL} is proportional to the optical intensity, the extent of $\delta\nu_{NL}$ will depend on how fast the intensity varies. For short pulses, new frequencies are generated and the power spectrum broadens during propagation. Since from (3.62) and (3.63) these new instantaneous frequencies are created at different parts of the pulse, the pulse shape evolution along the fiber, taking dispersion into account, is in general difficult if not impossible to evaluate analytically and numerical simulations must often be used. Nevertheless, a general picture of pulse evolution can be obtained when the launched pulse has a smooth envelope.

As an example, again consider a Gaussian pulse whose envelope is given by (3.53) with $C = 0$

$$A(0, \tau) = A_0 \exp\left\{-\frac{\tau^2}{\tau_0^2}\right\} \tag{3.64}$$

Inserting (3.64) into (3.62), we obtain immediately from (3.63)

$$\delta\nu_{\mathrm{NL}}(L, \tau) = \frac{n_2}{\lambda_0} L A_0^2 \frac{4\tau}{\tau_0^2} \exp\left\{-\frac{2\tau^2}{\tau_0^2}\right\} \tag{3.65}$$

Equation (3.65) shows that $\delta\nu_{NL}$ is negative in the leading edge ($\tau < 0$) and positive in the trailing edge ($\tau > 0$) of the pulse. Over the central region of the Gaussian pulse, (3.65) reduces to

$$\delta\nu_{\mathrm{NL}}(L, \tau) \cong \left(\frac{n_2}{\lambda_0} L A_0^2 \frac{4}{\tau_0^2}\right) \cdot \tau \tag{3.66}$$

so that the induced instantaneous phase variation (i.e., the induced chirp) is (quasi-)linear and positive.

It is noteworthy that SPM affects only the instantaneous carrier frequency but not the pulse shape; that is, if dispersion is negligible, SPM has little consequence on system performance even with amplitude modulated signals.

However, as we saw above, except for very short transmission distances and/or low bit rates, dispersion cannot be neglected, and when combined with SPM it will distort the pulse shape in different ways depending on the sign of β_0''.

With $\beta_0'' > 0$, high frequencies ($\nu > \nu_0$) propagate slower than low frequencies ($\nu < \nu_0$). As a result, during propagation along the fiber, the high frequencies generated by SPM in the trailing edge of the pulse will be retarded relative to the low frequencies generated in the leading edge. The pulse will thus broaden more rapidly than would be the case if SPM was absent. The net result is a reduction of the maximum achievable *BL* product. The amount of the *BL* product reduction depends on β_0'', the pulse shape, and its intensity, and has been analyzed in [Agrawal, 1989].

In contrast, with $\beta_0'' < 0$, SPM can be used to advantage to compensate for pulse broadening, as discussed in Section 3.4.1.1. For a specific initial pulse shape and a specific peak intensity, the pulse broadening can even be completely compensated, resulting in undistorted pulse propagation. Such pulses are referred to as fundamental solitons. The actual soliton pulse shape and the required peak intensity can be determined by solving the nonlinear differential equation (the so-called *nonlinear Schroedinger equation*) that governs the pulse propagation in nonlinear and dispersive single-mode fibers [Hasegawa, 1973].

The fundamental soliton solution of this equation is given by [Zakharov, 1972]

$$A(0, \tau) = A_0 \operatorname{sech}\left\{ 1.76 \frac{\tau}{T_0} \right\} \tag{3.67a}$$

with

$$A_0^2 = \frac{\lambda_0}{4n_2 Z_0} \qquad \text{where} \qquad Z_0 = 0.322 \frac{\pi}{2} \frac{|\beta_0''|}{T_0^2} \tag{3.67b}$$

T_0 being the intensity FWHM of the pulse.

The solution given by (3.67) is obtained using quite complicated mathematical methods explained in [Zakharov, 1972]. However, an approximate expression of the fundamental soliton peak intensity (3.67b) can be obtained with the help of very simple mathematics assuming initial Gaussian pulses instead of *sech* pulses.

Recalling (3.51), we see that when $\beta_0'' < 0$, the dispersion-induced chirp and the nonlinear self-induced chirp are of opposite signs. They can thus cancel each other provided that they have equal absolute values. To allow for soliton formation, they must do so from the beginning (i.e., when $L \to 0$ in (3.51)).

For $L \to 0$, (3.51) reduces to

$$\delta\nu(L \to 0, \tau) \cong \left(\frac{2}{\pi} \frac{\beta_0''}{\tau_0^4} L \right) \cdot \tau \tag{3.68}$$

Equating the absolute values of $\delta\nu_{NL}$ (3.66) and $\delta\nu$ (3.68), we obtain after straightforward arithmetic

$$A_0^2 = 0.44 \frac{\lambda_0}{2n_2} \frac{|\beta_0''|}{T_0^2} \tag{3.69}$$

where $T_0 = \sqrt{2 \ln 2}\,\tau_0$ is the intensity FWHM of the Gaussian pulse.

A_0 derived from (3.69) is about 2/3 that of the actual value derived from (3.67(b)).

The difference comes from the fact that the approximated value is obtained assuming Gaussian-shaped pulse (allowing very simple derivation) rather than *sech*-shaped pulses.

The peak power inside the fiber is obtained by

$$P_{\text{peak}} = \frac{\varepsilon_0}{2} A_0^2 \frac{c}{n} S_{\text{eff}} \tag{3.70}$$

where ε_0 is the vacuum permittivity, and S_{eff} the effective core area [Agrawal, 1989]. If the signal wavelength is slightly longer than the fiber cutoff wavelength, then $S_{\text{eff}} = S$, where S is the actual core area of the fiber. For typical values of β_0'' at 1.55-μm signal wavelength, P_{peak} is about 500 mW for T_0 = 10 ps, but reduces to about 0.5 mW for T_0 = 100 ps.

A number of experiments have demonstrated stable soliton transmission over distances of many thousands of kilometers [Mollenauer, 1988]. Since soliton transmission involves the interplay between dispersion and nonlinear refraction effects, periodic amplifications are required to compensate for signal attenuation due to fiber losses. Erbium-doped fiber amplifiers (Chapter 5) have proven particularly suitable for this purpose [Mollenauer, 1991].

Many engineering problems must still be solved before soliton-based systems become practical, but the rapid advances in this research field will probably lead to the use of soliton-like pulses, especially in long-distance submarine and transcontinental systems. Soliton pulses are also attractive candidates for ultrahigh-capacity multichannel fiber networks. Indeed, while nonsoliton systems may achieve the maximum bit rate distance product at only one wavelength close to the zero-dispersion wavelength, soliton systems can be wavelength multiplexed on the same fiber to transmit multiple channels that are each automatically dispersion-compensated [Mollenauer, 1991; Moores, 1992].

3.5.1.2 Cross-Phase Modulation

When more than one optical wave propagates inside the fiber (e.g., using wavelength division multiplexed systems as discussed in Chapter 7), the refractive index seen by a particular wave depends not only on the intensity of that wave but also on the intensity of the other copropagating waves.

It can be shown that, in this case, the nonlinear part of the refractive index seen by the wave labeled '*i*' is given by [Agrawal, 1989]

$$\Delta n_{\text{NL}}^{(i)} = n_2 \left\{ |E_i|^2 + 2\sum_{j \neq i}^{N} |E_j|^2 \right\} \tag{3.71}$$

where N is the total number of optical channels, and n_2 is the same as that of (3.61).

The first term in the right-hand side of (3.71) is responsible for SPM and has been discussed earlier. The second term is responsible for cross-phase modulation (XPM). For equal channel intensity, (3.71) shows that XPM is twice as effective as SPM.

Using (3.71) with (3.70), the nonlinear induced phase shift of lightwave 'i' after propagating through a fiber of length L is given by

$$\Delta\Phi_{\mathrm{NL}}^{(i)} = \frac{2\pi}{\lambda_i}\Delta n_{\mathrm{NL}}^{(i)} \cdot L = \gamma L \left\{ P_i + 2\sum_{j \neq i}^{N} P_j \right\} \tag{3.72}$$

where $\gamma = \dfrac{4\pi n_0 n_2}{\varepsilon_0 \lambda_i c S_{\mathrm{eff}}}$ whose typical value is $\approx 1\mathrm{W}^{-1}\mathrm{km}^{-1}$, and P_j is the optical power of the jth channel ($j = 1, \ldots N$).

Such phase changes can severely limit system performance when coherent optical demodulation techniques are employed. This is because coherent detection is a phase-sensitive technique (Chapter 4) and since the phase of a signal in a given channel will be shifted randomly through XPM due to power fluctuations in all other channels, the SNR at the fiber output is reduced. This will be discussed in more detail in Chapter 7.

3.5.2 Four-Photon Mixing

The same nonlinearity (3.57) that gives rise to the nonlinear refractive index is also responsible for four-photon mixing (FPM) in single-mode fibers. The nonlinear polarization, $\varepsilon_0 \tilde{\chi}^{(3)}$: $\boldsymbol{E}_i \cdot \boldsymbol{E}_j \cdot \boldsymbol{E}_k$ ($\boldsymbol{E}_i$ is the electric field of the wave at pulsation ω_i, $i = 1, 2, 3$), generates new optical waves at pulsations

$$\omega_{ijk} = \omega_i + \omega_j - \omega_k \tag{3.73}$$

where i, j, and k can be 1, 2, or 3.

These new optical waves copropagate with the initial wave and grow at their expense. In quantum-mechanical terms, four-photon mixing can be viewed as the annihilation of photons from some waves and the creation of new photons at different frequencies so that the total energy and momentum are conserved during the process.

It can be shown that the power $P_{ijk}(L)$ of the wave at pulsation ω_{ijk} exiting the fiber is given by [Shibata 1988, Chraplyvy 1990]

$$P_{ijk}(L) = \eta \left(\frac{1024\pi^6}{n_0^4 \lambda_{ijk}^2 c^2} \right) (6\chi_{xxxx}^{(3)})^2 \left(\frac{L}{S_{\mathrm{eff}}} \right)^2 \cdot P_i P_j P_k \cdot \exp\{-\alpha L\} \tag{3.74}$$

where η is the efficiency of the FPM process, which depends on the phase-matching condition (3.75) (see below).

Clearly, in multiwavelength systems with equally spaced channels the additional waves appear as undesirable crosstalk and degrade the system performances [Chraplyvy 1990, Inoue 1992].

Significant FPM occurs only if the phase-matching condition is satisfied (conservation of photon momentum). In mathematical terms, this phase-matching condition is expressed as [Butcher, 1991]

$$\beta(\omega_{ijk}) = \beta(\omega_i) + \beta(\omega_j) - \beta(\omega_k) \tag{3.75}$$

Because of dispersion in single-mode fibers, the four-wave mixing efficiency decreases with increasing channel spacing. Consequently, FPM will set a limit on the minimum channel spacing in order to keep the crosstalk below a certain level. This will be discussed in more detail in Chapter 7.

3.5.3 Stimulated Raman Scattering

The *Raman* effect results from inelastic scattering in which incident photons transfer part of their energy to mechanical vibration of the molecules of the propagating medium, while the remaining energy is radiated as light of a longer wavelength than the incident light (this radiated light is referred to as the *Stokes* light) [Bloembergen, 1971].

For intense light, the process can be stimulated by photons at the Stokes wavelength, resulting in most of the incident power being converted into the frequency down-shifted light. This process is referred to as stimulated Raman scattering (SRS).

In quantum mechanical terms, Raman scattering is explained as the annihilation of a photon of the incident field and the creation of a photon at the Stokes wavelength along with a quantum of vibrational energy (an optical phonon) in the scattering molecule.

It can be shown that, under continuous wave (CW) operation, the power $P_s(L)$ of the Stokes wave exiting the fiber of length L is given by [Smith, 1972]

$$P_s(L) = P_0 \cdot \exp\{g_R P_0 L / b S_{\text{eff}}\} \tag{3.76}$$

where P_0 is the launched power at the signal wavelength, g_R is the Raman gain coefficient, and b is a factor that accounts for the relative polarization of the launched and Stokes waves and the polarization properties of the fiber. In what follows, $b = 2$ will be assumed, which corresponds to a conventional fiber (i.e., not polarization maintaining fiber) [Chraplyvy, 1990].

In silica fibers, g_R extends over a large frequency range (up to 40 THz) with a peak near 13.2 THz.

The value of g_R increases more or less linearly with the wavelength difference between the injected light (sometimes called the pump) and the Stokes light up to a wavelength difference of about 100 nm (which corresponds to $\Delta\nu = 13.2$ THz at 1.55-μm pump wavelength, see Appendix B).

Equation (3.76) can be used to evaluate the power level at which the SRS effect becomes significant. It is customary to define the *Raman threshold* as the input power at which the Stokes power becomes equal to the pump power at the fiber output. This threshold power is given by [Smith, 1972]

$$P_0^{\text{th}} \cong 32\frac{S_{\text{eff}}}{L \cdot g_R} \tag{3.77}$$

As an example, if we use the typical values $S_{\text{eff}} = 50\ \mu\text{m}^2$, $L = 30$ km ($L_{eff} \cong 22$ km at 1.55 μm with $\alpha = 0.2$ dB/km), and $g_R = 6.5\ 10^{-14}$ m/W, (3.77) gives $P_0^{\text{th}} \cong 0.8$W (or 1.1W using L_{eff} in place of L).

Thus, it is clear that SRS is not likely to be a significant degradation effect in single-channel fiber communication systems.

In multichannel systems, SRS puts much more severe limitations on the maximum launched power per channel. The analysis is more complicated than in the single-channel situation because intermediate channels transfer part of their energy to the longer wavelength channels and simultaneously receive energy from the shorter wavelength channels by amounts determined by the Raman gain profile. This situation will be analyzed in more detail in Chapter 7. For the moment, to give an idea of the constraints imposed by SRS in multiaccess wavelength networks, let us consider the same values as those used in the example of (3.77). As will be discussed in Chapter 7, it can be shown that within a 10-channel WDM system with 10-nm channel spacing, the input power in each channel must be kept below 3 mW for a power penalty not exceeding 0.5 dB.

3.5.4 Stimulated Brillouin Scattering

Stimulated Brillouin scattering (SBS) is similar to SRS in that it essentially manifests itself through the generation of a Stokes wave of longer wavelength than the incident wave. The main difference (which leads to markedly different system consequences) is that SBS involves acoustic phonons rather than optical phonons, as is the case in SRS.

SBS can be intuitively understood as follows [Mestdagh, 1988]. Consider an incident optical wave propagating along the fiber, and suppose that this wave interferes with a second optical wave slightly downshifted in frequency and propagating in the opposite direction. Through the electrostrictive effect [Ippen, 1972], the coherent superposition of these two waves will give rise to variations of the medium density (i.e., the creation of a sound wave whose frequency is determined by the frequency difference of the two lightwaves). In SBS, one of the two lightwaves is the launched pump (or signal) wave. The other lightwave is created by scattering of this pump wave due to small material density fluctuations randomly distributed along the fiber (thermal fluctuations). When the frequency shift and the direction of the scattered wave are appropriate, the pump wave interferes with the scattered wave and the density variations are amplified through the electrostrictive effect. These variations will cause a small amount of reflection of the pump wave, and this reflected wave will again interfere with the pump, creating even stronger density fluctuations. These stronger density fluctuations will in turn cause stronger pump reflections, and so on: the reflections increase exponentially with the distance until the reflected lightwave (also called the Stokes wave by analogy with SRS) finally exits the fiber.

It is clear from this qualitative explanation that the Stokes wave propagates backward relative to the pump wave. This is in contrast with SRS for which the Stokes wave propagates in both directions down the fiber.

The power of the Stokes wave for a fiber of length L is also given by (3.77), except that g_R is replaced by the Brillouin gain g_B, which is over two orders of magnitude larger than g_R.

The *Brillouin threshold* is defined similarly to the Raman threshold and is given by [Smith, 1972]

$$P_0^{\mathrm{th}} = 42 \cdot \frac{S_{\mathrm{eff}}}{L \cdot g_B} \tag{3.78}$$

If we use the same typical values as those used in the example of Eq.(3.77), and with $g_B = 5.10^{-11}$ m/W, (3.78) gives $P_0^{\mathrm{th}} \cong 1.4\ m$W (or 1.9 mW with $L_{\mathrm{eff}} \cong 22$ km).

Thus, there are many indications that SBS is one of the dominant nonlinear effects in single-mode optical-fiber systems. However, the optical gain bandwidth $\Delta\nu_B$ of SBS is only about 20 MHz at 1.55 μm and varies as λ^{-2} [Ippen 1972, Cotter 1982]. Thus, efficient SBS occurs only for pump lightwaves whose linewidths are less than 20 MHz. For larger pump linewidths $\Delta\nu_{\mathrm{Laser}}$, SBS efficiency decreases as the ratio $\Delta\nu_B/\Delta\nu_{\mathrm{Laser}}$; that is, $g = g_B \cdot \dfrac{\Delta\nu_B}{\Delta\nu_{\mathrm{Laser}}}$ [Agrawal, 1989].

Since high modulation rates produce broad spectra, SBS is expected to be significantly reduced for signal bit rates usually encountered in fiber-optic communication systems. This has indeed been experimentally observed and it has been shown that the SBS threshold can be increased by an order of magnitude for bit rates $B \geq 10$ Gbps. The amount by which the Brillouin threshold power increases depends also on the signal modulation format chosen (i.e., ASK, FSK, or PSK) [Aoki, 1988; Lichtman, 1989]. The maximum reduction factor is 4 for ASK and FSK. For PSK, the SBS gain decreases linearly with B.

Furthermore, the Brillouin shift (i.e., the carrier frequency difference between the pump and the Stokes waves) is given by $\delta\nu_B = 2nV_s / \lambda$, where n is the refractive index (linear), and V_s the sound velocity in the fiber. At 1.55-μm pump wavelength,

$\delta\nu_B \cong 11$ GHz. Thus, for significant SBS effect to occur, the signal wave must have exactly an 11-GHz offset in frequency (i.e., $\delta\lambda \cong 0.1$ nm at 1.55 μm).

In multichannel systems, SBS can lead to undesirable channel crosstalk. It turns out that this occurs only in very densely spaced multiwavelength networks in which the fibers support channels in both directions. The small Brillouin shift $\delta\nu_B$ and its very narrow bandwidth $\Delta\nu_B$ impart a very selective frequency effect to the SBS.

As an overview of the results to be discussed in Chapter 7, nonlinear effects in multichannel fiber-optic systems limit the launched power to a range of 100 mW to a few milliwatts. The dominant limiting nonlinear effect depends, among other parameters, on the number of channels. For a relatively small number of channels (≤ 10) the dominant effects are SBS and FWM (although SBS is negligible for unidirectional channels). For $N > 10$, XPM begins to dominate, and for $N \geq 500$ SRS becomes the limiting effect.

3.6 SUMMARY

Of all transmission media used in today's communication systems, single-mode optical fibers are by far the most attractive in terms of attenuation and bandwidth.

The losses of a fiber link are due to various factors including intrinsic fiber losses, whose values range from 0.2 dB/km to 0.5 dB/km with a minimum occurring at 1.55-μm wavelength, and connecting losses like laser-fiber coupling loss, connector and splice losses.

The intrinsic bandwidth of a single-mode fiber exceeds 20,000 GHz. However, due to dispersion effects, the bit rate distance product associated with a signal propagating along the fiber is limited to much lower values ranging from about 50 Gbps · km up to about 1,000 Gbps · km depending upon the type of light source used. The best performance in terms of the *BL* product is obtained with a signal wavelength of 1.3 μm where the interplay between material and waveguide dispersions can produce a nearly zero overall dispersion.

At first glance, the *BL* product can be improved by increasing the launched optical power into the single-mode fiber. However, we have seen that by trying to do so, a point can be reached where nonlinear optical effects can no longer be neglected. Some of these nonlinear effects can be used profitably to compensate for fiber dispersion (e.g., the formation of optical solitons) or to compensate for fiber attenuation (e.g., optical amplification by stimulated Raman or Brillouin scatterings). On the other hand, these nonlinear effects can also lead to degradation of system performance, especially for multichannel communication systems (a subject which we will discuss in detail in Chapter 7).

REFERENCES

[Bloembergen, 1967] Bloembergen, N. "The stimulated Raman effect," *Am. J. Physics*, Vol. 35, 1967, p. 989.
[Gloge, 1971] Gloge, D. "Dispersion in weakly guiding fibers," *Applied Optics*, Vol. 10, 1971, pp. 2252–2258.

Zakharov, 1972] Zakharov, V. E., et al. "Exact theory of two-dimensional self-focusing and one-dimensional self-modulation of waves in nonlinear media," *Soviet Physics JEPT*, Vol. 34, No. 1, Jan. 1972.

Ippen, 1972] Ippen, E. P., et al. "Stimulated Brillouin scattering in optical fibers," *Applied Phys. Letters*, Vol. 21, 1972, p. 539.

Smith, 1972] Smith, R. G. "Optical power handling capacity of low loss optical fibers as determined by stimulated Raman and Brillouin scattering," *Applied Optics*, Vol. 11, No. 11, 1972, p. 2489.

Hasegawa, 1973] Hasegawa, A., et al. "Transmission of stationary nonlinear optical pulses in dispersive dielectric fibers. Anomalous dispersion," *Applied Phys. Letters*, Vol. 23, No. 3, 1973, pp. 142–144.

Clemmow, 1973] Clemmow, P. C. *An Introduction to Electromagnetic Theory*, New York, NY: Cambridge University Press, 1973.

Stolen, 1978] Stolen, R. H., et al. "Self-phase modulation in silica optical fibers," *Phys. Rev. A*, Vol. 17, 1978, p. 1448.

CRC, 1978] Standard Mathematical Tables, CRC Press, 1978, p. 417.

Stolen, 1979] Stolen, R. H. "Nonlinear properties of optical fibers," in *Optical Fiber Telecommunications*, Edited by S. E. Miller and A. G. Chynoweth, New York, NY: Academic Press, 1979.

Jenkins, 1981] Jenkins, F. A., and H. E. White. *Fundamental of Optics*, New York, NY: McGraw-Hill, 1981, p. 482.

Kaminow, 1981] Kaminow, I. P. "Polarization in optical fibers," *IEEE J. Quant. Electron.*, Vol. QE-17, No. 1, 1981, pp. 15–22.

Marcuse, 1982] Marcuse, D. *Light Transmission Optics*, Van Nostrand Reinhold, 1982.

Cotter, 1982] Cotter, D. "Observation of stimulated Brillouin scattering in low-loss silica fiber at 1.3 *μm*," *Electron. Letters*, Vol. 18, 1982, p. 495.

Ogawa, 1982] Ogawa, K. "Analysis of mode partition noise in laser transmission systems," *IEEE J. Quant. Electron.*, Vol. QE-18, No. 5, May 1982, pp. 849–855.

Senior, 1985] Senior, J. M. *Optical Fiber Communications: Principles and Practice*, New York, NY: Prentice Hall, 1985.

Tajima, 1986] Tajima, K. "Self-amplitude modulation in *PSK* coherent optical transmission systems," *IEEE J. Light. Technology* , Vol. 4, No. 7, July 1986, pp. 900–904.

Geckeler, 1987] Geckeler, S. *Optical Fiber Transmission Systems*, Norwood, MA: Artech House, 1987.

Agrawal, 1988] Agrawal, G., et al. "Dispersion penalty for 1.3 *μm* lightwave systems with multimode semiconductor lasers," *IEEE J. Light. Technology*, Vol. 6, No. 5, May 1988, pp. 620–625.

Heidemann, 1988] Heidemann, R. "Investigations on the dominant dispersion penalties occurring in multigigabit direct detection systems," *IEEE J. Light. Technology*, Vol. 6, No. 11, Nov. 1988, pp. 1693–1697.

Koch, 1988] Koch, T. L., et al. "Semiconductor laser chirping-induced dispersion distortion in high-bit-rate optical fiber communication systems," *Proc. IEEE International Conference on Communications '88 (ICC88)*, June 12–15, 1988, pp. 19.4.1–4.

Elrefaie, 1988] Elrefaie, A. F., et al. "Chromatic Dispersion Limitations in Coherent Lightwave Transmission Systems," *IEEE J. Light. Technology*, Vol. 6, No. 5, May 1988, pp. 704–709.

Miller, 1988] Miller, S. E., and I. P. Kaminow. *Optical Fiber Telecommunications II*, New York, NY: Academic Press, 1988.

Mollenauer, 1988] Mollenauer, L. F., et al. "Demonstration of soliton transmission over more than 4000*km* in fiber loss periodically compensated by Raman gain," *Optics Letters*, Vol. 13, 1988, pp. 675–677.

Mestdagh, 1988] Mestdagh, D. "Les effets non lineaires dans les fibres optiques: applications et limitations en télécommunication," *Nouvelles des Sciences et des Technologies*, Vol. 6, No. 3, Sept. 1988, pp. 73–79.

Shibata, 1988] Shibata, N., et al. "Experimental verification of efficiency of wave generation through four-wave mixing in low-loss dispersion-shifted single-mode optical fiber," *Electron. Letters*, Vol. 24, 1988, p. 1528.

Aoki, 1988] Aoki, Y., et al. "Input power limits of single-mode optical fibers due to stimulated Brillouin scattering in optical communication systems," *IEEE J. Light. Technology*, Vol. 6, 1988, p. 710.

Agrawal, 1989] Agrawal, G. *Nonlinear Fiber Optics*, New York, NY: Academic Press, 1989.

[Lichtman, 1989] Lichtman, E., et al. ''Stimulated Brillouin scattering excited by a modulated pump wave in single-mode fibers,'' *IEEE J. Light. Technology*, Vol. 7, 1989, p. 1.

[Kikushima, 1990] Kikushima, K., et al. ''Statistical dispersion budgeting method for single-mode fiber transmission systems,'' *IEEE J. Light. Technology*, Vol. 8, No. 1, Jan. 1990, pp. 11–15.

[Gnauck, 1990] Gnauck, A. H., et al. *IEEE Photon. Technology Letters*, Vol. 2, 1990, p. 585.

[Cimini, 1990] Cimini, L. J., et al. ''Optical equalization to combat the effects of laser chirp and fiber dispersion,'' *IEEE J. Light. Technology*, Vol. LT-8, No. 5, May 1990, pp. 649–659.

[Chraplyvy, 1990] Chraplyvy, A. R. ''Limitations on lightwave communications imposed by optical fiber nonlinearities,'' *IEEE J. Light. Technology*, Vol. 8, No. 10, 1990, pp. 1548–1557.

[Winters, 1990] Winters, J. H., et al. ''Electrical signal processing techniques for fiber optic communication systems,'' *IEEE J. Light. Technology*, Oct. 1990.

[Benedetto, 1991] Benedetto, S., et al. ''Applications of Trellis coding to coherent optical communication employing polarization shift keying modulation,'' *Electron. Letters*, Vol. 27, 1991, pp. 1061–1063.

[Butcher, 1991] Butcher, P. N., and D. Cotter. *The Elements of Nonlinear Optics*, New York, NY: Cambridge University Press, 1991.

[Lin, 1991] Lin, C. ''Optical Fiber Transmission Technology,'' in *HandBook of Microwave and Optical Components*, Edited by K. Chang, Vol. 4, 1991, pp. 1–35.

[Mollenauer, 1991] Mollenauer, L. F., et al. ''Wavelength division multiplexing with solitons in ultra-long distance transmission using lumped amplifiers,'' *IEEE J. Light. Technology*, Vol. 9, No. 3, March 1991, pp. 362–367.

[Gnauck, 1991] Gnauck, A. H., et al. ''8*Gb/s*-130*km* transmission experiment using *Er*-doped fiber preamplifier and optical dispersion equalization,'' *IEEE Photon. Technology Letters*, Vol. 3, No. 2, Dec. 1991, pp. 1147–1149.

[Gitlin, 1992] Gitlin, R. D., F. H. Jeremiah, and B. W. Wienstein. *Data Communications Principles*, New York, NY: Plenum, 1992 pp. 697–718.

[Wentworth, 1992] Wentworth, R. H., et al. ''Laser mode partition noise in lightwave systems using dispersive optical fiber,'' *IEEE J. Light. Technology*, Vol. 10, No. 1, Jan. 1992, pp. 84–89.

[Inoue, 1992] Inoue, K., et al. ''Influence of fiber four-wave mixing on multichannel *FSK* direct detection transmission systems,'' *IEEE J. Light. Technology*, Vol. 10, No. 3, March 1992, pp. 350–360.

[Moores, 1992] Moores, J. D. ''Ultra-long distance wavelength-division multiplexed soliton transmission using inhomogenously broadened fiber amplifiers,'' *IEEE J. Light. Technology*, Vol. 10, No. 4, April 1992, pp. 482–487.

[Hill, 1992] Hill, M. H., et al. ''Optical polarization division multiplexing at 4*Gb/s*,'' *IEEE Photonics Technology Letters*, Vol. 4, No. 5, May 1992, pp. 500–502.

[Hakki, 1992] Hakki, B. W. ''Evaluation of transmission characteristics of chirped *DFB* lasers in dispersive optical fiber,'' *IEEE J. Light. Technology*, Vol. 10, No. 7, July 1992, pp. 964–970.

Chapter 4
Optical Receivers

The role of the optical receiver is to recover the data that was originally transmitted by the light source and which has been sent through the optical-fiber communication system. Its main components are a photodetector, an electronic preamplifier, and some electronic signal-processing elements. Figure 4.1 shows a block diagram of a representative optical digital data receiver.

The photodetector converts the modulated optical input into an electronic signal through the photoelectric effect. The ideal photodetector would have a high optoelectrical conversion efficiency, add no noise to the signal, not limit the system's operating bit rate, be highly reliable and inexpensive, have a size comparable to the fiber-core size, and, possibly, be electronically compatible with integrated circuits. These requirements are best met with semiconductor photodiodes, which are at present universally employed within optical receivers for fiber systems. Two types of semiconductor photodiodes are commonly used: the so-called PIN photodiode and the avalanche photodiode (APD), which will both be described in the present chapter.

Because the received optical signal impinging on the photodiode is typically weak, the photodetected signal is first amplified by a preamplifier. The main goal of the preamplifier is to provide a signal level compatible with the following circuitry used for further processing. The combination of the photodiode and the preamplifier is referred to as the *receiver front-end*. It is crucial that the noise added by the preamplifier be minimized because the receiver performance is essentially determined by the noise introduced within the front-end. Since there is a tradeoff between noise and bandwidth, the lowest noise preamplifier designs suffer from bandwidth limitation and an equalizer is sometimes used to alleviate the problem of pulse spreading within the front-end. This equalizer is shown as an optional element in Figure 4.1 because there are receiver front-end designs that partially avoid the problem of bandwidth limitation (for instance, the transimpedance front-end discussed in Section 4.2.2.2).

The next step following the (optional) equalizer is to boost the signal with a

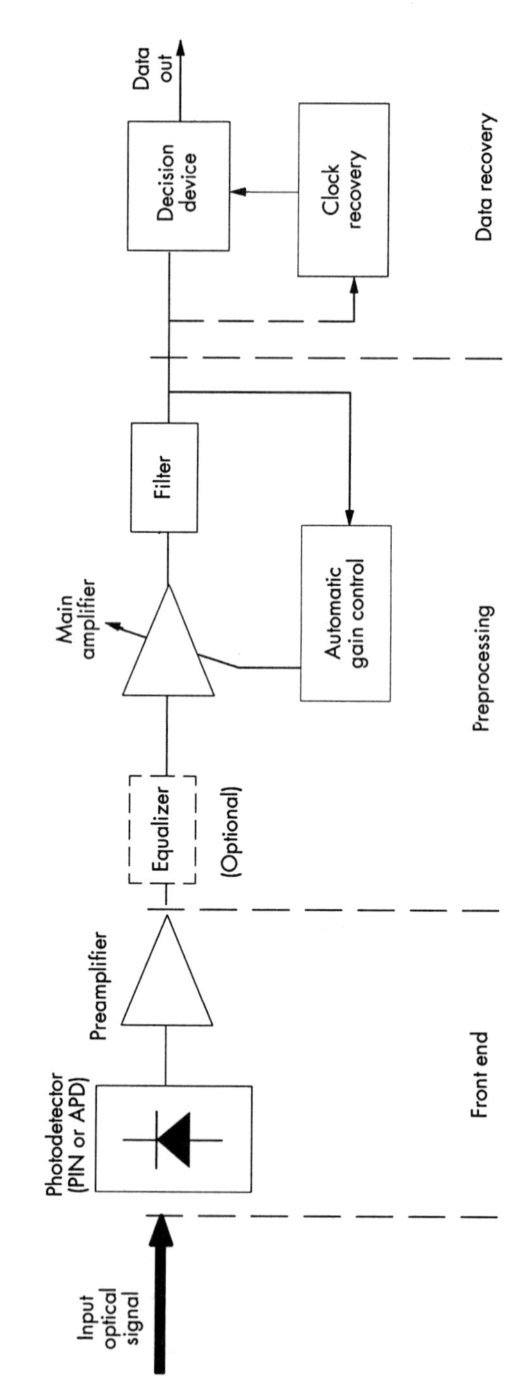

Figure 4.1 Block diagram of a direct detection optical receiver for a digital data link.

postamplifier, usually with an automatic gain control (AGC) that adjusts the gain as a function of the input signal strength. The average output voltage of the AGC is fixed at an adequate level irrespective of the incident average optical power at the receiver. The filter following the postamplifier removes unwanted frequency components that might have been generated to this point.

The last step in the receiver is the actual data recovery. In some low-cost (but poorly performing) optical receivers, data recovery is done asynchronously by merely using a comparator to decide whether a pulse is present or not. This simplified data recovery scheme is practical only for low data rate signals for which the signal pulses have sharp rise and fall times compared to the bit duration. For high data rate systems, a *clock recovery circuit* is necessary to achieve optimum performance. The recovered clock is fed onto the data decision circuit, which samples the input signal at optimum times corresponding to the situation where the signal level difference between bits "1" and bits "0" is maximum (usually in the middle of the bit slot). The output of the decision circuit is the recovered data stream, which may contain some errors.

The present chapter is intended to review the characteristics and performances of optical receivers used in optical-fiber communication systems. Broadly viewed, it is divided into two separate parts. The first part is devoted to the so-called direct detection (DD) optical receivers, while the second part discusses the coherent detection (COH) optical receivers.

With direct detection, the receiver recovers the original data by deciding on the presence or absence of light in each bit slot. Therefore, the only way the original data stream can be imposed on the optical carrier at the transmitter is by modulating the intensity of the emitted light. Systems based on such schemes are referred to as intensity modulation with direct detection (IM/DD). They do not use the optical carrier as a means to support the original data stream and the detection process is essentially a photon counting process.

By contrast, in COH schemes the original data is imposed on the optical carrier by modulating its amplitude, frequency, or phase and it is detected at the receiver by using homodyne or heterodyne techniques, well-known in the context of radio or microwave communication systems.

Although most current systems usually employ a IM/DD scheme, there are several motivations for using COH schemes when dealing with multiaccess optical-fiber networks. Indeed, they not only allow for improved receiver sensitivity, which results in an enhanced network power budget (thereby increasing the number of nodes connected to the network), but also for more efficient use of the available fiber network bandwidth. This issue will be explored in Chapter 7 where we will look at wavelength division multiple access (WDMA) networks.

The present chapter is organized as follows. Section 4.1 introduces the basic concepts behind the two types of photodetectors; namely, the PIN and the avalanche photodiodes. Section 4.2 is devoted to DD receivers. First the basic detection theory that introduces the concept of bit error rate (BER) is considered. This is followed by an analysis of the various noise sources that limit the performance of DD receivers. Performance degradations under nonideal conditions such as a nonzero extinction ratio, fiber dispersion effects and timing jitter are discussed in the succeeding sections. Finally, a review of the performance of DD optical receivers in actual fiber communication systems and experiments is presented.

Section 4.3 is devoted to COH receivers. First the basic concept and the improved performance of coherent optical detection as compared to DD are highlighted. Next, modulation and demodulation techniques are discussed, and the receiver sensitivities obtained with various modulation/demodulation schemes are presented. This is followed by a discussion of distinct receiver performance degradation mechanisms such as the finite laser linewidth, the intensity noise, and the polarization mismatch. The section ends with a review of recent system experiments and field trials using coherent detection. The important aspects of optical receivers within multiaccess networks are summarized in Section 4.4.

4.1 SEMICONDUCTOR PHOTODETECTORS

There are essentially two types of semiconductor photodetectors commonly used in optical-fiber communication systems: the PIN and the APD photodiodes. They are described separately in the following two sections.

4.1.1 PIN Photodiodes

The most prevalent photodetectors used in early lightwave communication systems were reverse-biased *p-n* (homo-) junctions [Smith 1980, Personick 1981]. In these devices, photons which have an energy greater than or equal to the bandgap of the semiconductor compound (i.e., $h\nu \geq E_g$) are absorbed and electron-hole pairs are generated. The pairs generated inside the depletion region are separated and drift rapidly, under the influence of the large electric field created across the junction. This effect produces the useful photodetected current in the external circuit used for further signal processing and eventually data recovery.

A problem, however, with a reverse-biased *p-n* junction is that incident light is also absorbed outside the depletion region. Since there is no voltage gradient outside the depletion region, electrical carriers generated there will stay there until they either recombine via nonproductive processes or diffuse towards the depletion region boundary. Diffusion is an inherently slow process ($v_g \leq 1$ μm/ns) and limits the operating speed of a *p-n* junction to about 1 Gbps at the most.

Although the diffusion problem can to some extent be overcome by increasing the applied voltage to widen the depletion region and to narrow the outside regions, another approach has been proven to be much more effective.

The idea consists of introducing a third layer of very lightly doped (i.e., nearly intrinsic) semiconductor material between the p- and n-doped regions. Such a structure is referred to as the PIN (P-doped, Intrinsic, N-doped) photodiode and is shown in Figure 4.2. It is seen that the width W of the depletion region, where absorbed photons create useful electron-hole pairs, is considerably enlarged so that most of the incident optical power is absorbed inside it, thereby eliminating the major drawback of a single p-n junction.

A further improvement is obtained by using a double-heterostructure design similar to the case of semiconductor lasers discussed in Chapter 2. In this way, the p region through which the incident light travels before reaching the depletion region can be made transparent to the light wavelength (wider bandgap), so that most of the power arrives in the middle i layer. In addition, the depletion region can be made of a semiconductor compound having a large absorption coefficient at the wavelength of the incident light. This allows to narrow the i layer so that the device's response time is as short as possible (because it takes less time for the carriers to sweep across the depletion region). At the same time, high quantum efficiency is achieved. Typical values of the response time of PIN photodiodes are in the range of 30 to 50 ps, to which corresponds a bandwidth of about 3 to 5 GHz (a record of up to 70 GHz has been obtained with optimized PIN photodiodes [Tucker, 1986]).

4.1.1.1 *Quantum Efficiency and Responsivity*

An ideal PIN photodiode should produce one electron-hole pair in the depletion region for every photon within the incident optical signal. In practice, however, a departure from this ideal case is observed. The efficiency of optical-to-electrical conversion is characterized by the fundamental quantity η, called the *quantum efficiency*, and defined as [Senior, 1985]

$$\eta = \frac{\text{Electron-hole pair generation rate}}{\text{Incident photon rate}} \tag{4.1}$$

In PIN photodiodes, η is always less than unity since not all of the incident photons are absorbed to create electron-hole pairs.

The value η is strongly dependent on the wavelength of the incident signal (λ) and on the material used to make the photodiode. The dependence of η on the wavelength of the incident light λ enters through the absorption coefficient $\alpha(\lambda)$ of the depletion region. Assuming antireflection coating on the input facet of the PIN photodiode, the amount of power that passes through the depletion region of depth W without being absorbed is $P_{\text{tr}} = \exp\{-\alpha(\lambda)\text{W}\}P_{\text{i}}$, where P_i is the incident optical power. The absorbed power within the depletion region is thus given by

$$P_{\text{abs}} = P_i - P_{\text{tr}} = [1 - e^{-\alpha(\lambda)W}] \cdot P_i \tag{4.2}$$

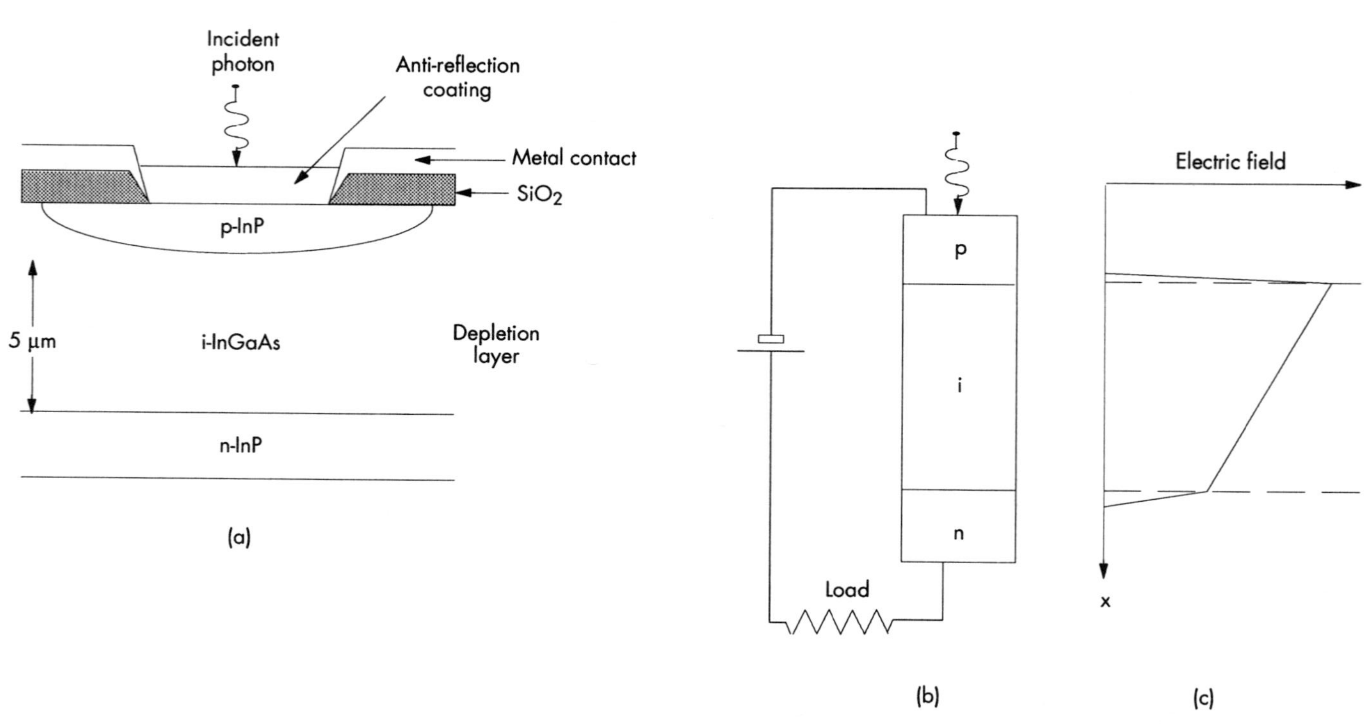

Figure 4.2 (a) Layer structure of a 1.3-/1.55-µm PIN photodiode, (b) reverse-biased with a load resistor, and (c) the internal electric field distribution.

and using (4.1), we have

$$\eta = \frac{P_{\text{abs}}}{P_i} = 1 - e^{-\alpha(\lambda)W} \tag{4.3}$$

Equation (4.3) shows that η becomes zero when $\alpha = 0$, but approaches unity when $\alpha W >> 1$.

Figure 4.3(a) shows the absorption coefficient $\alpha(\lambda)$ as a function of the wavelength for several of the major photodiode materials commonly used in lightwave communication systems. It is seen that α drops to zero for wavelengths $\lambda > \lambda_c$ simply because the energy of photons with such long wavelengths becomes less than the bandgap of the semiconductor compound and therefore cannot be absorbed. The wavelength λ_c is referred to as the *cutoff wavelength* of the photodiode.

As it appears from (4.1), the quantum efficiency does not involve the wavelength of the incident optical signal. In order to directly relate the generated photocurrent with the wavelength of the incident optical signal, another quantity, called the *responsivity* $\Re$, is often of more use in characterizing the performance of the photodiode. It is defined as

$$\Re = \frac{I}{P_i} \ [A/W] \tag{4.4}$$

where I is the generated photocurrent.

The relationship between $\Re$ and η can be obtained by remarking that

$$\eta = \frac{\text{Carrier generation rate}}{\text{Incident photon rate}} = \frac{(I/e)}{(P_i/h\nu)} = \frac{h\nu}{e} \cdot \Re \tag{4.5}$$

where (4.4) was used.

The responsivity is thus given by

$$\Re = \eta \frac{e}{h\nu} \cong \eta \frac{\lambda}{1.24} \tag{4.6}$$

where $\lambda = c/\nu$ is expressed in μm.

It may be noted that $\Re$ is directly proportional to η and λ. The linear dependency on η is obvious. The linear dependency on λ results simply from the fact that the same photocurrent can be generated with photons of reduced energy.

Figure 4.3(b) shows the responsivity of the materials of Figure 4.3(a) as a function of the wavelength. It is seen that the most efficient semiconductor compounds for the 1.3- and 1.5-μm wavelength bands are Ge ($\Re \approx 0.7$ A/W) and InGaAs ($\Re \approx 0.8$ A/W), while Si ($\Re \approx 0.5 - 0.6$A/W) is the best material for the 0.8-μm band.

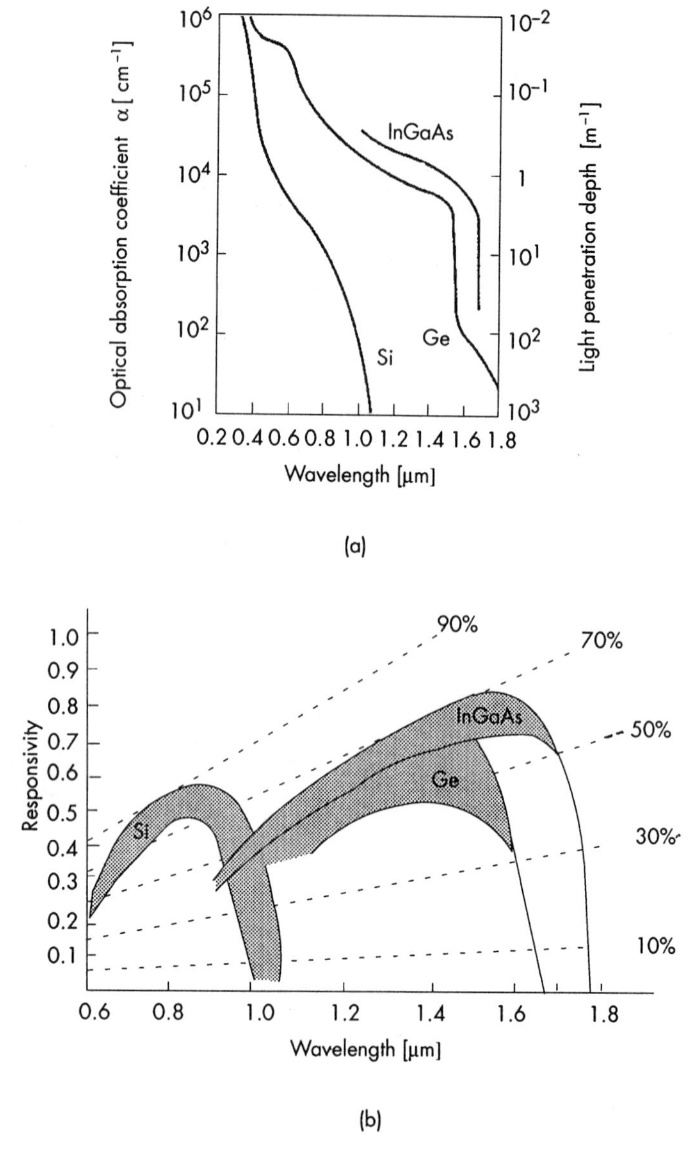

Figure 4.3 (a) The absorption coefficient $\alpha(\lambda)$ and the light penetration depth as a function of the wavelength and (b) the corresponding responsivity $\Re$. [After Green, 1993.]

4.1.2 Avalanche Photodiodes (APDs)

The second major type of photodetector for optical communication systems is the avalanche photodiode (APD). The idea behind the APD is to amplify the generated photocurrent inside the photodetector itself before it reaches the electrical preamplifier input. The internal current gain is obtained by what is known as *impact ionization*: whereby an electron (or hole) generated by the absorption of an incident photon can acquire sufficient kinetic energy under a strong applied reverse-bias voltage to generate a new carrier through impact ionization. This new carrier can be further accelerated to produce still other carriers, resulting in an avalanche process. To avoid avalanche breakdown with the current growing to extremely large values irrespective of the incident power level, the applied voltage is typically set to 10% below that which would produce the breakdown condition.

The structure of an APD differs from that of a PIN in that an additional layer is added where the impact ionization process occurs. This layer is referred to as the gain region or the avalanche multiplication region.

Figure 4.4 shows the structure of an InGaAsP/InP APD and the internal electric field, which is produced by a reverse-bias voltage of typically tens to hundreds of volts (this is to be compared with PINs that require only few volts). It is seen that the i layer still acts as the depletion region where most of the incident photons are absorbed and generate the primary electron-hole pairs. The strong electric field in the gain region results from suitable doping of both the n^+- and the added p layers, which gives rise to a strong charge density of opposite polarity at the junction. It is the resulting strong electric field inside the gain region that allows the generation of secondary electron-hole pairs responsible for the current gain.

Referring to (4.4), it is clear that the responsivity of an APD is enhanced by the current gain corresponding to the number of carriers produced by a single incident photon. However, this is only valid in the sense of ensemble averaging. In fact, the avalanche process is intrinsically noisy and results in an instantaneous gain $M(t)$ that fluctuates around its average value $<M>$ ($<M>$ is typically between 30 and 100).

The instantaneous responsivity of an APD can be written as

$$\Re_{\text{APD}}(t) = \Re_1 \cdot M(t) \tag{4.7}$$

where $\Re_1$ is the device's responsivity for unity avalanche gain.

Therefore, the instantaneous photocurrent generated by the APD can be written as

$$I_{\text{APD}}(t) = \Re_1 M(t) \cdot P_i \tag{4.8}$$

This current fluctuates around an average value given by

$$<I_{\text{APD}}> = \Re_1 \cdot <M> \cdot P_i \tag{4.9}$$

The implications of the APD gain noise on system performance will be considered in Section 4.2.2.

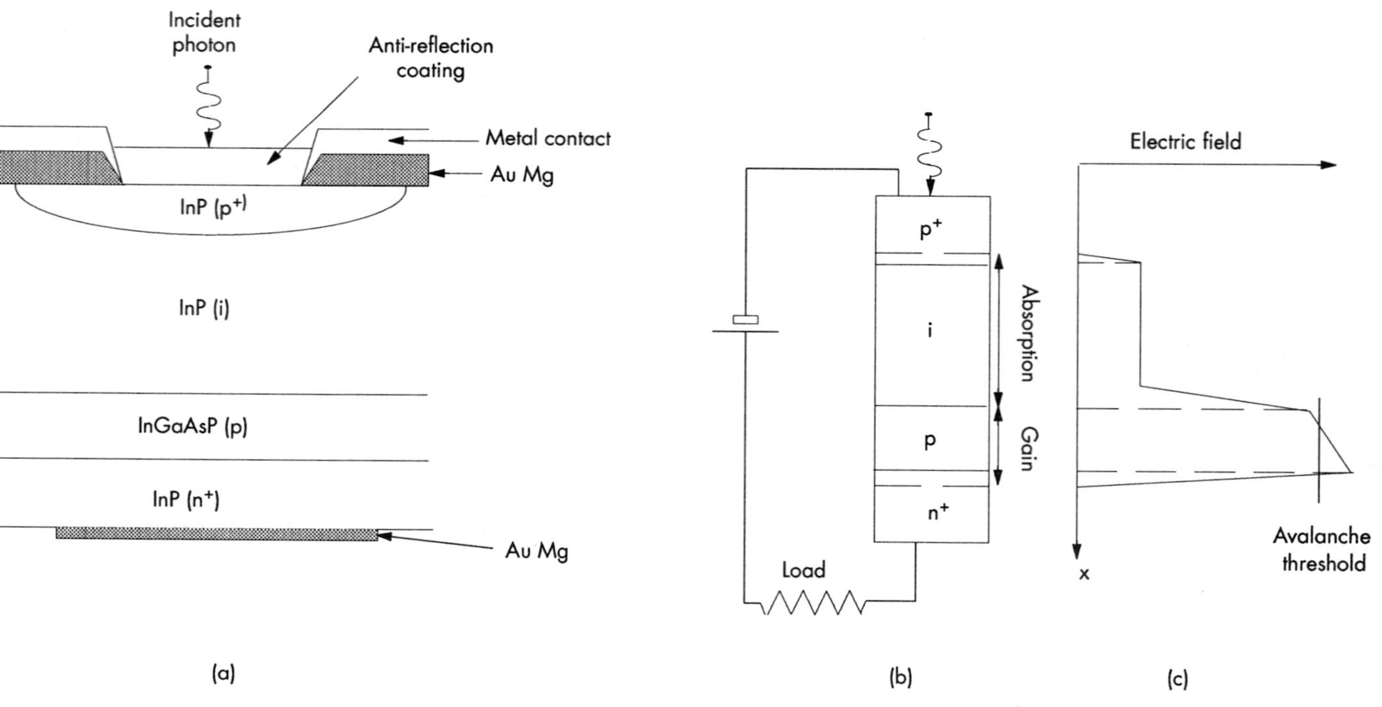

Figure 4.4 (a) Layer structure of an APD for 1.3-/1.55-μm wavelength detection, (b) reverse-biased with a load resistor and (c) the internal electric field distribution.

4.2 DIRECT DETECTION (DD) RECEIVERS

Direct detection (DD) receivers are called so because the incident optical signal is directly converted into an electrical form by one of the photodiode types discussed in the previous section, without signal preprocessing in the optical domain. This is to be contrasted with the coherent (COH) receivers discussed in Section 4.3. (Notice that optical amplifiers, discussed in Chapter 5, can be used in combination with direct detection). In essence, DD receivers are based on direct signal photon counting as bits "0" and "1" are decoded, depending whether or not there is optical power within the allocated bit slot.

Equations (4.4) and (4.9) show that, in principle, if the responsivity of the photodetector has a finite value, then an electrical current is always generated even if the incident optical power approaches to zero. For example, a binary digital signal could (ideally) be recovered by DD without error provided that each bit "1" carries at least η photons while each bit "0" corresponds to the absence of any light. However, this is never the case even for a perfect photodetector, and all receivers require more "photons per bit" than this hypothetical limit to operate reliably.

The reasons are the quantum nature of light and the noise mechanisms that lead to fluctuations in the generated photocurrent. These mechanisms prevent error-free data recovery because a bit "1" can be decided at the receiver while a bit "0" was originally transmitted, and vice versa. The optical receiver will thus produce only an "estimate" of the originally transmitted message.

The performance of an optical receiver is specified by the minimum received optical power (or equivalently, the minimum number of photons per bit) required to achieve a given bit error rate (BER) between the original message at the transmitter and its estimate at the receiver. In telecommunication networks, the BER is typically set at 10^{-9}, or one error per billion bits on average, while for computer networks a BER of 10^{-15} (i.e., six orders of magnitude lower) is often highly desirable. The minimum optical power required to achieve the specified BER is referred to as the *receiver sensitivity*.

Since the receiver sensitivity depends on the BER, let us first derive a general expression of the BER of a digital receiver as a function of the noise levels.

4.2.1 BER

The possible errors in the decoded bit stream will occur at the decision device which compares the input signal level to a preset threshold level (I_D) and issues a bit "1" or "0" depending on whether the sampled level I is above or below the threshold respectively.

Figure 4.5 represents the generated electrical signal, corrupted by noise, at the input of the decision device.

Errors occur if either $I < I_D$ while a bit "1" was originally transmitted, or conversely, if $I > I_D$ while a bit "0" was originally transmitted. The total probability of making an error is the sum of the probability of deciding a "0" given that a "1" was sent,

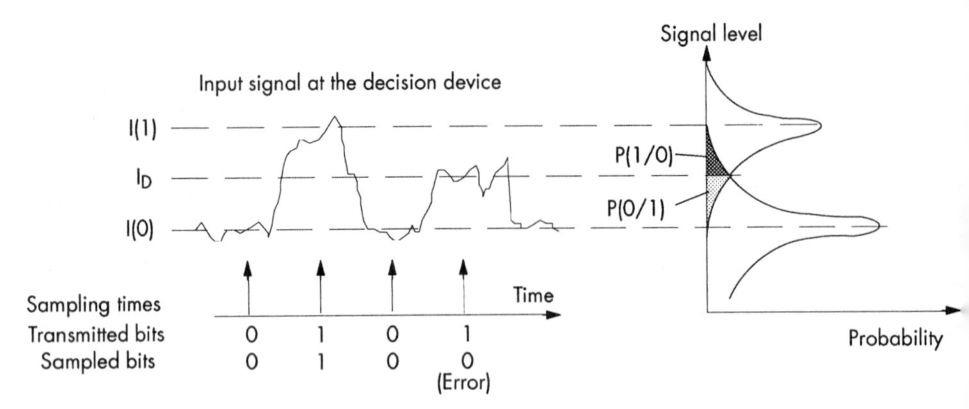

Figure 4.5 Electrical signal at the input of the decision device and the conditional probability distribution of bits "0" and "1."

plus the probability of deciding a "1" given that a "0" was sent. The BER is therefore given by

$$\text{BER} = p(1)\text{Prob}(0/1) + p(0)\text{Prob}(1/0) \tag{4.10}$$

where $p(0)$ and $p(1)$ are the a priori probabilities that a "0" or a "1," respectively, is sent, Prob(0/1) is the conditional probability of deciding a "0" when a "1" is sent, and Prob(1/0) is the conditional probability of deciding a "1" when a "0" is sent. The values $p(0)$ and $p(1)$ depend on the information source and the line code used to encode the data.

Prob(0/1) and Prob(1/0) depend on the noise statistics at the input of the decision device. While various assumptions can be made about these noise statistics, we will assume Gaussian noise statistics. This can be proven to be a reasonable assumption for practical optical receivers [Personick 1973, Yariv 1991].

If $\bar{I}(0)$ and $\bar{I}(1)$ refer to the average input level for bits "0" and "1" respectively, and if σ_0^2 and σ_1^2 are the corresponding variances of their Gaussian distributions, then the conditional probabilities are obtained by

$$\text{Prob}(0/1) = \frac{1}{\sigma_1\sqrt{2\pi}}\int_{-\infty}^{I_D} \exp\left\{-\frac{[\bar{I}(1) - I]^2}{2\sigma_1^2}\right\}dI = \frac{1}{2}\,\text{erfc}\left\{\frac{\bar{I}(1) - I_D}{\sigma_1\sqrt{2}}\right\} \tag{4.11}$$

$$\text{Prob}(1/0) = \frac{1}{\sigma_0\sqrt{2\pi}}\int_{I_D}^{+\infty} \exp\left\{-\frac{[I - \bar{I}(0)]^2}{2\sigma_0^2}\right\}dI = \frac{1}{2}\,\text{erfc}\left\{\frac{I_D - \bar{I}(0)}{\sigma_0\sqrt{2}}\right\} \tag{4.12}$$

where erfc(x) is the complementary error function, defined as

$$\text{erfc}(x) = \frac{2}{\sqrt{\pi}}\int_{x}^{+\infty} exp\{-y^2\}dy \cong \frac{\exp\{-x^2\}}{x\sqrt{\pi}} \tag{4.13}$$

The approximation given by the last term in (4.13) is accurate within one percent for $x \geq 3.0$.

For most messages, the bits "0" and "1" are equally likely, so $p(0) = p(1) = 1/2$. Therefore, by substituting (4.11) and (4.12) in (4.10), the BER is given by

$$\text{BER} = \frac{1}{4}\left\{\text{erfc}\left(\frac{\bar{I}(1) - I_D}{\sigma_1\sqrt{2}}\right) + \text{erfc}\left(\frac{I_D - \bar{I}(0)}{\sigma_0\sqrt{2}}\right)\right\} \tag{4.14}$$

The BER must be minimized to operate the receiver reliably. (4.14) is minimized by choosing I_D so that the arguments of the erfc functions are equal. This gives

$$\frac{\bar{I}(1) - I_D}{\sigma_1} = \frac{I_D - \bar{I}(0)}{\sigma_0} = Q \; or \; I_D = \frac{\sigma_0\bar{I}(1) + \sigma_1\bar{I}(0)}{\sigma_0 + \sigma_1} \tag{4.15}$$

When $\sigma_0 = \sigma_1$, (4.15) provides $I_D = (\bar{I}(0) + \bar{I}(1))/2$, which corresponds to a decision threshold at midpoint. As will be seen in the next section, this situation is valid for PIN receivers whose noise is dominated by thermal noise such that $\sigma_0 \cong \sigma_1$ is independent of the average currents corresponding to bits "0" or "1." By contrast, in the case of APD receivers as well as in the case of coherent receivers discussed in Section 4.3, σ_0 and σ_1 are signal-dependent and the optimum threshold level I_D must be set according to (4.15).

In all cases, the BER with optimum threshold level is obtained by using (4.14) and (4.15) and can be written as

$$\text{BER} = \frac{1}{2}\,\text{erfc}\left(\frac{Q}{\sqrt{2}}\right) \cong \frac{\exp\{-Q^2/2\}}{Q\sqrt{2\pi}} \tag{4.16}$$

where Q is given by (4.15), or also

$$Q = \frac{\bar{I}(1) - \bar{I}(0)}{\sigma_0 + \sigma_1} \tag{4.17}$$

By using the approximate expression of (4.16), it is seen that the BER improves as Q increases, and becomes lower than 10^{-9} for $Q \geq 6$ and lower than 10^{-15} for $Q \geq 8$. Notice that the BER decreases rapidly with a small increase of Q: a change from 6 to 8 in Q produces six orders of magnitude improvement in BER. The next section will provide explicit expressions of σ_0 and σ_1 for DD receivers.

4.2.2 Receiver Noise

The receiver noise is essentially determined by the characteristics of the photodetector and the electrical preamplifier. The role of the photodetector is obvious. The specific role played by the preamplifier is simply due to the fact that it is at the input of the preamplifier that the detected signal is weakest and therefore most subject to contamination by noise.

The purpose of the present section is to describe the two fundamental noise mechanisms; namely, *shot noise* and *thermal noise*, which determine the sensitivity of DD receivers. For each noise mechanism, the cases of PIN and APD photodiodes will be considered separately because APDs suffer from the additional avalanche gain noise.

4.2.2.1 Shot Noise

The shot noise results from the quantum nature of light. This characteristic can be observed in a photon counting experience in which a laser beam impinges on an ideal photodetector that generates one electron-hole pair each time a photon is incident on it. Such experience [Hong, 1986] reveals that within a laser beam of constant mean intensity, the number of photons arriving at the detector follows a *discrete Poisson probability distribution*, so that the probability to have n photons within the laser beam during the time interval $\Delta\tau$ is given by

$$\text{Prob}(n) = \frac{e^{-<n>} \cdot <n>^{n}}{n!} \tag{4.18}$$

where $<n>$ is the mean number of photoelectrons detected during $\Delta\tau$, and is related to the mean intensity of the photocurrent by

$$\bar{I} = <i(t)> = \frac{e<n>}{\Delta\tau} \tag{4.19}$$

where we assumed an ideal photodetector such that $\eta = 1$.

Figure 4.6 shows the Poisson distribution (4.18) for two different values of $<n>$. It is seen qualitatively that as $<n>$ grows larger, the Poisson distribution approaches a Gaussian distribution.

The quantum nature of optical beams sets the fundamental bounds on the performance of any optical communication system. (4.18) can be used to determine σ_0 and σ_1 for both types of photodetector (i.e., PIN and APD). Let us first consider the case of PIN photodiodes.

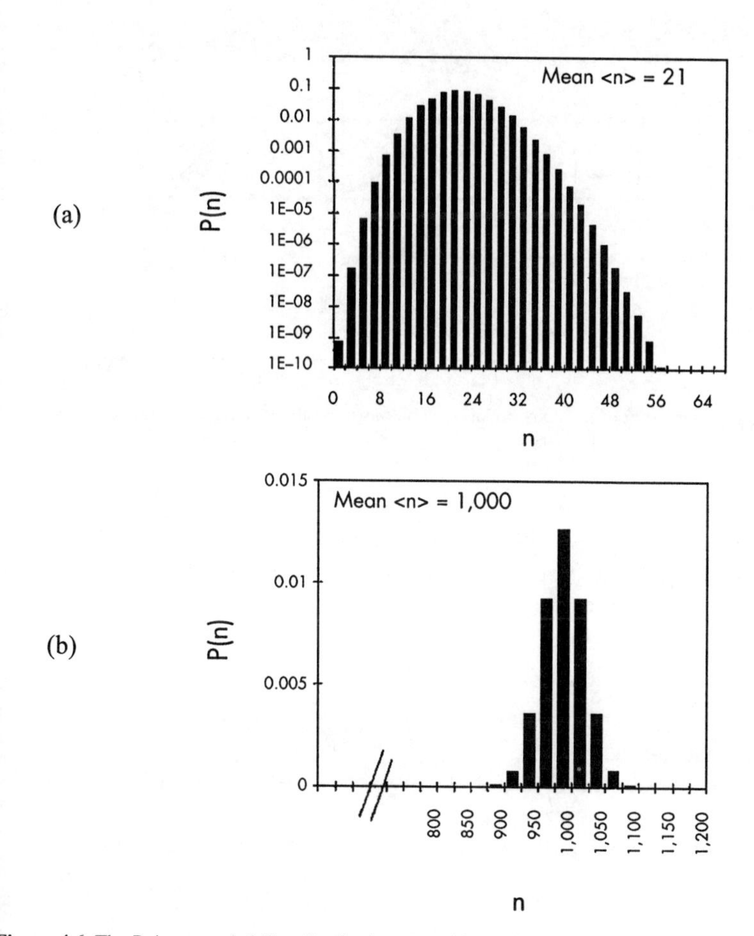

Figure 4.6 The Poisson probability distribution: (a) with small mean $<n> = 21$ and (b) with large mean $<n> =$ 1,000.

PIN Photodiodes

Figure 4.7 qualitatively represents the consequence of the shot noise on the photocurrent generated by a PIN photodiode (considering that $\eta<1$). Every photoelectron generates a current pulse of charge (area) e and time duration τ_p in the external circuit. These current pulses together constitute the instantaneous electric current $i(t)$.

Because of the randomness of the arriving photon stream, this instantaneous current fluctuates. In order to characterize the statistical properties of these fluctuations (i.e., the mean and the variance), let us assume, for instance, a nonreturn-to-zero (NRZ) optical

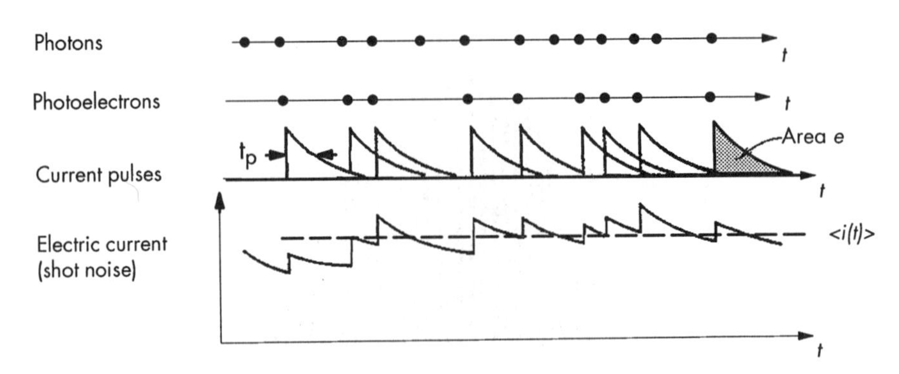

Figure 4.7 Schematic representation of the consequence of shot noise on the electrical current generated by a PIN photodiode. [After Saleh, 1991.]

signal with bit duration T. This case will allow us to derive σ_0 and σ_1 in a straightforward manner by using the remarkable property of the Poisson probability distribution, which states that the variance is equal to the mean, such that

$$<[n - <n>]^2> = <n> \tag{4.20}$$

where the brackets indicate the average.

A more rigorous treatment of the shot noise power is reported in Appendix D [Marcuse, 1980].

Assuming an ideal PIN photodiode, the instantaneous current measured during the bit duration T is

$$i_{0,1}(t) = \frac{en_{0,1}}{T} \tag{4.21}$$

where the indices 0 and 1 refer to bits "0" and "1," respectively.

The average current for bits "0" and "1," $\bar{I}_{0,1}$, is given by (4.19) with $\Delta\tau = T$. The mean-square current fluctuations for bits "0" and "1" are obtained as

$$\sigma^2_{\text{shot }0,1} = <[i_{0,1}(t) - \bar{I}_{0,1}]^2> = \frac{e^2}{T^2}<[n_{0,1} - <n_{0,1}>]^2> = \frac{e^2<n_{0,1}>}{T^2} = \frac{e}{T}\bar{I}_{0,1} \tag{4.22}$$

where (4.19), (4.20), and (4.21) were used.

Since the incident NRZ optical signal has a bit period of T, its bandwidth Δf is $1/2T$.

Therefore, (4.22) becomes

$$\sigma^2_{\text{shot }0,1} = 2e\bar{I}_{0,1}\Delta f = 2e\Re\bar{P}_{0,1}\Delta f \tag{4.23}$$

Equation (4.23) is the well-known expression for the shot noise power in vacuum tubes. Notice that the bandwidth $\Delta f = 1/2T$ in (4.23) corresponds also to the bandwidth of the matched filter associated with rectangular pulses of width T. It can be shown (see Appendix D) that (4.23) is valid in general provided that Δf is understood as the receiver noise bandwidth (i.e., the bandwidth that effectively controls the receiver noise).

It is also noteworthy that all photodiodes generate some current even in the absence of incident light because electron-hole pairs are generated randomly by thermal effect. This unwanted resulting current is referred to as the dark current $\bar{I}_{\text{dark}}$. It can be shown that the shot noise contribution of the dark current can be included in (4.23) by replacing $\bar{I}_{0,1}$ by $\bar{I}_{0,1} + \bar{I}_{\text{dark}}$ [Saleh, 1991].

Therefore, (4.23) becomes

$$\sigma^2_{\text{shot }0,1} = 2e(\Re\bar{P}_{0,1} + \bar{I}_{\text{dark}})\Delta f \tag{4.24}$$

which is the general expression for the shot noise in PIN receivers.

APD Photodiodes

The internal gain in an APD photodiode increases the generated photocurrent. APDs are therefore expected to be much more sensitive than their PIN counterparts. Unfortunately, as we have seen in Section 4.1.2, the avalanche process in APD contributes to additional noise and the actual improvement in receiver sensitivity is considerably reduced. The physical reason behind this added noise is that secondary electron-hole pairs are generated at random times through the process of impact ionization.

Figure 4.8 qualitatively represents the consequence of the shot noise on the photocurrent generated by an APD. This figure must be compared to Figure 4.7: it is seen that, because of the random gain fluctuations, the variance of the APD's shot noise is enhanced.

It can be shown that the shot noise introduced by an APD is given by [Saleh, 1991]

$$\sigma^2_{\text{shot }0,1} = 2e\langle M\rangle^2 F(M)(\Re\bar{P}_{0,1} + \bar{I}_{\text{dark}})\Delta f \tag{4.25}$$

where $F(M)$ is the excess noise factor, which is a measure of the extra noise added by the avalanche process. An explicit expression of $F(M)$ is given by

$$F(M) = k\langle M\rangle + (1-k)\left[2 - \frac{1}{\langle M\rangle}\right] \tag{4.26}$$

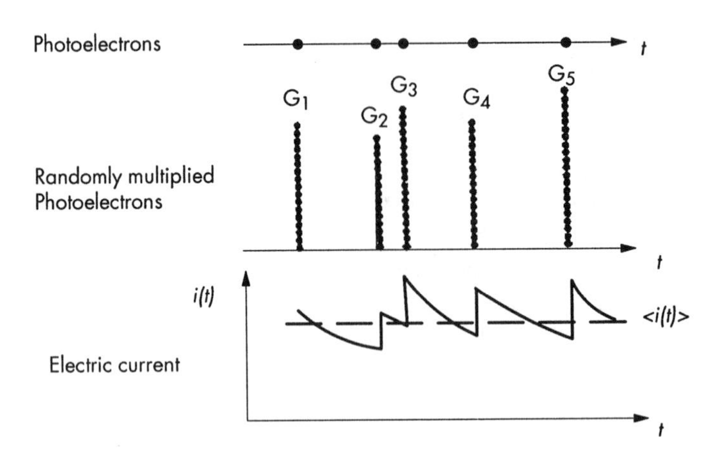

Figure 4.8 Schematic representation of the consequence of shot noise on the photocurrent generated by an APD. [After Saleh, 1991.]

The factor k is referred to as the *ionization coefficient ratio* and is a measure of the ability of a hole or an electron to generate other carrier pairs in the avalanche process. Typical ranges of values of k are found in Table 4.1. The factor k should be as small as possible to achieve the best APD performance.

An approximate form of $F(M)$ is often used in place of (4.26):

$$F(M) = \langle M \rangle^{x} \tag{4.27}$$

with typical ranges of x shown in Table 4.1.

4.2.2.2 Thermal Noise

The current generated by the photodiode is generally converted to a voltage by means of a load resistor R_L. This voltage is amplified by the preamplifier whose output serves as input to the electrical signal processing circuitry that actually recovers the transmitted

Table 4.1
The Ionization Coefficient k and the x Parameter for Various APD Compounds

Semiconductor Compound	k	x
Silicon	0.02–0.04	0.3–0.5
Germanium	0.7–1.0	1.0
InGaAs	0.3–0.5	0.5–0.8

data. The load resistor introduces its own noise (known as *thermal noise*, but sometimes also called *Johnson noise* or *Nyquist noise* after the two scientists who first studied it) [Agrawal, 1992].

The thermal noise is a manifestation of the random motion of electrons in any conductor whose temperature is above absolute zero. This random thermal motion of electrons in a resistor or in any device with an associated resistance leads to a current that fluctuates around a zero mean but with a variance of

$$\sigma_{\text{th}}^2 = \langle i_{\text{th}}^2 \rangle = \frac{4kT\Delta f}{R_L} \tag{4.28}$$

where k is the Boltzmann constant (not to be confused with the ionization coefficient of APDs in (4.26)), T is the absolute temperature in kelvin, Δf is the receiver bandwidth, and R_L is the load resistor.

This thermal noise adds to the useful signal current and degrades the performance of practical DD receivers.

As can be seen from (4.28), the thermal noise can be reduced by using a load resistor R_L as large as possible. This is often applied in the so-called *high-impedance* front-end receiver, as shown in Figure 4.9(a). Although the high-impedance front-end design provides the best receiver sensitivity by reducing the thermal noise, its main drawback is its limited bandwidth, given by $\Delta f = (2\pi R_L C)^{-1}$ where C is the total capacitance that includes contributions from the photodiode, parasitic, and preamplifier. To overcome this limitation, an equalizer that attenuates low-frequency components of the signal and boosts high-frequency components is sometimes used.

Another disadvantage of the high-impedance front-end, besides requiring an equalizer, is that the dynamic range of the receiver is limited. This is because the low-frequency components that are integrated can quickly saturate the preamplifier output.

Where the receiver sensitivity is not a major concern, the receiver bandwidth limitation can be overcome by simply decreasing R_L (e.g., $R_L = 50\Omega$). Such front-end designs are referred to as *low-impedance* front-ends.

A distinct design approach, particularly useful in multiaccess networks, to achieve high receiver sensitivity as well as large bandwidth is the *transimpedance front-end* as shown in Figure 4.9(b). Its dynamic range is also improved compared with high-impedance front-ends. This feature is of particular importance in multiaccess optical-fiber networks since a particular receiver in the network must be able to support connections to a variety of other transmitters. The optical signals emanating from different transmitters will, in general, experience a wide range of network losses, so that the received power level will vary considerably.

In the transimpedance front-end, the photocurrent I_{in} is converted to an output voltage V_{out} using the load resistor connected as a feedback resistor around an inverting amplifier. The transfer function $H(f)$ for the transimpedance configuration is readily obtained as [Senior, 1985]

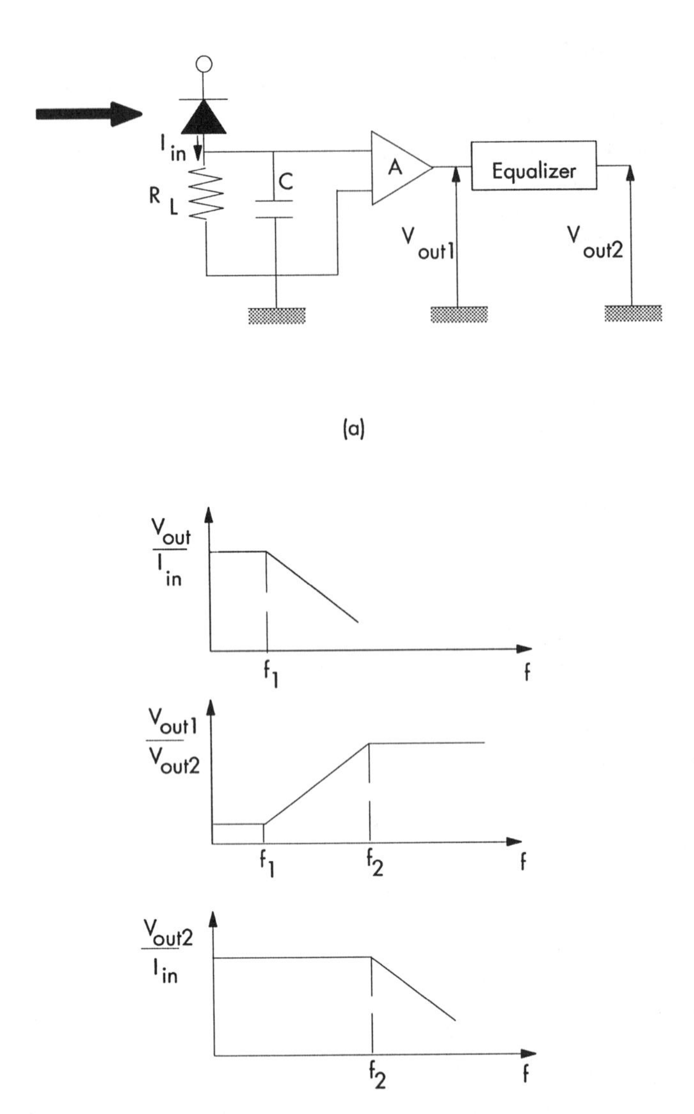

Figure 4.9 Receiver front-end configurations: (a) the high- and low-impedance front-end and (b) the transimpedance front-end.

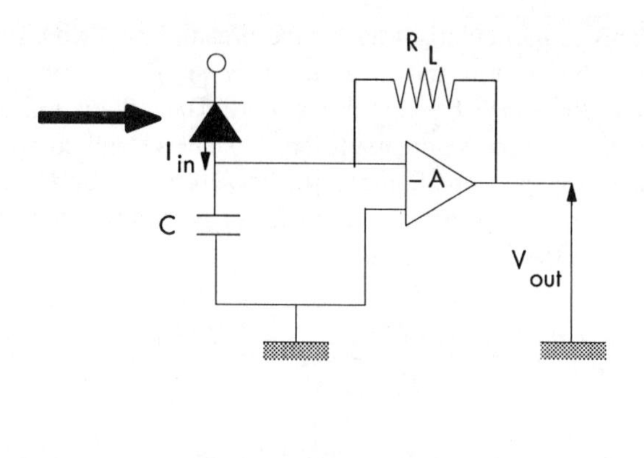

(b)

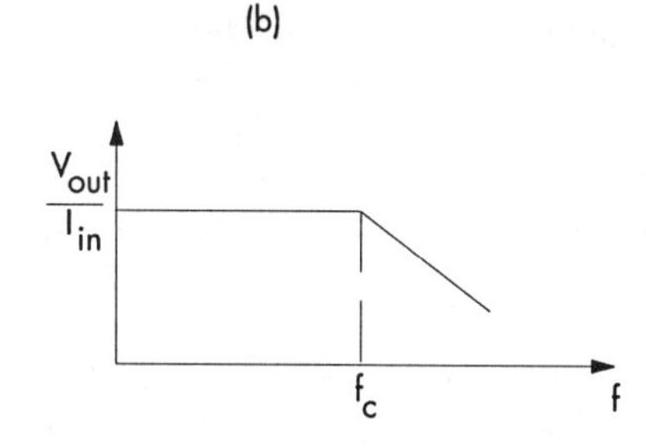

Figure 4.9 (continued)

$$H(f) = \frac{V_{\text{out}}(f)}{I_{\text{in}}(f)} \cong \frac{-R_L}{1 + j\dfrac{fR_LC}{A}} \text{ [V/A]} \tag{4.29}$$

where A is the amplifier gain.

Hence, from (4.29), the permitted electrical bandwidth of the transimpedance configuration (without equalization) is given by

$$\Delta f \leq \frac{A}{R_LC} \tag{4.30}$$

Therefore, by making A and R_L as large as possible (while maintaining the stability of the closed-loop transimpedance configuration), the bandwidth can be enhanced while the thermal noise can be reduced to approach that of the high-impedance front-end.

Equation (4.28) expresses the thermal noise generated by the load resistor alone. In actual receivers, however, several other electrical components may contribute to additional noise. This is particularly true for the preamplifier, which invariably adds its proper noise. This noise is referred to as the *preamplifier noise* and depends on the front-end design and the type of preamplifier used. The calculation of the actual noise introduced by a particular type of preamplifier is extensive and complex. A comprehensive treatment of the subject can be found in [Miller, 1988]. A simple approach to account for a particular amplifier noise is to introduce a factor F_n, referred to as the *amplifier noise figure*, and modify (4.28) so as to obtain

$$\sigma_{th}^2 = \frac{4kT\Delta f}{R_L} + \langle i_{amp}^2 \rangle = \frac{4kTF_n\Delta f}{R_L} \tag{4.31}$$

where $\langle i_{amp}^2 \rangle$is the mean-square noise current introduced by the preamplifier.

Physically, F_n can be understood as the factor by which the thermal noise generated in the load resistor is enhanced by the preamplifier. The value F_n is usually specified in units of dB. Typically, for a good low-noise preamplifier $F_n = 3$ dB, implying a total noise that is twice the noise of the load resistor alone.

As will be seen in the next section, the thermal noise given by (4.31) by far dominates the shot noise and is therefore the major limiting factor for the sensitivity of practical direct detection receivers.

In actual design, the frequency dependence of F_n must be carefully considered to obtain optimum performance. The frequency dependence of F_n can be determined by noting from (4.31) that F_n can be expressed as

$$F_n = 1 + \frac{R_L}{4kT\Delta f}\langle i_{amp}^2 \rangle \tag{4.32}$$

A general expression for $\langle i_{amp}^2 \rangle$ can be written as (see Appendix E) [Personick 1973, Smith 1982, Kasper 1988]

$$\langle i_{amp}^2 \rangle = k_1\Delta f + k_2(\Delta f)^2 + k_3(\Delta f)^3 \tag{4.33}$$

where k_1, k_2, and k_3 are constants that depend on the amplifier technology (e.g., FET or bipolar transistors) and the pulse shape entering and leaving the fiber. The actual values of k_1, k_2, and k_3 for several amplifier types and pulse shapes are summarized in Appendix E.

Figure 4.10 shows a comparison of the mean-square noise currents $\langle i_{amp}^2 \rangle$ for several typical field-effect transistor (FET) front-ends and one bipolar junction transistor (BJT) front-end. It is seen that GaAs metal-semiconductor FETs (MESFETs) have superior noise properties over a large range of bandwidth from about 10 Mbps up to 10 Gbps and more. Silicon BJTs and metal-oxyde-semiconductor FETs (MOSFETs) have, however, comparable noise performance from about 100 Mbps. Silicon JFETs are not suitable

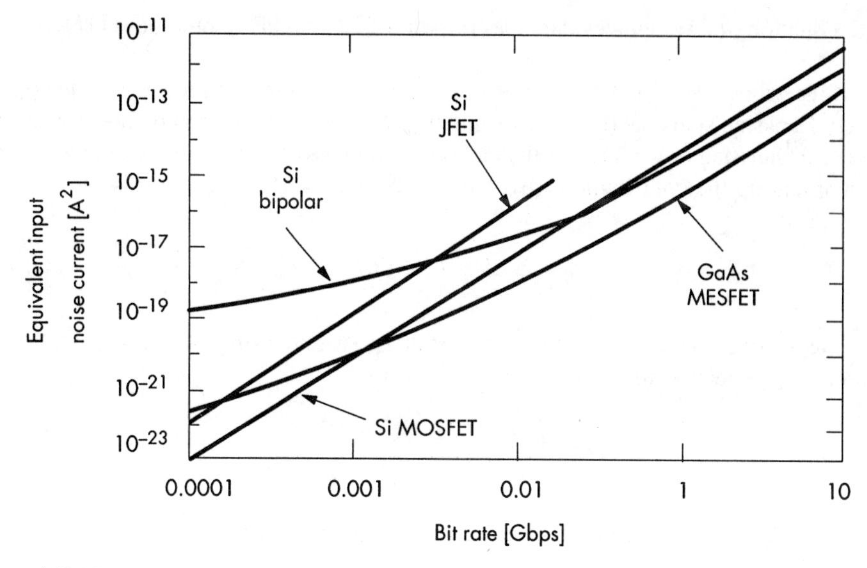

Figure 4.10 Mean-square noise currents for several front-end amplifiers.

for high-bit-rate systems and are practical only for systems with bit rates ≤100 Mbps. At low bit rates, the silicon MOSFET is slightly advantageous over all amplifier types [Powers, 1993].

By combining (4.32) and (4.33) with (4.31), the frequency dependency of the receiver thermal noise is obtained as

$$\sigma_{\text{th}}^2 = K_1\Delta f + K_2(\Delta f)^2 + K_3(\Delta f)^3 \tag{4.34}$$

where K_1, K_2, and K_3 can be deduced from the values given in Appendix E.

4.2.3 Sensitivity of Direct Detection Receivers

Having determined the shot noise and thermal noise as a function of the receiver parameters, we can now proceed to the calculation of the minimum optical power a receiver needs to operate with a BER below a specified value. Since the BER is related to the factor Q by (4.16), a simple way to obtain the receiver sensitivity is to express Q as a function of the incident optical power and to include this expression in (4.16).

For instance, consider the case in which the bit "0" carries no optical power ($\overline{P}_0 = 0$) and bit "1" carries the power $\overline{P}_1$. Hence, $\overline{I}(0) = 0$ and $\overline{I}(1)$ is given by

$$\overline{I}(1) = \Re\langle M\rangle\overline{P_1} = 2\Re\langle M\rangle\overline{P} \tag{4.35}$$

where $\overline{P}$ is the average received power defined as $\overline{P} = (\overline{P}_0 + \overline{P}_1)/2 = \overline{P}_1/2$.

Equation (4.35) includes the cases of both PIN and APD photodiodes (<M> = 1 for PINs) .

Since shot noise and thermal noise are independent random processes with approximately Gaussian statistics (i.e., we assume that there is enough light to consider the shot noise as a Gaussian process rather than a Poisson process), the total noise can be obtained by summing up the contributions given by (4.25) and (4.31). The result is

$$\sigma_{0,1}^2 = \sigma_{\text{shot } 0,1}^2 + \sigma_{\text{th}}^2 = 2e{<}M{>}^2 F(M)(\Re \overline{P}_{0,1} + \bar{I}_{\text{dark}})\Delta f + \frac{4kTF_n\Delta f}{R_L} \tag{4.36}$$

Neglecting the dark current to the shot noise and using (4.17) and (4.35), the parameter Q is related to $\bar{P}$ by

$$Q = \frac{\bar{I}(1)}{\sigma_0 + \sigma_1} = \frac{2{<}M{>}\Re \overline{P}}{\sigma_{\text{th}} + (\sigma_{\text{shot } 1}^2 + \sigma_{\text{th}}^2)^{1/2}} \tag{4.37}$$

Solving (4.37) for a given value of Q, we obtain [Smith, 1980]

$$\overline{P} = \frac{Q}{\Re}\left(eQ \cdot \Delta f \cdot F(M) + \frac{\sigma_{\text{th}}}{{<}M{>}}\right) \tag{4.38}$$

Equation (4.38) shows how $\overline{P}$ depends on various receiver parameters. We will now discuss how $\overline{P}$ can be optimized to operate the receiver with a given BER.

Consider first the case of PIN receivers by setting $M = 1$. With an ideal PIN receiver for which $\sigma_{th} = 0$, the receiver sensitivity is directly obtained from (4.38) as

$$\overline{P}_{\text{PIN}}\,(\text{shot} - \text{noise}) = \frac{eQ^2\Delta f}{\Re} \tag{4.39}$$

Equation (4.39) shows the linear dependency of $\overline{P}_{\text{PIN}}$ on the bandwidth Δf, which is a general feature of shot-noise-limited receiver ($\sigma_{th} = 0$). Using $Q = 6$ in order to obtain a BER of 10^{-9}, $\Re$ = 1A/W and Δf = 1 GHz, (4.39) provides $\overline{P}_{\text{PIN}}$ (shot − noise) = −52dBm.

In practice, the shot-noise limit is never achieved with direct detection receivers. This is because thermal noise always by far dominates the shot noise for such receivers. Consider the following receiver parameters as an example: $\Re$ = 1 A/W, Δf = 1 GHz, R_L = 1 kΩ, F_n = 3dB and $\overline{P}$ = 1μW (−30 dBm). Equations (4.23) and (4.31) provide σ_{shot} = 18 nA and σ_{th} = 180 nA. In all practical cases $\sigma_{shot} \ll \sigma_{th}$ and (4.38) gives the thermal noise receiver sensitivity of a PIN photodiode as

$$\overline{P}_{\text{PIN}}\,(\text{thermal} - \text{noise}) = \frac{\sigma_{\text{th}} Q}{\Re} \tag{4.40}$$

Equation (4.38) can be used to evaluate how APDs improve the receiver sensitivity in the usual case of thermal-noise limit. Since the shot noise for APD is enhanced by a factor $<M>^2 F(M)$, the thermal noise and the shot noise may be of comparable magnitude and (4.38) must be optimized in its general form. Using (4.26) giving the explicit expression of $F(M)$ and (4.38), it can easily be verified that $\overline{P}_{\text{APD}}$ is minimized for an optimum gain $<M>_{\text{opt}}$ given by

$$<M>_{\text{opt}} = k^{-1/2}\left(\frac{\sigma_{\text{th}}}{eQ\Delta f} + k - 1\right)^{1/2} \cong \left(\frac{\sigma_{\text{th}}}{keQ\Delta f}\right)^{1/2} \tag{4.41}$$

to which corresponds a receiver sensitivity given by

$$\overline{P}_{\text{APD}} = [k<M>_{\text{opt}} + 2(1-k)]\frac{eQ^2\Delta f}{\Re} \cong k <M>_{\text{opt}}\frac{eQ^2\Delta f}{\Re} \tag{4.42}$$

where we neglected (1—k) compared to $k<M>_{\text{opt}}$, which is justifiable for 1.3 to 1.55-μm wavelength optical receivers (see Table 4.1 for typical values of k). Equation (4.42) reveals that with InGaAs APDs for which $k = 0.3$–0.5, the receiver sensitivity is typically improved by 5 to 10 dB compared to PIN-based receivers.

The receiver sensitivity is sometimes expressed in terms of the received signal-to-noise ratio (SNR) or in terms of the average number of photons per bit, rather than in terms of Q as in (4.38), (4.39), (4.40), and (4.42).

In the shot-noise limit, $\sigma_0 = 0$ and the factor Q given by (4.17) becomes $Q = \bar{I}(1)/\sigma_1$. Since the SNR is given by $\text{SNR} = \bar{I}^2(1)/\sigma_1^2$, we have the simple relation $Q = \sqrt{\text{SNR}}$, which can be included in (4.38) to express the receiver sensitivity in terms of the SNR.

The SNR can also be expressed in terms of the average number of photons per bit, $\bar{n}_p$, by observing that $\bar{n}_p = \overline{P}/h\nu B$ where $h\nu$ is the energy of each photon in the incident signal and B is the signal bit rate. Since $\text{SNR} = \bar{I}^2(1)/\sigma_1^2 = \bar{I}(1)/2eB$ (recall that $\sigma_1^2 = 2e\bar{I}(1)B$) and $\bar{I}(1) = 2\eta e\bar{n}_p B$, the SNR of DD receivers simply becomes

$$\text{SNR}_{\text{DD}} = \eta\bar{n}_p \tag{4.43}$$

Using the above expressions relating Q, SNR and $\bar{n}_p$, the BER in the shot-noise limit can be expressed as

$$\text{BER} = \frac{1}{2}\operatorname{erfc}\left(\sqrt{\frac{\text{SNR}}{2}}\right) = \frac{1}{2}\operatorname{erfc}\left(\sqrt{\frac{\eta\bar{n}_p}{2}}\right) \cong \frac{e^{-\eta\bar{n}_p/2}}{\sqrt{2\pi\eta\bar{n}_p}} \tag{4.44}$$

Equation (4.44) shows that in order to achieve a BER of 10^{-9} with an ideal receiver (i.e., $\eta = 1$ and $\sigma_{\text{th}} = 0$), $\bar{n}_p = 18$, (or SNR = 12.5 dB).

Notice that the required average number of photons per bit to achieve a given BER in the shot-noise limit can also be obtained directly from (4.18): assume an ideal photodetector that records a "1" when at least one photon is received, and a "0" if not. Then, the only way an error can occur is when an optical pulse is actually transmitted but no photon is detected. With (4.10) and (4.18), we immediately find

$$\mathrm{BER} = p(1)\mathrm{Prob}(0/1) = \frac{1}{2}e^{-2\bar{n}_p} \tag{4.45}$$

For BER $\leq 10^{-9}$, (4.45) provides $\bar{n}_p \leq 10.5$. The discrepancy between $\bar{n}_p = 18$ obtained from (4.44) and $\bar{n}_p = 10.5$ obtained from (4.45) is due to the different assumptions made for the noise statistics. In (4.44), Gaussian noise statistics are assumed, while in (4.45) the exact Poisson statistics for the shot noise are used. As can be seen from Figure 4.6, the Poisson probability distribution approaches a Gaussian distribution when $\bar{n}_p$ ($\langle n \rangle$) is large. For the shot-noise limit, which involves rather small values of $\bar{n}_p$, (4.45) provides the exact value of $\bar{n}_p$. Since this value results from the quantum nature of light, $\bar{n}_p = 10.5$ is often referred to as the *quantum limit* of DD receivers.

In practice, as already mentioned, most DD receivers operate far away from the quantum limit by 20 dB or more because their performance is severely limited by thermal noise. In the thermal noise limit, $\sigma_0 \cong \sigma_1$ and (4.17) provides $Q = \bar{I}(1)/2\sigma_1$. Since the SNR is still given by $\mathrm{SNR} = \bar{I}^2(1)/\sigma_1^2$, we have $\mathrm{SNR} = 4Q^2$. Therefore, to achieve a BER = 10^{-9} in the thermal-noise limit, the SNR must be at least 144 or 21.5 dB. In terms of the number of photons per bit, $\bar{n}_p$ typically ranges between 1,000 and 6,000 to achieve a BER of 10^{-9} with practical DD receivers.

4.2.4 Sensitivity Degradation Mechanisms

The previous section considered the receiver noise as the only basic mechanism that determines receiver sensitivity. In practice, other degradation mechanisms must be taken into account to determine the actual sensitivity of DD receivers. Some of these mechanisms, such as the nonzero extinction ratio, may occur already at the transmitter side. Others may occur when the optical signal propagates through the fiber (e.g., dispersion effects). Still others may occur at the decision circuit itself (e.g., timing jitter).

Because of these additional mechanisms, the receiver sensitivity degrades compared with the value of $\bar{P}$ derived in the previous section. This degradation in the receiver sensitivity is referred to as the *power penalty*.

In what follows, we will discuss the receiver performance degradation and the power penalty associated with the three most prevalent degradation mechanisms; namely, the extinction ratio, the fiber dispersion, and the timing jitter.

4.2.4.1 Extinction Ratio

The extinction ratio r is defined as the ratio of the optical power transmitted for a bit "0" to the power transmitted for a bit "1":

$$r = \frac{\bar{P}_0}{\bar{P}_1} \tag{4.46}$$

With semiconductor lasers, the off-state power depends on the relative values of the bias and the threshold currents. For high-speed operation, the bias current is typically set just above the threshold, resulting in an extinction ratio of $r \approx 0.03$ to 0.1 (i.e., −15 dB to −10 dB).

For a PIN receiver dominated by thermal noise, the receiver sensitivity due to a nonzero extinction ratio can be obtained by modifying (4.37) according to (4.17) and (4.46), and by including the result in (4.40). The result is

$$\overline{P}_{\text{PIN}}(r) = \frac{1+r}{1-r} \cdot \frac{\sigma_{\text{th}} Q}{\Re} \tag{4.47}$$

This equation clearly shows that $\overline{P}_{\text{PIN}}$ $(r \neq 0)$ is increased by a factor $(1 + r)/(1 - r)$ with respect to $\overline{P}_{\text{PIN}}$ $(r = 0)$. This factor is the extinction ratio power penalty and is commonly expressed in decibel units as

$$\text{Penalty}(r) = 10 \cdot \log_{10}\left(\frac{1+r}{1-r}\right) \tag{4.48}$$

Figure 4.11 shows the power penalty as a function of r. To restrict the power penalty to less than about 1 dB, the extinction ratio must satisfy $r \leq 0.1$ (i.e., -10 dB).

For the APD receiver, the extinction ratio affects the optimum APD gain. By including the shot noise contribution to σ_0, it can be shown that the optimum APD gain decreases when $r \neq 0$ [Muoi 1984, Kasper 1988]. As a result, the receiver sensitivity degrades. Compared to the PIN case, the power penalty is larger by about a factor of 2 for the same value of r.

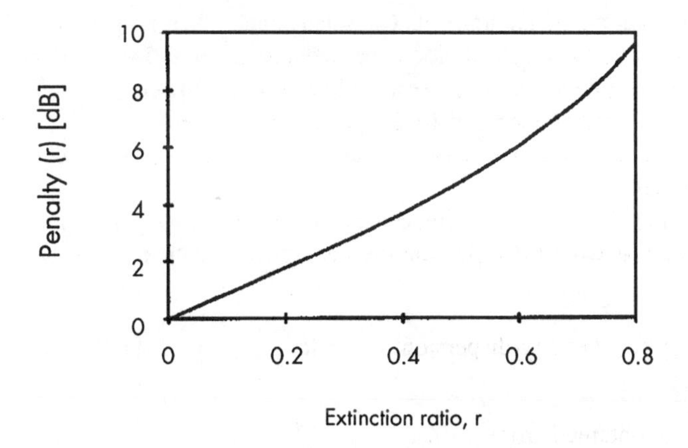

Figure 4.11 Power penalty of DD receivers as a function of the extinction ratio.

4.2.4.2 Fiber Dispersion

As discussed in Section 3.4, the maximum bit rate distance product, BL, achievable with single-mode fibers is limited because of dispersion-induced pulse broadening. Equations (3.55) and (3.56) provide the maximum achievable BL products when single-longitudinal mode (SLM) lasers (chirped pulses) and multilongitudinal-mode (MLM) lasers are used, respectively. These equations are obtained by imposing a maximum dispersion-induced pulse broadening factor (20% of the original optical pulse width) such that a limited amount of the received pulse energy spreads beyond the allocated bit slot.

Systems designed according (3.55) or (3.56) limit the effect of intersymbol interference (ISI). Notice that small ISI can be reduced through (adaptive) equalization [Cimini 1990, Gitlin 1993].

However, by allowing a certain amount of pulse broadening, we implicitly degrade the receiver sensitivity. This is because optical pulse broadening is accompanied by a reduction of pulse energy within the allocated bit slot. Such a reduction decreases the SNR at the decision point. In order to maintain the same SNR as in the absence of pulse broadening, and thereby maintain the same system performance, the receiver requires more average optical power. This increase in the required optical power is referred to as the *dispersion-induced power penalty.*

A general analysis of this power penalty is not tractable analytically as it depends on many details about the transmitter, the fiber, and the receiver. Nevertheless, rough estimations can be obtained by separately considering the case of chirped Gaussian pulses (Section 3.4.1.1) and the case of multilongitudinal-mode pulses exhibiting mode partition noise (Section 3.4.1.2).

Chirped-Gaussian Pulses

A chirped-Gaussian pulse maintains its Gaussian shape when propagating along the fiber but broadens by a factor $T(z)/T_0$ that can be deduced from (3.54). This pulse broadening is accompanied by a decrease of the received pulse peak power. The factor by which the peak power is reduced is identical to the pulse broadening factor, as can be seen from (3.48) for the case of unchirped pulses (it can be shown that the same conclusion is valid for chirped pulses).

By defining the dispersion-induced power penalty as the increase (in dB) in the received power that would compensate the peak-power reduction, we have

$$\text{Penalty(dispersion)} = 10 \cdot \log_{10}\left(\frac{T(z)}{T_0}\right) \quad [\text{dB}] \tag{4.49}$$

where $T(z)/T_0$ is obtained from (3.54).

Applying (3.54) to (4.49), it is seen that the power penalty is negligible (i.e., < 0.1

dB) as long as $T(z)/T_0 < 1.023$. In other words, only 2.3% of pulse broadening is allowed to keep the receiver sensitivity independent of fiber dispersion effects. When 20% pulse broadening is allowed, the dispersion-induced power penalty raises to 0.8 dB.

Mode Partition Noise

Mode partition noise (MPN) has been briefly discussed in Section 2.3.3.2 and the resulting maximum *BL* product has been discussed in Section 3.4.1.2 on the basis of (3.56).

The effect of MPN on receiver sensitivity has been extensively studied in the literature [Okano 1980, Ogawa 1982, Agrawal 1988, Wenworth 1992]. It can be shown that the dispersion-induced power penalty obtained with multilongitudinal-mode laser sources is given by

$$\text{Penalty(MPN)} = 5 \cdot \log_{10}\left(\frac{1}{1 - Q^2\sigma_{\text{MPN}}^2}\right) \text{ [dB]} \tag{4.50}$$

where σ_{MPN}^2 is the relative noise level of the received power in the presence of MPN, and can be expressed as [Agrawal, 1988]

$$\sigma_{\text{MPN}} = \frac{\varkappa}{\sqrt{2}}\{1 - \exp[-(\pi BLD\Delta\lambda)^2]\} \tag{4.51}$$

The value $\varkappa$ is known as the mode partition coefficient, whose typical values range between 0.6 and 0.8.

In (4.51), D is the dispersion coefficient defined by (3.38) (see also Appendix C) and $\Delta\lambda$ is the rms width of the averaged longitudinal-mode power distribution (assumed to be Gaussian).

Figure 4.12 shows the MPN-induced power penalty at $Q = 6$ (BER = 10^{-9}) and $Q = 8$ (BER = 10^{-15}) as a function of the normalized dispersion parameter (BLD$\Delta\lambda$) for two typical values of $\varkappa$.

As can be clearly seen from this figure, penalty (MPN) increases rapidly with an increase in (BLD$\Delta\lambda$). Equation (4.50) shows that the power penalty becomes infinite for

$$\sigma_{\text{MNP}} = \frac{1}{Q} \tag{4.52}$$

With $Q = 6$ (BER = 10^{-9}), (4.52) provides $\sigma_{\text{MNP}} = 0.167$ or, using (4.51), $BLD\Delta\lambda = 0.2$ with $\varkappa = 0.7$. For example, with $D = 2$ ps/km·nm and $\Delta\lambda = 3$ nm, this last condition shows that the MPN-induced power penalty becomes infinite if $BL \geq 30$ Gbps·km. At $B = 1$ Gbps, the distance is then limited to 30 km. It can be shown that the MPN effect induces an error floor (i.e., a BER of 10^{-9} cannot be achieved) when the *BL* product exceeds 30 Gbps·km.

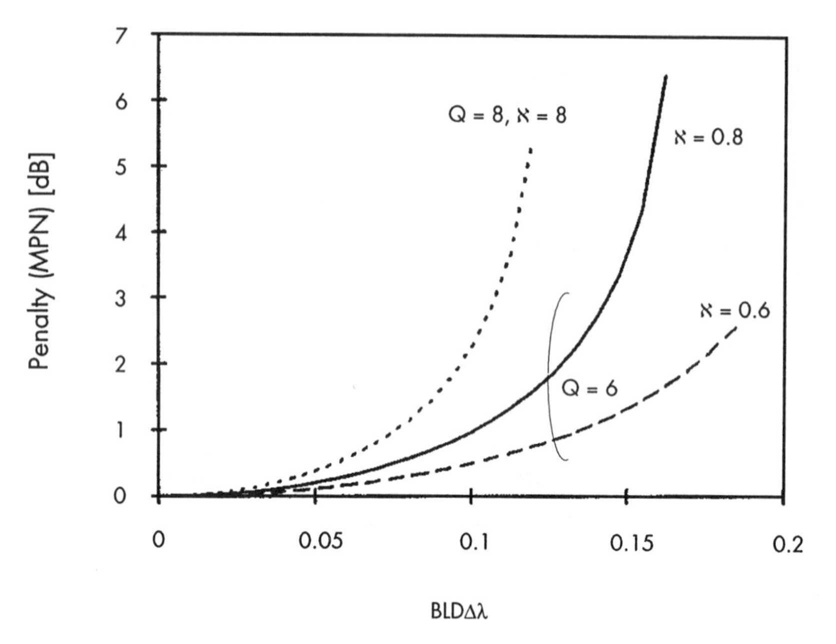

Figure 4.12 Power penalty of DD receivers due to mode partition noise (MPN).

For the dispersion effect to be negligible (i.e., Penalty (MPN) ≤ 0.1 dB), (4.50) shows that $\sigma_{MPN} \leq 0.212$ with $Q = 6$. Using (4.51), this gives $BL \leq 22$ Gbps·km, assuming $\varkappa = 0.7$, $D = 2$ ps/km·nm, and $\Delta\lambda = 3$ nm.

The two above examples show that the MPN-induced power penalty is very sensitive to σ_{MPN}.

It is important to stress that the analytical results expressed by (4.49), (4.50) and (4.51) provide only a rough estimation of the power penalty. In practice, the dispersion-induced power penalty depends on many system parameters and must be evaluated by means of numerical simulations.

4.2.4.3 Timing Jitter

The output of the clock-recovery circuit provides the sampling times used by the decision device. Since this timing information is recovered from the noisy input signal, the sampling times fluctuate around a mean value usually centered at the middle of the bit slot. Such fluctuations are referred to as *timing jitter*.

The consequence of timing jitter is that different portions of the received signal waveform are sampled from bit to bit. Therefore, the sampled values fluctuate by an amount that depends on both the received signal waveform and the timing jitter. These unwanted fluctuations translate into a reduction of the SNR at the sampling instant and result in performance degradation.

The power penalty induced by timing jitter is defined as the increase of the received optical power required to maintain the same SNR as that expected in the absence of timing jitter.

The evaluation of this power penalty is complicated by the fact that it depends not only on the statistical properties of the timing jitter, but also on the exact waveform of the received signal. Since the received waveform depends on details of the transmitter (through relaxation oscillations, frequency chirping, etc.), the fiber (through dispersion phenomena), and the receiver (through equalizer, filters, etc.), the power penalty must generally be computed on a per-case basis by means of numerical simulations.

In general, the rms value of the timing jitter should be below 10% of the bit slot for a negligible power penalty [O'Reilly 1985, Agrawal 1986, Shen 1986, Schumacker 1987].

4.2.5 Performance of Practical DD Receivers

Figure 4.13 represents the measurement results for direct detection receiver sensitivities obtained with either PINs or APDs as a function of the bit rate. The theoretical quantum limit is also shown for comparison.

We note that practical PIN receiver sensitivities are worse by about 25 dB or more compared with the quantum limit. The use of APDs improves the receiver sensitivity, but to a limited extent only because of the excess-noise factor. APD receivers typically operate 20-dB away from the quantum limit. This advantage of APDs over PINs is,

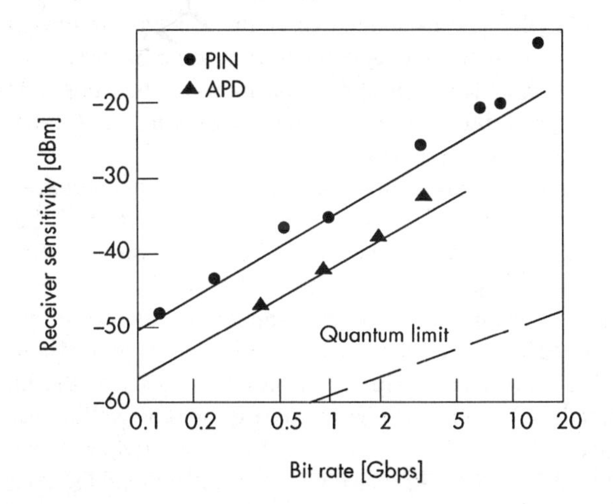

Figure 4.13 DD receiver sensitivities obtained with PIN and APD photodiodes. The quantum limit is also shown for comparison.

however, restricted to bit rates less than about 3 to 4 Gbps due to the limited gain-bandwidth product of APDs.

Moreover, APDs are considerably more expensive than PIN diodes (or even packaged PIN-FETs) and require a high-power voltage supply that must be carefully controlled to avoid avalanche breakdown. For these reasons, PINs are often preferred to APDs, except in systems where the few decibels of improvement in receiver sensitivity is really advantageous.

From the mid-1980s, considerable effort has been made to develop optoelectronic integrated circuits (OEICs) that monolithically integrate several receiver components on the same chip. InP-based OEIC receivers have been demonstrated for lightwave systems operating in the 1.3 to 1.6-μm wavelength range.

A 10-Gbps single-channel InP OEIC PIN receiver with −21.4 dBm sensitivity has been recently demonstrated [Kuebart, 1993].

It is expected that in the near future single-channel as well as multichannel OEIC receivers will be routinely produced [Soole, 1993]. Such a breakthrough is particularly useful for multiaccess networks, which involve several concurrent channels.

Notice that receiver sensitivities approaching the quantum limit can be achieved by using a PIN diode in combination with an optical preamplifier or with a laser diode that serves as the local oscillator in a coherent optical detector. Optical amplifiers will be described in Chapter 5, while coherent optical detectors are discussed in the next section.

4.3 COHERENT DETECTION (COH)

So far, we have been concerned with intensity-modulated optical signals followed by DD receivers. These IM/DD systems do not exploit the wave nature of light since the original information is basically recovered by a photon counting process.

By contrast, coherent optical communication systems use lightwave modulation/demodulation (i.e., amplitude, phase, or frequency modulation of the optical carrier wave) instead of merely intensity modulation.

The term *coherent* stems from the fact that the phase coherence of the optical wave carrier plays an important role in the design of such systems. The endeavor towards coherent optical systems started in the late 1970s, and many experiments and field trials established their feasibility in the late 1980s [DeLange 1968, Linke 1987, Kimura 1987].

The motivation behind using coherent optical techniques in multiaccess networks lies in the improved receiver sensitivity and the improved frequency selectivity. The improvement of the receiver sensitivity is the main focus of the present section. The improved frequency selectivity will be briefly explored as well, but more details on this issue will be given in Chapter 7 when we will look at WDMA networks.

In Section 4.3.1 we will introduce the basic concept behind coherent detection and the resulting improvement of the receiver sensitivity. Section 4.3.2 describes various techniques that can be used to modulate the amplitude, the phase, or the frequency of the optical carrier. Section 4.3.3 discusses the two classes of (electrical) demodulation schemes; namely, the synchronous and the asynchronous demodulators, used in coherent receivers. The three principal receiver sensitivity degradation mechanisms in practical coherent receivers will be discussed in Section 4.3.4. These include the effect of laser linewidth (or laser-phase noise), the effect of laser-intensity fluctuation and the effect of polarization mismatch. Section 4.3.5 will review the performance of coherent optical communication experiments and field trials that have been reported so far.

4.3.1 Basic Concept

In coherent optical communication systems, the original information is imposed on the optical field by either amplitude, frequency or phase modulation. Since photodiodes are responsive to the photon flux only (and not to the phase of the optical carrier), preprocessing of the optical signal is required prior to optoelectronic conversion by the photodiode.

The basic idea behind coherent detection is to mix the input signal optical field with a locally generated optical wave and to detect the resulting interference (i.e., *beating*) with a conventional photodiode (usually a PIN). The locally generated optical wave is commonly referred to as the local oscillator (LO) wave because its role is equivalent to that of the local oscillator in superheterodyne radio receivers.

As a result of the beating between the two optical fields, the photocurrent generated by the photodiode contains information about the amplitude, the frequency or the phase of the signal field. This information can be extracted by further electronic processing. Figure 4.14 shows a block diagram of a coherent receiver.

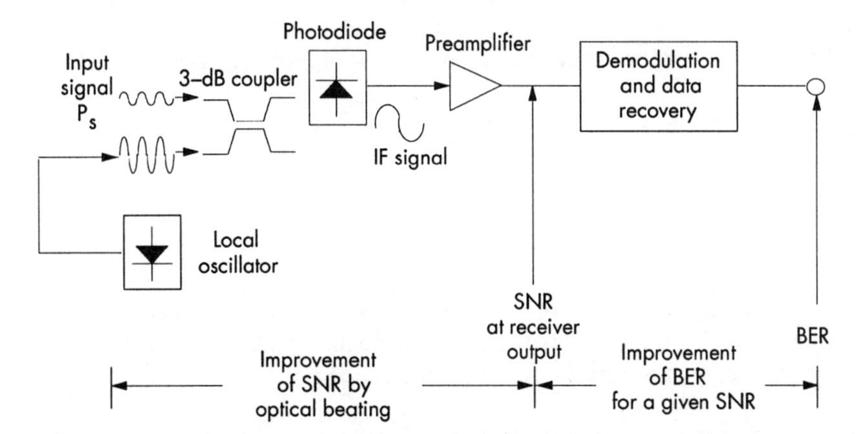

Figure 4.14 Block diagram of a coherent optical receiver. The signal wave at frequency ν_s is mixed with a local oscillator wave of frequency ν_{LO}. The photocurrent generated by the photodiode oscillates at the frequency difference $\nu_{IF} = \nu_s - \nu_{LO}$.

Coherent receivers are more sensitive than DD receivers. This results from two distinct effects. The first effect, due to optical mixing, provides an improvement of the SNR at the output end of the receiver preamplifier. The second effect is the improvement of the BER for a given SNR brought about by the use of different modulation/demodulation formats (ASK, FSK, PSK, DPSK, etc.) [Okoshi, 1988].

The SNR improvement provided by optical mixing will first be discussed in the present section. The improvement due to the second effect will be discussed in Section 4.3.3.

To see how optical mixing can improve the received SNR, let us write the field of the incident optical signal in complex notation as

$$E_s(t) = |A_s| \cdot \exp\{j(2\pi\nu_s t + \varphi_s)\} \tag{4.53}$$

where $|A_s|$ is the magnitude, ν_s is the carrier frequency and φ_s is the phase of the signal lightwave. Depending on the modulation format used, either $|A_s|$ (ASK), ν_s (FSK), or φ_s (PSK) are modulated with the information signal.

The local oscillator field is written similarly as

$$E_{\mathrm{LO}}(t) = |A_{\mathrm{LO}}| \cdot \exp\{j(2\pi\nu_{\mathrm{LO}} t + \varphi_{\mathrm{LO}})\} \tag{4.54}$$

Equations (4.53) and (4.54) assume that the two fields are perfectly parallel plane waves and have aligned linear polarization. The effect of polarization mismatch on the performance of coherent receivers will be discussed in Section 4.3.4.3.

The two fields are combined by the optical coupler (see Chapter 6 for a detailed discussion of optical couplers) and the total field incident upon the photodiode is the sum of the two constituent fields; $E = E_s + E_{\mathrm{LO}}$. Since the optical power is proportional to the squared modulus of the associated optical field [Saleh, 1991], the total power incident upon the photodiode is given by (K is a constant of proportionality)

$$P(t) = K \cdot |E_s + E_{\mathrm{LO}}|^2 = P_s + P_{\mathrm{LO}} + 2\sqrt{P_s P_{\mathrm{LO}}} \cdot \cos(2\pi\nu_{\mathrm{IF}} t + \varphi_s - \varphi_{\mathrm{LO}}) \tag{4.55}$$

where

$$P_s = K \cdot |A_s|^2 \quad P_{\mathrm{LO}} = K \cdot |A_{\mathrm{LO}}|^2 \quad \text{and} \quad \nu_{\mathrm{IF}} = \nu_s - \nu_{\mathrm{LO}} \tag{4.56}$$

P_s and P_{LO} are the powers of the signal and the local oscillator beams, respectively. The frequency ν_{IF} is known as the intermediate frequency (IF). By inserting (4.55) into (4.4), the photocurrent generated by the photodiode is given as

$$\begin{aligned} I(t) &= \Re(P_s + P_{\mathrm{LO}}) + 2\Re\sqrt{P_s P_{\mathrm{LO}}} \cdot \cos(2\pi\nu_{\mathrm{IF}} t + \varphi_s - \varphi_{\mathrm{LO}}) \\ &\cong \Re P_{\mathrm{LO}} + 2\Re\sqrt{P_s P_{\mathrm{LO}}} \cdot \cos(2\pi\nu_{\mathrm{IF}} t + \varphi_s - \varphi_{\mathrm{LO}}) \end{aligned} \tag{4.57}$$

where $\Re$ is the photodiode's responsivity and where we assumed that the local oscillator power is much stronger than the arriving signal power. Notice that the amount of signal power and local oscillator power that actually reaches the photodiode is half that indicated in (4.57) because one of the two 3-dB coupler outputs is not used. The values P_s and P_{LO} must therefore be understood as the optical powers that effectively reach the photodiode. We will return to this issue in Section 4.3.4.2 when we will discuss the balanced receiver.

The last term in (4.57), which oscillates at the intermediate frequency ν_{IF}, carries the useful information. Two main coherent detection schemes can be distinguished, depending on whether ν_{IF} equals zero or not. They are known as homodyne ($\nu_{IF} = 0$) or heterodyne ($\nu_{IF} \neq 0$, typically $\nu_{IF} = 0.1$ to 5.0 GHz) detections.

Let us first consider the heterodyne case and see how heterodyne coherent detection offers improved receiver sensitivity as compared to direct detection.

From (4.57), the average power of the photocurrent at frequency ν_{IF} is given by (recall that the average of $\cos^2 \theta$ over θ is 1/2)

$$\bar{S}_{\text{Het}} = \frac{1}{2}(2\Re\sqrt{P_s P_{\text{LO}}})^2 = 2\Re^2 P_s P_{\text{LO}} \tag{4.58}$$

If we note that the average electrical signal power for direct detection is $\Re^2 P_s^2$, we immediately see from (4.58) that the received signal power is increased by a factor of $2P_{LO}/P_s$ by using heterodyne coherent detection. Since $2P_{LO}/P_s$ can be made considerably larger than one, the enhancement can be by orders of magnitude. In other words, the local oscillator acts as a sort of amplifier in the optical domain since it amplifies the signal seen by the photodiode.

However, the receiver noise is also enhanced because of the additional shot noise due to the strong local oscillator power, and the improvement in SNR is reduced accordingly.

The total noise variance is obtained by adding the contributions from shot noise due to the signal power, dark current, local oscillator power, and thermal noise due to the electrical circuitry (load resistor and amplifier noise). Therefore, we have

$$\begin{aligned}\sigma_{\text{Tot}}^2 &= \sigma_{\text{shot}}^2 + \sigma_{\text{th}}^2 = 2e(\bar{I}_s + \bar{I}_{\text{dark}} + \bar{I}_{\text{LO}})\Delta f + \frac{4kTF_n}{R_L}\Delta f \\ &\cong 2e\bar{I}_{\text{LO}}\Delta f = 2e\Re P_{\text{LO}}\Delta f\end{aligned} \tag{4.59}$$

The last approximation in (4.59) becomes more and more accurate as P_{LO} grows larger. In other words, the local oscillator is made so strong that the thermal electronic noise becomes negligible compared to the shot noise. Although the resulting overall noise is enhanced compared to the noise of a DD receiver, we have seen from (4.58) that the signal power is also enhanced by a factor $2P_{LO}/P_s$, so that the heterodyne detection provides a SNR given by

$$\mathrm{SNR_{Het}} = \frac{2\Re^2 P_s P_{\mathrm{LO}}}{2e\Re P_{\mathrm{LO}}\Delta f} = \frac{\eta P_s}{h\nu_s \Delta f} \quad (4.60)$$

where we used $\Re = \dfrac{\eta e}{h\nu_s}$ from (4.6).

Equation (4.60) can be expressed in terms of the average number of photons per bit, $\bar{n}_p$. If we note that $\bar{n}_p = P_s/2h\nu_s\Delta f$, we immediately have from (4.60)

$$\mathrm{SNR_{Het}} = 2\eta\bar{n}_p \quad (4.61)$$

The principal advantage of heterodyne coherent detection now becomes evident by comparing (4.61) with (4.43). In the shot-noise limit, direct detection provides $\mathrm{SNR_{DD}} = \eta\bar{n}_p$; therefore, the heterodyne coherent receiver offers an improvement in SNR by a factor of 2. This is often referred to as the 3-dB advantage of heterodyning over direct detection.

In practice, the advantage is even more substantial because a DD receiver always operates far away from the shot-noise limit since its performance is dominated by thermal noise. By contrast, the receiver sensitivity of coherent detection can approach the shot-noise limit provided that the power of the local oscillator is sufficiently large.

For the case of homodyne coherent detection, the photocurrent generated by the photodiode is given by (4.57) with $\nu_{\mathrm{IF}} = 0$

$$I(t) \cong \Re P_{\mathrm{LO}} + 2\Re\sqrt{P_s P_{\mathrm{LO}}} \cdot \cos(\varphi_s - \varphi_{\mathrm{LO}}) \quad (4.62)$$

Therefore, the optical signal is directly converted to the baseband. Assuming that the local oscillator is locked to the signal phase ($\varphi_{\mathrm{LO}} = \varphi_s$), the average power of the signal photocurrent is now given by $\bar{S}_{\mathrm{Hom}} = 4\Re^2 P_s P_{\mathrm{LO}}$ (the factor 1/2 due to the averaging of $\cos^2\theta$ over θ disappears for the homodyne case), so that the SNR power ratio becomes

$$\mathrm{SNR_{Hom}} = 4\eta\bar{n}_p \quad (4.63)$$

A direct comparison of (4.63) with (4.61) shows that homodyne detection offers a 3-dB advantage over heterodyne detection.

At first glance, this additional 3-dB advantage in receiver sensitivity would turn the preference towards the homodyne scheme. However, a major disadvantage of homodyne detection results from its larger sensitivity to laser-phase noise. This sensitivity imposes stringent requirements on the laser linewidth for both the transmitter and the local-oscillator optical sources to be used in the system. These requirements are significantly relaxed in the case of heterodyne detection. In addition, the design of homodyne coherent optical receivers requires the implementation of a complex optical phase-lock loop to keep the phase difference $\varphi_s - \varphi_{\mathrm{LO}}$ in (4.62) constant [Kazovsky, 1990]. Since these drawbacks can be partially avoided by the use of heterodyne schemes, heterodyne detection ... implemented in all practical coherent optical systems.

4.3.2 Modulation Techniques

The modulation of either the amplitude, the phase, or the frequency of the optical carrier is achieved using distinct techniques and physical mechanisms. In general, the modulation tends to be applied externally to the laser since this does not add noise to the high-purity light sources that are required (see Section 4.3.4.1): the semiconductor laser is pumped with a constant current and its output is modulated by using the external modulator.

The most familiar external modulators make use of titanium-diffused $LiNbO_3$ waveguides, but other modulators based on semiconductor compounds have also been demonstrated.

In what follows, implementation techniques for amplitude, phase, and frequency modulations in practical coherent lightwave systems will be described.

4.3.2.1 Amplitude Modulation (ASK)

The electrical field associated with an amplitude shift keyed (ASK) optical signal can be written as

$$E_s(t) = |A_s(t)| \cdot \cos[2\pi\nu_s t + \varphi_s] \tag{4.64}$$

where $|A_s(t)|$ carries the information by taking one of the two fixed values during each bit period depending on whether a bit "1" or a bit "0" is being transmitted ($|A_s(t)|$ can take more than two distinct values if *multilevel* ASK modulation is used).

The most obvious way to modulate the amplitude of the optical carrier is to modulate the output power of the laser source by varying the current applied directly to the device, similarly to the on-off keying (OOK) modulation commonly used for IM/DD systems. However, as we have seen in Chapter 2, direct modulation of semiconductor lasers gives rise to phase changes (chirp) in the output optical field. Such phase noise is particularly detrimental in coherent detection receivers since the detector response depends on the phase of the received signal. This problem can be overcome by the use of an external modulator.

Mach-Zender interferometers or directional couplers designed with titanium-diffused $LiNbO_3$ are commonly used as ASK modulators. Mach-Zender external modulators will be discussed in detail in Chapter 7 in the context of tunable optical filters. When used as ASK modulators, they can provide an ON-OFF ratio (also called extinction ratio) in excess of 20 (i.e., 13 dB) and can be modulated at speeds of up to 20 GHz [Korotky, 1987].

It is noteworthy that all external modulators introduce a power penalty because of insertion loss. This insertion loss can be reduced to below 1 dB for well-designed modulators.

External ASK modulation can also be achieved by using semiconductors and the electroabsorption effect. All semiconductors begin to absorb incident light when the wavelength λ is shorter than the wavelength λ_c that corresponds to the semiconductor bandgap. The term λ_c is referred to as the absorption edge and is the equivalent of the cutoff wavelength of photodiodes. Its value can be varied through the electroabsorption effect by applying an external voltage. By decreasing λ_c below the wavelength of the incident light, the semiconductor becomes transparent. Semiconductor modulators based on the electroabsorption effect have been demonstrated with conventional double heterostructure and multiple-quantum-well (MQW) designs. ON-OFF ratios of 15 to 20 dB have been achieved with voltage of about 5V and modulation bandwidths in excess of 20 GHz [Kotaka, 1989; Mak, 1990].

The main advantage of semiconductor modulators is that they can be monolithically integrated with the semiconductor laser on the same chip [Goto, 1990].

4.3.2.2 Phase Modulation (PSK)

In the case of phase shift keying (PSK) modulation, the electrical field associated with the optical signal can be written as

$$E_s(t) = |A_s| \cdot \cos\{2\pi\nu_s t + \varphi_s(t)\} \tag{4.65}$$

For binary data source, the phase φ_s takes two values, commonly chosen to be 0 and π, while the amplitude $|A_s|$ of the carrier is kept constant.

External modulators based on the electrorefractive effect can be used for PSK modulation. The design of $LiNbO_3$-based PSK modulators is much simpler than for ASK since a Mach-Zender interferometer or directional coupler configuration is no longer needed. Instead, a single waveguide whose refractive index is controlled by the applied voltage can be used. The same performances as for ASK Mach-Zender modulators have been demonstrated.

Semiconductor PSK modulators have also been developed using MQW structures. An MQW device has been operated at 10 Gbps with 2.5V for π phase shift at a wavelength of 1.55 μm [Wakita, 1987].

4.3.2.3 Frequency Modulation (FSK)

The optical field for frequency shift keying (FSK) can be written as

$$\begin{aligned} E_s(t) &= |A_s| \cdot \cos\{2\pi(\nu_s \pm \Delta\nu)t + \varphi_s\} \\ &= |A_s| \cdot \cos\{2\pi\nu_s t + (\varphi_s \pm 2\pi\Delta\nu t)\} \end{aligned} \tag{4.66}$$

The binary information is encoded on the optical carrier by shifting its frequency between the two values $\nu_s + \Delta\nu$ and $\nu_s - \Delta\nu$ depending on whether a bit "1" or a bit "0" is being transmitted. The shift $\Delta\nu$ is called the frequency deviation, while $2\Delta\nu$ is often referred to as the tone spacing. The second equality in (4.66) indicates that FSK modulation can be achieved with a PSK modulator by varying the phase of the optical carrier linearly over the bit duration.

However, a simpler and more commonly used method for FSK modulation is obtained by varying the injection current applied directly to the semiconductor laser. As discussed in Chapter 2, variations in the injection current lead to changes in both the amplitude and the frequency of the emitted lightwave. Typical values of frequency shifts are 0.1 to 1.0 GHz/mA. This is efficient enough to produce significant frequency shifts while keeping the amplitude nearly unchanged from bit to bit.

It should be noticed that most semiconductor lasers typically exhibit a dip in their frequency modulation (FM) response in the low-frequency range 0.1 to 10 MHz, as shown in Figure 4.15. For long strings of "1" or "0" (i.e., low-frequency components of the data spectrum), the frequency deviation is reduced, resulting in performance degradation of the FSK-based communication system.

In order to overcome this problem, precoding techniques that reduce the low-frequency components of the data spectrum can be used [Emura, 1984]. Alternative techniques consist of designing semiconductor lasers with flat FM response over the entire data spectrum. For example, flat FM response from 100 kHz up to 15 GHz has been demonstrated with multisection DFB lasers [Yamazaki, 1985].

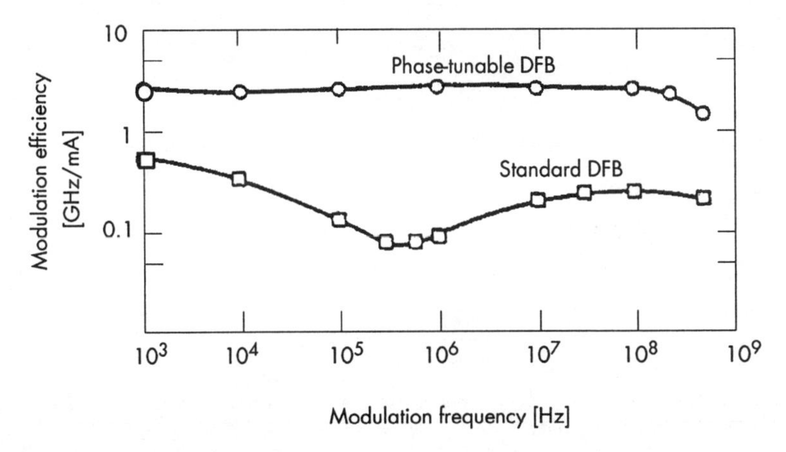

Figure 4.15 Frequency modulation response of a standard DFB laser diode and a multisection DFB laser diode. [After Yamazaki, 1985.]

4.3.3 Electrical Demodulation Schemes and BER

Homodyne or heterodyne optical detection schemes convert the received optical signal into an electrical signal which must be processed by further electrical circuits to recover the original information. In the case of homodyne detection, the optical signal is converted directly to the baseband and the same electrical signal-processing circuitry as that used in IM/DD receivers can be used to recover the original data.

Heterodyne optical detection provides an electrical signal in the form of a modulated RF/microwave carrier whose center frequency is the intermediate frequency ν_{IF}. The demodulation of such signals can be carried out either synchronously or asynchronously using well-known techniques developed for RF/microwave communication systems.

The BER of the communication system, which eventually determines the receiver sensitivity, is dependent upon the combination of the modulation format (ASK, PSK, FSK), the type of coherent optical detection (homodyne, heterodyne), as well as the type of electrical demodulation (synchronous, asynchronous).

In what follows, we will discuss the synchronous and asynchronous RF/microwave demodulation schemes and summarize the BER obtained with various combinations of modulation formats/optical detection/electrical demodulation schemes.

Detailed calculations of the results presented here can be found in [Okoshi, 1988; Agrawal, 1992].

4.3.3.1 Synchronous Demodulation

In synchronous demodulators, the RF/microwave carrier at the intermediate frequency ν_{IF} is first recovered and then used to downconvert the RF/microwave signal to baseband. A schematic block diagram of a synchronous heterodyne receiver is shown in Figure 4.16.

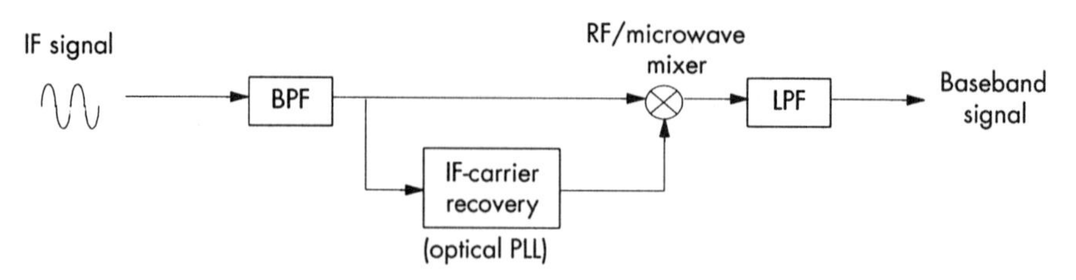

Figure 4.16 Synchronous demodulator with carrier recovery of the intermediate frequency ν_{IF}. BPF = bandpass filter, LPF = lowpass filter.

The current generated by the receiver front-end is filtered by a bandpass filter (BPF) centered at ν_{IF}. From (4.57), this filtered current is given by

$$I_f(t) = 2\Re\sqrt{P_s P_{LO}} \cdot \cos\{2\pi\nu_{IF}t + \varphi_s - \varphi_{LO}\} = I_s \cdot \cos\{2\pi\nu_{IF}t - \Delta\varphi\}$$
$$= I_s(t) \cdot \cos\Delta\varphi \cdot \cos(2\pi\nu_{IF}t) + I_s(t) \cdot \sin\Delta\varphi \cdot \sin(2\pi\nu_{IF}t) \quad (4.67)$$

where $I_S(t) = 2\Re\sqrt{P_S(t)P_{LO}}$ and $\Delta\varphi(t) = \varphi_{LO} - \varphi_s(t)$.

In addition to this signal current, the receiver front-end generates noise which is also filtered by the BPF. Assuming Gaussian statistics for the filtered noise, its in-phase (i_P) and quadrature-phase (i_Q) components can be included in (4.67) to represent the noisy filtered signal current

$$I_f(t) = (I_s(t)\cos\Delta\varphi(t) + i_P) \cdot \cos(2\pi\nu_{IF}t) + (I_s(t)\sin\Delta\varphi(t) + i_Q) \cdot \sin(2\pi\nu_{IF}t) \quad (4.68)$$

where i_Q and i_P are Gaussian random variables with zero mean and variance given by (4.59).

The RF/microwave mixer mixes the recovered carrier at ν_{IF} with $I_f(t)$ given by (4.68), and the output is filtered by the lowpass filter (LPF) to reject the ac component at frequency $2\nu_{IF}$. The resulting baseband signal is then given by

$$I_b(t) = \langle I_f \cdot \cos(2\pi\nu_{IF}t)\rangle = \frac{1}{2}(I_s(t)\cos\Delta\varphi(t) + i_P) \quad (4.69)$$

where the brackets denote averaging over a period of $1/\nu_{IF}$ (i.e., the effect of the LPF).

Equation (4.69) shows that the quadrature-phase noise component has been eliminated from the baseband signal. Only the in-phase noise component remains which, on average, is half of the total noise.

The data, contained in the term $I_s(t)\cos\Delta\varphi(t)$, must be recovered from the baseband signal. More specifically, for ASK modulation the data is contained in $I_s(t)$ while $\Delta\varphi$ is ideally kept constant. For PSK or FSK modulations, the data is contained in the variations of $\Delta\varphi$ while I_s is kept constant.

The BERs obtained with synchronous demodulation for various modulation formats have been calculated in [Okoshi 1988, Agrawal 1992] and are summarized in Table 4.2.

4.3.3.2 Asynchronous Demodulation

Asynchronous demodulators do not require the recovery of the RF/microwave carrier at ν_{IF}. Instead, an envelope detector is used to convert the filtered signal $I_f(t)$ given by (4.68)

Table 4.2
BER and Sensitivity of Coherent Optical Receivers for Various Combinations of Modulation Formats/Optical Detection/Electrical IF Signal Processing

	Demodulation	*BER*	n_p	$\bar{n}_p$
	IM/DD	$1/2 \cdot \exp(-\eta n_p)$	*20*	*10*
Synchronous	ASK heterodyne	$1/2 \cdot \text{erfc}(\sqrt{\eta n_p/4})$	72	36
	ASK homodyne	$1/2 \cdot \text{erfc}(\sqrt{\eta n_p/2})$	36	18
	PSK heterodyne	$1/2 \cdot \text{erfc}(\sqrt{\eta n_p})$	18	18
	PSK homodyne	$1/2 \cdot \text{erfc}(\sqrt{2\eta n_p})$	9	9
	FSK heterodyne	$1/2 \cdot \text{erfc}(\sqrt{\eta n_p/2})$	36	36
Asynchronous	ASK heterodyne	$1/2 \cdot \exp(-\eta n_p/4)$	80	40
	FSK heterodyne	$1/2 \cdot \exp(-\eta n_p/2)$	40	40
	DPSK heterodyne	$1/2 \cdot \exp(-\eta n_p)$	20	20

to baseband, resulting in a much simpler receiver design. A block diagram of an asynchronous demodulator for ASK signals is shown in Figure 4.17(a). The signal output of the envelope detector is given by

$$I_b(t) = |I_f(t)| = [(I_s\cos\Delta\varphi + i_P)^2 + (I_s\sin\Delta\varphi + i_Q)^2]^{1/2} \tag{4.70}$$

The main difference compared to the case of synchronous demodulation is that the baseband signal is corrupted by both the in-phase and quadrature-phase noise components. The SNR at the input of the decision device is therefore degraded, resulting in less sensitive receivers. Nevertheless, the receiver sensitivity degradation is quite small ($\cong$ 0.5 dB) and asynchronous heterodyne receivers have been widely used, not only because of their design simplicity but also (and this is the main reason) because the laser linewidth requirements for both the transmitter and the local oscillator are much more relaxed compared to synchronous receivers. This can be qualitatively understood by comparing (4.69) with (4.70). In (4.69), the term $\cos\Delta\varphi$ plays a vital role in determining the BER. Ideally, $\Delta\varphi$ should remain constant ($\Delta\varphi = 0$) except for intentional modulation by the data as in PSK or FSK. Small random fluctuations of $\Delta\varphi$ directly result in significant changes of $I_b(t)$. The impact on the BER is obvious. In contrast with asynchronous demodulation, the baseband signal contains both $\cos\Delta\varphi$ and $\sin\Delta\varphi$ terms. A decrease of $\cos\Delta\varphi$ due to random fluctuation of $\Delta\varphi$ is balanced by an increase of $\sin\Delta\varphi$, thereby reducing the sensitivity of asynchronous receivers to laser-phase noise. This will be discussed in more quantitative terms in Section 4.3.4.1.

Figure 4.17(b, c) shows asynchronous heterodyne receivers for FSK and DPSK demodulation, respectively. FSK demodulation is achieved by separately processing the "0" and the "1" bits whose carrier frequencies, and hence intermediate frequencies, are different. The BPFs are centered at the intermediate frequencies corresponding to bits "0"

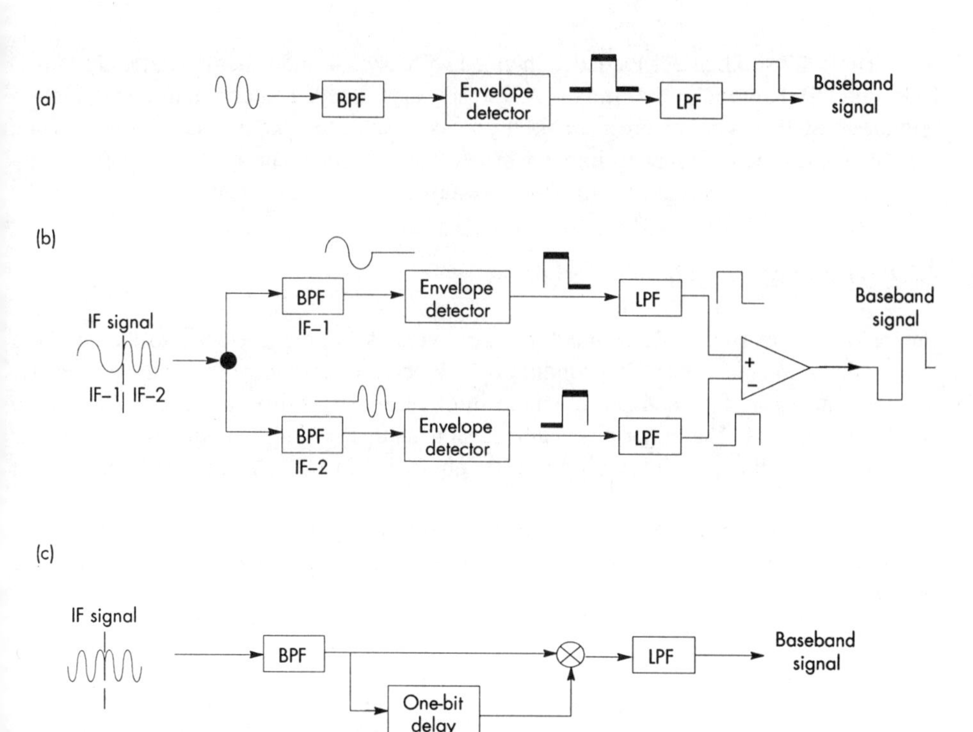

Figure 4.17 Block diagram of asynchronous heterodyne demodulators: (a) ASK, (b) FSK and (c) DPSK.

and "1." The dual-filter FSK demodulator can be thought of as two ASK demodulators in parallel whose outputs are combined before reaching the decision device [Agrawal, 1992]. This scheme works well provided that the spectra of "0" and "1" bits have negligible overlap (i.e., wide-deviation FSK for which the tone spacing is much larger than the bit rate). With narrow-deviation FSK, for which the tone spacing is less than or comparable to the bit rate, synchronous demodulation is generally applied.

For phase-modulated signals, some modifications are required when asynchronous demodulation is used. Since PSK signals have constant amplitude and single-carrier frequency, envelope detectors or BPFs are of little help to demodulate the signal. However, differential PSK (DPSK) format can be demodulated asynchronously by using the delay-demodulation scheme shown in Figure 4.17(c). With DPSK format, the information is encoded in the phase difference ($\varphi_k - \varphi_{k-l}$) between two consecutive bits. The demodulation uses the IF signal as the reference, but with a delay equal to the bit slot duration T. The signal output of the mixer shown in Figure 4.17(c) has a component of the form $\cos(\varphi_k - \varphi_{k-1})$, which contains all the information about the data. An advantage of DPSK, besides allowing asynchronous demodulation, is that the laser-phase stability is required only for a relatively short duration (two bit periods).

Table 4.2 summarizes the BER obtained with asynchronous demodulators for ASK, FSK, and DPSK modulation formats. As already mentioned, asynchronous demodulation provides slightly worse receiver sensitivity as compared to synchronous schemes. They are, however, much simpler to implement because they alleviate practical problems associated with laser-phase noise, to which we turn to in the next section.

4.3.4 Sensitivity Degradation Mechanisms

The receiver sensitivities determined in the previous section correspond to the quantum limit and assume ideal operating conditions with perfect components. Actual devices are, however, never perfect and the quantum-limit receiver sensitivity is very difficult to achieve in practice. The major physical mechanisms that lead to departure of the receiver sensitivity from the quantum limit are: laser phase and intensity noises, and polarization mismatch.

This section discusses the corresponding sensitivity degradations and some of the performance requirements for the system components to be used in coherent lightwave systems.

4.3.4.1 Laser Phase Noise

As we have seen in Section 2.3.4.2, the output from a SLM laser diode contains phase noise, which results in a lightwave output that is not purely monochromatic. This has important consequences on the performance of coherent lightwave systems as can be seen from (4.57) and (4.62): both the signal phase φ_s and the local-oscillator phase φ_{LO} must remain relatively stable to be able to recover the original signal reliably.

Considerable attention has been paid to evaluating the allowed amount of phase noise in coherent systems. Since the phase noise is characterized by the laser linewidth $\Delta\nu$ through $<\varphi^2(t)> = 2\pi\Delta\nu \cdot t$ (where t is the time) [Saleh, 1991], it is usual to specify the maximum tolerable value of $\Delta\nu$. This value depends on the modulation format and on the optical demodulation scheme (i.e., homodyne or heterodyne), as well as on the type of processing applied to the IF signal (i.e., synchronous or asynchronous).

In all cases, the tolerable laser linewidth $\Delta\nu$, often defined so that the induced power penalty is below 1 dB, can be expressed as a fraction of the data bit rate B [Kazovsky 1986, Kimura 1987]. This is summarized in Table 4.3. Notice that since both φ_s and φ_{LO} fluctuate independently, $\Delta\nu$ in Table 4.3 is actually the sum of the linewidths $\Delta\nu_s$ and $\Delta\nu_{LO}$ associated with the transmitter and the local oscillator, respectively.

Some general conclusions can be drawn from Table 4.3:

- Since the tolerable $\Delta\nu$ is proportional to B, high-bit-rate systems are less demanding on laser spectral purity.
- The linewidth requirements are most stringent for homodyne receivers. The combi-

Table 4.3
Laser Linewidth Requirements for Different Combinations of Modulation Formats/Optical Detection/Electrical IF Signal Processing ($\Delta\nu = \Delta\nu_s + \Delta\nu_{LO}$)

IF Processing	*Detection*	*Modulation*	$\Delta\nu/B$
Synchronous	Homodyne	ASK	$< 3.10^{-4}$
		PSK	$< 5.10^{-4}$
	Heterodyne	ASK	$< 5.10^{-3}$
		PSK	$< 2.10^{-3}$
		FSK	$< 5.10^{-3}$
Asynchronous	Heterodyne	ASK	$< 10^{-1}$
		DPSK	$< 10^{-2}$
		FSK	$< 10^{-1}$

nation leading to the best receiver sensitivity (i.e., PSK/homodyne/synchronous) also requires the smallest linewidth.
The linewidth requirements are considerably relaxed for heterodyne receivers, especially when ASK and FSK formats and asynchronous demodulation are used.

The requirements on laser linewidths (and stability) are quite difficult to obtain in practice and even in laboratory experiments (especially for homodyne detection). Fortunately, an alternative approach, known as *phase-diversity*, significantly relaxes these requirements [Davis 1987, Kazovsky 1989].

The basic idea of phase-diversity receivers consists of using two or more photodetectors whose outputs are combined to produce an electrical signal that is independent of the phase difference $\varphi_s - \varphi_{LO}$. Figure 4.18 shows a two-port phase-diversity receiver. It uses a 90-deg hybrid whose output fields E_1 and E_2 are given by

$$E_1 = C_1 \cdot (E_s + E_{LO}) \tag{4.71}$$

$$E_2 = C_1 (E_s + e^{j\pi/2} E_{LO}) \tag{4.72}$$

where C_1 is a constant of proportionality.

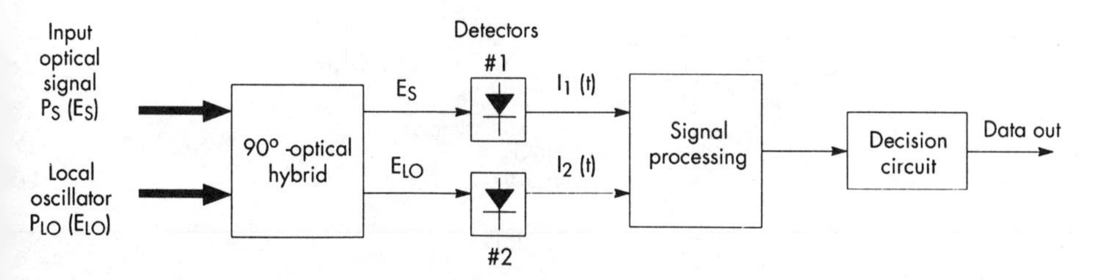

Figure 4.18 A two-port phase-diversity receiver.

The 90-deg hybrid can be implemented as shown in Figure 4.19: the output fields E_1 and E_2 have orthogonal polarization and hence the phase relationship between the signal and the LO components of E_1 and E_2 can be controlled independently by the input LO polarization controller.

When the 90-deg hybrid-output fields are detected, the following signal currents are generated (we assume, for simplicity, ASK modulation format):

$$I_1(t) = C_2(P_s + P_{\mathrm{LO}}) + 2\sqrt{P_s P_{\mathrm{LO}}} \cdot \cos\{2\pi\nu_{\mathrm{IF}}t + \varphi_s - \varphi_{\mathrm{LO}}\} \quad (4.73)$$

$$I_2(t) = C_2(P_s + P_{\mathrm{LO}}) + 2\sqrt{P_s P_{\mathrm{LO}}} \cdot \sin\{2\pi\nu_{\mathrm{IF}}t + \varphi_s - \varphi_{\mathrm{LO}}\} \quad (4.74)$$

where C_2 is another constant of proportionality that depends on the 90-deg hybrid losses and the photodetector responsivity.

After bandpass filtering (to eliminate the dc terms), the currents I_1 and I_2 are squared and summed. The resulting signal becomes independent of the laser-phase noise. The values of $\Delta\nu/B$ can approach unity without introducing significant power penalty, even in

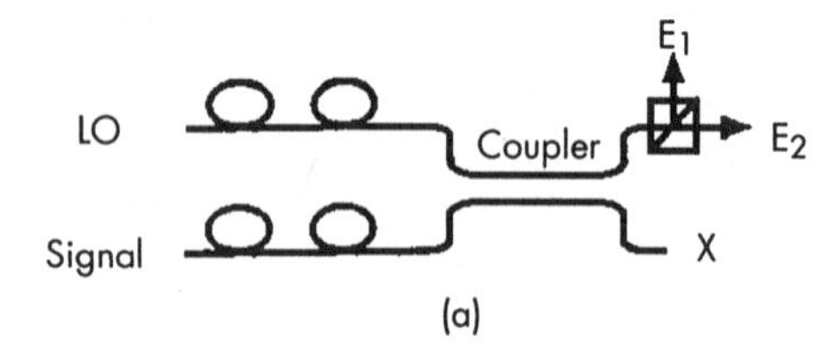

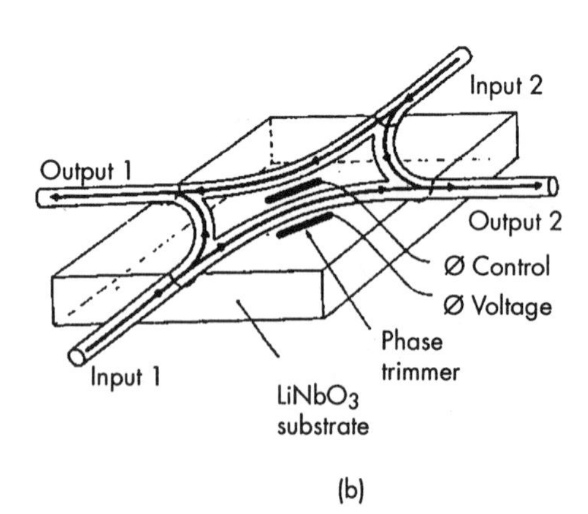

Figure 4.19 Two distinct implementations of a 90-deg optical hybrid: (a) using bulk optic components and (b) using integrated optics. (© 1989 IEEE [After Kazovsky, 1989.])

the case of homodyne receivers. The concept has been extended to receivers with three or more branches. However, because of the increasing complexity, the number of ports in practical phase-diversity receivers has been limited to two or three. Experiments have been carried out with ASK, FSK and DPSK formats [Welter 1988, Hodgkinson 1985, Pettit 1987].

4.3.4.2 Laser-Intensity Noise

In Section 2.3.4.1, we have seen that the output power of any semiconductor laser is always accompanied by some random amplitude fluctuations, called laser-intensity noise. This noise, characterized by the RIN parameter defined in (2.34), is converted by the photodiode into current fluctuations which add to the shot noise and thermal noise, thereby reducing the SNR at the input of the decision device.

To see how laser-intensity noise affects the performance of coherent receivers, let us modify (4.59) by adding a term associated with the laser RIN so as to obtain [Agrawal, 1992]

$$\sigma_{\text{Total}}^2 \cong \sigma_{\text{shot (LO)}}^2 + \sigma_{\text{RIN}}^2 = 2e\Re P_{\text{LO}}\Delta f + \sigma_{\text{RIN}}^2 \tag{4.75}$$

Equation (4.75) implicitly assumed that laser-intensity noise follows Gaussian statistics, so that the total noise variance is obtained by adding the variances of the independent noise sources. An explicit expression of σ_{RIN}^2 can be obtained by noting that

$$\sigma_{\text{RIN}}^2 = <(\Re\delta P_{\text{LO}}(t))^2> = \Re^2\overline{P}_{\text{LO}}^2 r_I^2 \tag{4.76}$$

where $\Re$ is the photodiode responsivity and $\delta P_{\text{LO}} = (P_{\text{LO}}(t) - \overline{P}_{\text{LO}})/\overline{P}_{\text{LO}}$ represents the normalized fluctuations of the local-oscillator output power ($\overline{P}_{\text{LO}}$ is the mean power).

The parameter r_I, defined by $r_I^2 = <(\delta P_{\text{LO}}(t))^2> /\overline{P}_{\text{LO}}^2$ is related to the RIN parameter defined by (2.34) as

$$r_I^2 = \int_{-\infty}^{+\infty} \text{RIN}(f)df \cong 2 \cdot \text{RIN} \cdot \Delta f \tag{4.77}$$

The last approximation in (4.77) is obtained by assuming a flat RIN spectrum up to the receiver bandwidth Δf. The r_I parameter can be understood as a measure of the local-oscillator intensity noise level since, following its definition, it is simply the inverse of the SNR of the light emitted by the LO.

The SNR_{Het} given by (4.60) must be modified to include the effect of the RIN. According to (4.58), (4.75), (4.76), and (4.77), we have

$$\mathrm{SNR_{Het}} = \frac{\Re P_s}{e\Delta f + \Re \overline{P}_{\mathrm{LO}} \cdot \mathrm{RIN} \cdot \Delta f} \tag{4.78}$$

Equation (4.78) clearly shows the degradation of $\mathrm{SNR_{Het}}$ due to the LO intensity noise. In order to maintain the same SNR as in the absence of LO intensity noise, the signal power P_s must be increased to offset the increase of receiver noise. This increase ΔP in P_s is the corresponding power penalty. From (4.78), the power penalty (in dB) is given by the simple expression

$$\text{Penalty (RIN)} = 10 \cdot \log_{10}\left(\frac{P_s + \Delta P}{P_s}\right) = 10 \cdot \log_{10}\left(1 + \frac{\eta \overline{P}_{\mathrm{LO}} \cdot \mathrm{RIN}}{h\nu}\right) \tag{4.79}$$

where $\Re = \eta e/h\nu$ was used.

Figure 4.20 shows Penalty(RIN) from (4.79) as a function of the local oscillator intensity noise (RIN) for several values of $\overline{P}_{\mathrm{LO}}$. It is seen that the power penalty is unacceptably large when $\overline{P}_{\mathrm{LO}} \geq 1\mathrm{mW}$ even for RIN values as low as -160 dB/Hz. In order to keep the power penalty below 1 dB with RIN of -150 dB/Hz, $\overline{P}_{\mathrm{LO}}$ should be less than about 0.3 mW.

Lower values of $\overline{P}_{\mathrm{LO}}$ would make the power penalty negligible. However, too low values of $\overline{P}_{\mathrm{LO}}$ would also reduce the coherent detection advantage since thermal noise

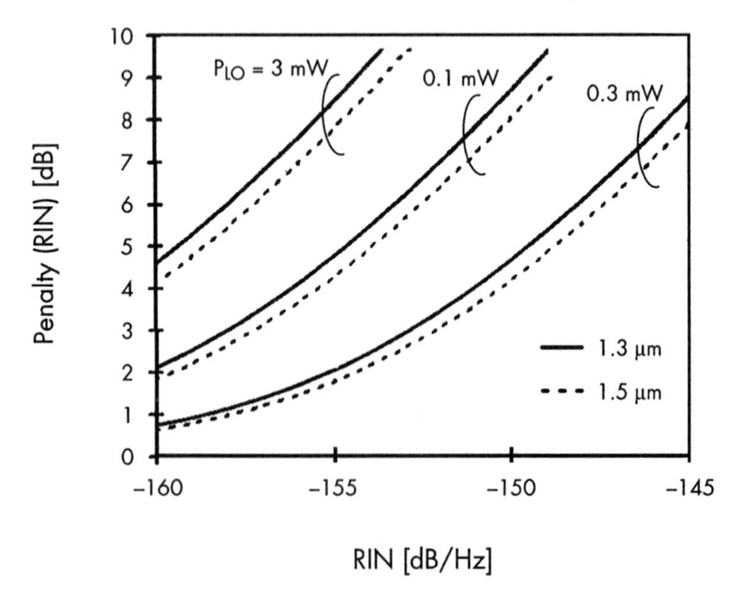

Figure 4.20 LO intensity-noise power penalty of heterodyne coherent receivers as a function of the RIN for various values of the LO output power.

would become increasingly significant compared to the LO shot noise. (Notice that $\overline{P}_{\mathrm{LO}}$ should be higher than a few milliwatts to make the thermal noise negligible—in practice therefore, low-noise preamplifiers are still required in coherent receivers.)

Unless further improvements in the semiconductor laser technology are achieved to reduce the RIN, intensity noise will continue to be a limiting factor in coherent receivers for $\overline{P}_{\mathrm{LO}} \geq 0.1\mathrm{mW}$.

Fortunately, the intensity-noise problem can be solved almost completely by the clever artifice of using two back-to-back photodiodes in the so-called *balanced-receiver* shown in Figure 4.21.

The operation of the balanced receiver can be understood by considering the 2×2 coupler analyzed in Section 6.1. This analysis (6.11) shows that at output #1 (Figure 4.21) the received signal appears unshifted, while the LO appears with a $\pi/2$ phase shift. At output #2, the received signal appears with a $\pi/2$ phase shift, while the LO appears unshifted. In other words, a signal that crosses over a 2×2 coupler exhibits a $\pi/2$ phase shift.

Returning to Section 4.3.1 and following the same analysis as that carried out to obtain (4.57), it can be shown that [Abbas, 1985]

$$I_{\mathrm{sig}\,1} = \frac{\Re}{2}(P_s + P_{\mathrm{LO}}) + \Re\sqrt{P_s P_{\mathrm{LO}}} \cdot \cos\{2\pi\nu_{\mathrm{IF}}t + \varphi_s - \varphi_{\mathrm{LO}}\} \tag{4.80}$$

$$I_{\mathrm{sig}\,2} = \frac{\Re}{2}(P_s + P_{\mathrm{LO}}) - \Re\sqrt{P_s P_{\mathrm{LO}}} \cdot \cos\{2\pi\nu_{\mathrm{IF}}t + \varphi_s - \varphi_{\mathrm{LO}}\} \tag{4.81}$$

where $I_{\mathrm{sig}\,1}$ and $I_{\mathrm{sig}\,2}$ are the photocurrents generated by the two photodiodes. Equations (4.80) and (4.81) assume a perfect 3-dB coupler with a 50% power-splitting ratio.

The back-to-back photodiode configuration provides a current

$$I(t) = I_{\mathrm{sig}\,1} - I_{\mathrm{sig}\,2} = 2\Re\sqrt{P_s P_{\mathrm{LO}}} \cdot \cos\{2\pi\nu_{\mathrm{IF}}t + \varphi_s - \varphi_{\mathrm{LO}}\} \tag{4.82}$$

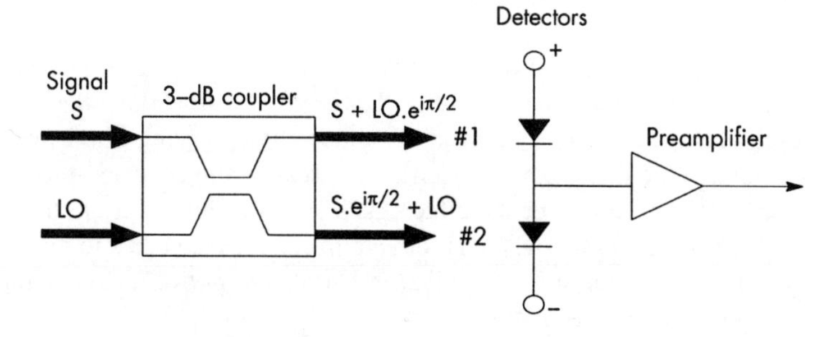

Figure 4.21 A two-port balanced coherent optical receiver.

Therefore, the dc terms of $I_{\text{sig 1}}$ and $I_{\text{sig 2}}$ (i.e., the first terms in the right-hand side of (4.80) and (4.81)) are completely canceled out. Since it is precisely those terms that have the most significant contributions to the noise induced by LO intensity noise, the RIN-induced power penalty has been eliminated. It should be noticed that balanced-receivers do not cancel intensity noise in the ac signal component (at ν_{IF}). However, its impact on system performance is much less severe owing to the square-root optical power dependency of this term.

Although the balanced-receiver configuration increases the complexity of the coherent receiver, and hence its cost, it has been widely used in all coherent lightwave systems. This is because, besides the ability to balance out the intensity fluctuations, all the signal and the local-oscillator powers are used effectively (recall what has been said in Section 4.3.1). This is to be contrasted with the case of single-photodetector coherent receivers where half the optical power coming out from the second output port of the 2×2 coupler is wasted (i.e., not used).

4.3.4.3 Polarization Mismatch

The polarization of the received signal and the local oscillator must be aligned to produce a photocurrent with maximum strength. If polarizations are orthogonal, no beating occurs and no signal photocurrent is produced.

The effect of polarization mismatch can be understood more quantitatively from the analysis of Section 4.3.1, where the polarizations of $\vec{E}_s$ and $\vec{E}_{\text{LO}}$ have been assumed linear and parallel. If $\vec{u}_s$ and $\vec{u}_{\text{LO}}$ denote the unit vector indicating the direction of polarization of $\vec{E}_s$ and $\vec{E}_{\text{LO}}$ respectively, then (4.57) becomes

$$I(t) = \Re\,(P_s + P_{\text{LO}}) + 2\Re\sqrt{P_s P_{\text{LO}}} \cdot \cos\theta \cdot \cos\{2\pi\nu_{\text{IF}}t + \varphi_s - \varphi_{\text{LO}}\} \qquad (4.83)$$

where θ is defined by $\cos\theta = \vec{u}_s \cdot \vec{u}_{\text{LO}}$ (the "·" indicates the vector product).

It is clear from (4.83) that any change of θ from its ideal value (i.e., $\theta = 0$) reduces the signal and degrades the receiver performance. This phenomenon is called *polarization fading*.

The polarization of the local oscillator is determined by the LO laser and remains fixed in time. The polarization state of the signal after propagation along the fiber network will, in general, change with time because of random time-varying birefringence of the fiber (Section 3.2.2). Such changes occur on a time scale, ranging from microseconds to minutes primarily due to environmental effects such as temperature variations [Imai 1988, Calvani 1989, Nicholson 1989]. They lead to random changes of $\cos\theta$ and render coherent receivers unusable unless measures are taken to make the performance of the receiver independent of polarization fluctuations.

There are five approaches to deal with the polarization mismatch problem.

1. *Installing special polarization-maintaining fibers*. There are two prices to be paid in applying this approach: the fibers already installed are not of the polarization-maintaining type, which is more expensive and has higher losses than conventional fibers.
2. *Polarization tracking*: $\vec{u}_s$ is first measured and then corrected to align it with $\vec{u}_{LO}$. This approach has been implemented (in laboratory) using different techniques to correct the received signal polarization (e.g., passive waveplates, optical-fiber coils, Faraday rotator, or electro-optic devices) [Okoshi 1985, Walker 1990]. The main drawback of such an approach is the resulting bulkiness of the receiver.
3. *Polarization scrambling*: this forces $\vec{u}_s$ to change rapidly and randomly during one bit duration, thus ensuring that $\vec{u}_s$ and $\vec{u}_{LO}$ will never be orthogonal for an entire bit [Zhou, 1993].
4. *Polarization shift keying* (POLSK): in this modulation scheme one state of polarization, $\vec{u}_{s,0}$, represents a bit "0" and the orthogonal state of polarization, $\vec{u}_{s,1}$, represents a bit "1." Since this orthogonality survives during propagation along the fiber, a differential receiver independent of the absolute state of polarization can be used.
5. *Polarization diversity:* this is shown in Figure 4.22 [Kazovsky, 1989]. At the output of the 3-dB coupler, a polarization beam splitter provides two outputs with orthogonal polarizations, which are then processed separately. The polarization of the local oscillator is adjusted to produce equal LO powers at the two outputs of the polarization beam splitter; it does not have to be readjusted when the signal polarization varies in time. The photocurrents generated by the two photodetectors are squared and added such that the output signal becomes polarization-independent.

The polarization-diversity technique has been combined with balanced receivers to avoid the limitations imposed by intensity noise as well as those imposed by polarization fluctuations. A balanced polarization-diversity receiver is shown in Figure 4.23. It requires four photodiodes, but nevertheless has the potential to be integrated on the same chip to provide a compact receiver.

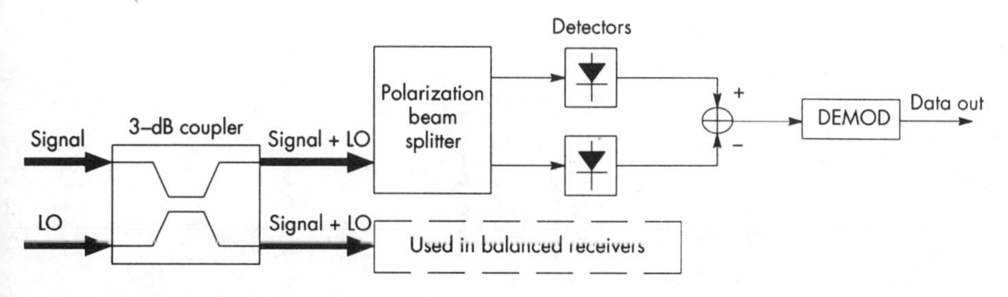

Figure 4.22 Polarization-diversity receiver.

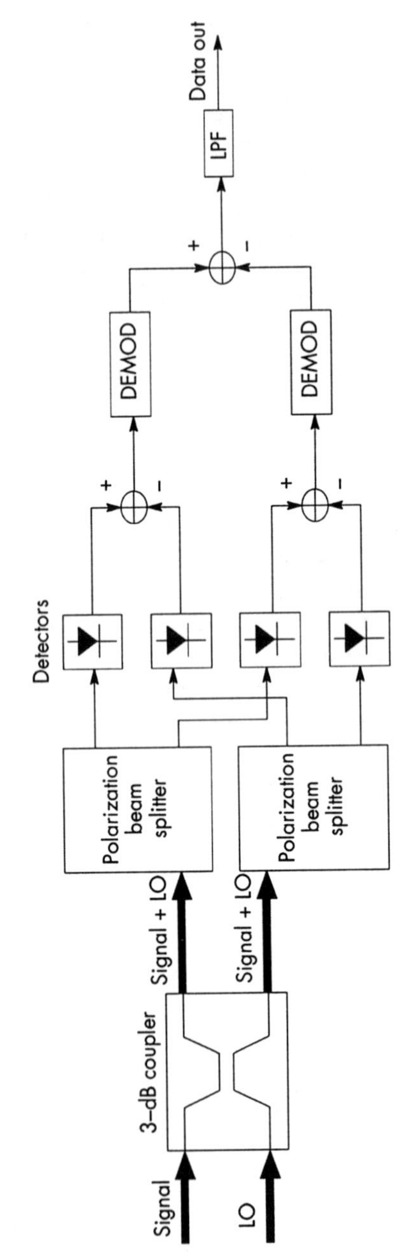

Figure 4.23 A balanced-polarization diversity coherent receiver.

4.3.5 Demonstrated Coherent Systems

Coherent communication system experiments have been carried out using all possible combinations of modulation format (ASK, FSK, PSK, DPSK)/optical detection technique (heterodyne or homodyne)/IF processing. The major goal of these experiments was to demonstrate the improved sensitivity obtained with coherent detection receivers as compared to DD receivers.

The general conclusion of these experiments is that receiver sensitivities of 3.5 to 7 dB above the quantum limit can be achieved in practice. This represents an improvement of more than 15 to 20 dB compared to DD receivers operating under the same system conditions (i.e., same bit rate, fiber length, etc.).

Several field trials using coherent lightwave technology have also been carried out in order to test this new technology in a life environment before its commercialization. Both land-based and submarine point-to-point systems have been tested [Van Bochove, 1993].

In all cases, a balanced polarization-diversity heterodyne receiver with asynchronous demodulation was used. The modulation format chosen was the so-called CPFSK (continuous phase FSK) [Iwashita, 1987], although the first field trial in 1988 used DPSK at 565 Mbps. The bit rate of the most recent field trials has been raised up to 2.48 Gbps.

In spite of these successful experiments and field trials, coherent lightwave systems have not as yet reached the commercial stage. Two main reasons can be given for this:

1. Coherent transmitters and receivers are more complex than their DD counterparts. This leads to increased system cost and decreased reliability. Although efforts are under way to solve these problems by designing optoelectronic integrated circuits (OEICs) [Deri, 1992], the same trend occurs for DD systems.
2. The sensitivity of DD receivers with optical preamplifiers (discussed in Chapters 5 and 6) can approach that of coherent detection receivers.

Nevertheless, coherent techniques reveal to be attractive for multichannel communication systems, such as shown in Figure 4.24 [Kazovsky, 1987]. The output of a bank of N transmitters, each operating at its own wavelength, is combined in the optical domain and the multiplexed signal is launched into a single fiber. At the receiver end, a bank of N receivers, each tuned to a distinct channel wavelength, are used to recover the original data.

For identical channel bit rate B, the system capacity (defined as the product of the bit rate times the fiber length) is enhanced by a factor N. Such a point-to-point multichannel transmission system can be implemented using either direct detection or coherent detection receivers.

When direct detection is used, each receiver must be equipped with an optical filter tuned to one of the N distinct wavelengths used in the system. Since practical optical filters have a rather broad frequency response, they can only be used to separate fairly widely

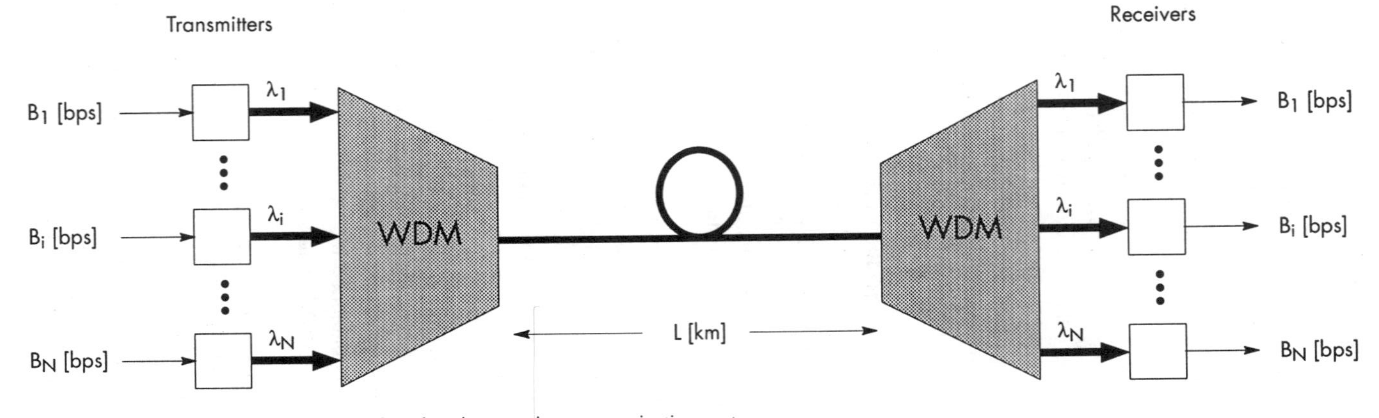

Figure 4.24 A multichannel multiwavelength point-to-point communication system.

spaced channel wavelengths of the order of one nanometer (i.e., $\Delta\nu$ = 180 GHz at λ = 1.55-μm wavelength).

A point-to-point transmission capacity of BL = 1.37 Tbps·km has been demonstrated with DD receivers by transmitting 10 channels at 2 Gbps over about 70 km of single-mode fiber. The channel spacing was 1.35 nm [Olsson, 1985].

When coherent detection is used, the channel spacing can be as small as a few times the channel bit rate (e.g., a channel spacing $\Delta\nu$ of 10 GHz has been proven sufficient for a channel bit rate of 2.5 Gbps). This results from the ability to design IF amplifiers and filters with a frequency bandpass response that is superior to any response of an optical filter. In particular, the steepness of the IF filter cutoff is much higher, allowing the channels to be spaced closely together without significant degradation from interchannel crosstalk.

In a coherent detection experiment, 100 channels at 622 Mbps with a 10-GHz separation have been transmitted over 50 km of fiber [Toba, 1990]. The transmission capacity for this experiment exceeded 3 Tbps·km. Another coherent detection experiment transmitted 4×2.5 Gbps channels over 300 km of fiber, resulting in a similar system capacity of 3 Tbps·km [Ichihashi, 1993].

Notice that such BL products would not be possible for single-channel time-division multiplexed (TDM) systems, as fiber dispersion limits the maximum BL product to values considerably below 1 Tbps·km.

Although the resulting system capacities of the above examples are similar for direct detection and coherent detection, coherent lightwave systems have the potential to fully fill the available fiber bandwidth. The realization of such very high capacity links is, however, well beyond the current state of the art.

4.4 SUMMARY

We have seen that the most prevalent photodetectors for lightwave fiber communication systems are the PIN and APD semiconductor photodiodes. For the 1.3 to 1.6-μm wavelength band where optical fibers provide the lowest attenuation and dispersion, InGaAs-based detectors exhibit the best characteristics.

These optoelectronic converters can be used in two distinct configurations; namely, the DD receiver and the coherent detection receiver.

With DD receivers, the thermal-noise contributions of the load resistor and the preamplifier essentially determine the receiver sensitivity and much design effort goes into the design of the receiver front-end, with emphasis on the preamplifying stage. Because of the capacitive nature of the photodetector, high-impedance preamplifiers, while providing the best optical receiver sensitivity, require the incorporation of an equalizer to operate at high bit rates. The transimpedance alternative, which sacrifices some noise performance for increased bandwidth and dynamic range, is often used as it is much more attractive, especially for multiaccess networks. With current designs, the sensitivity of DD receivers is 20 to 25 dB above the fundamental quantum limit set by the inherent shot noise, which results from the quantum nature of light.

The quantum-limit receiver sensitivity can be better approached with coherent detection, but at the expense of increased complexity and stringent requirements on the components to be used. This improved receiver sensitivity of coherent detectors is very advantageous in multiaccess networks since such networks generally suffer from a limited power budget, which limits the maximum number of nodes that can be connected to the network.

However, improved receiver sensitivity is currently fading away as optical amplifiers (discussed in the next chapter) make it possible to boost the lightwave signal whenever desired. Nevertheless, coherent detection holds the promise of allowing the full utilization of the available fiber network bandwidth. This will be explored in greater detail in Chapter 7.

REFERENCES

[Personick, 1973] Personick, S. "Receiver design for digital fiber optic communication systems," *Bell System Technical Journal*, Vol. 52, 1973, pp. 843–874.

[DeLange, 1968] DeLange, O. E. "Optical heterodyne detection," *IEEE Spectrum Magazine*, Oct. 1968, pp. 77–85.

[Marcuse, 1980] Marcuse, D. *Principles of Quantum Electronics*, New York, NY: Academic Press, 1980.

[Smith, 1980] Smith, R. G., and S. D. Personick. in *Semiconductor Devices for Optical Communications*, Edited by H. Kressel, New York, NY: Springer-Verlag, 1980.

[Okano, 1980] Okano, Y., et al. "Laser mode partition noise evaluation for optical fiber transmission," *IEEE Trans. on Communication*, Vol. COM-28, No. 2, Feb. 1980, pp. 238–243.

[Personick, 1981] Personick, S. "Photodetectors for fiber systems," in *Fundamentals of Optical Fiber Communications*, Edited by M. Barnovsky, New York, NY: Academic Press, 1981, pp. 295–328.

[Ogawa, 1982] Ogawa, K. "Analysis of mode partition noise in laser transmission systems," *IEEE J. Quantum Electron.*, Vol. QE-18, No. 5, May 1982, pp. 849–855.

[Smith, 1982] Smith, R., et al. "Receiver design for optical fiber communication systems," in *Topics in Applied Physics*, Vol. 39, Edited by H. Kressel, Berlin, Germany: Springer-Verlag, 1982, pp. 89–160.

[Kobayashi, 1982] Kobayashi, S., et al. *IEEE J. Quantum Electron.*, Vol. QE-18, 1982, p. 582.

[Emura, 1984] Emura, K., et al. "Novel optical *FSK* heterodyne single filter detection system using a directly modulated *DFB*-laser diode," *Electron. Letters*, Vol. 20, No. 24, 1984, pp. 1022–1023.

[Muoi, 1984] Muoi, T. V. "Receiver design for high-speed optical-fiber systems," *IEEE J. Light. Technology*, Vol. LT-2, No. 3, 1984, pp. 243–267.

[Senior, 1985] Senior, J. M. *Optical Fiber Communications: Principles and Practice*, Series Editor P. J. Dean, London, England: Prentice-Hall, 1985.

[Abbas, 1985] Abbas, G. L. "A dual-detector optical heterodyne receiver for local oscillator noise suppression," *IEEE J. Light. Technology*, Vol. LT-3, No. 5, Oct. 1985, pp. 1110–1122.

[Yamazaki, 1985] Yamazaki, S., et al. "Realization of a flat *FM* response by directly modulating a phase tunable *DFB* laser diode," *Electron. Letters*, Vol. 21, No. 7, 1985, pp. 283–285.

[O'Reilly, 1985] O'Reilly, J. J., et al. "Influence of timing errors on the performance of direct-detection optical-fiber communication systems," *IEEE Proc. J. Optoelectron.*, Vol. 132, 1985, pp. 309–313.

[Okoshi, 1985] Okoshi, T. "Polarization-state control schemes for heterodyne or homodyne optical fiber communications," *IEEE J. Light. Technology*, Vol. LT-3, No. 6, 1985, pp. 1232–1237.

Hodgkinson, 1985] Hodgkinson, T. G., et al. ''Demodulation of optical *DPSK* using in-phase and quadrature detection,'' *Electron. Letters*, Vol. 21, No. 19, Sept. 1985, pp. 867–868.

Olsson, 1985] Olsson, N. A., et al. ''68.3*km* transmission with 1.37*Tbit.km/s* capacity using wavelength division multiplexing of ten single-frequency lasers at 1.5 μm,'' *Electron. Letters*, Vol. 21, 1985, pp. 105–106.

Hong, 1986] Hong, C. K., et al. *Phys. Rev. Letters*, Vol. 56, 1986, p. 58.

Agrawal, 1986] Agrawal, G. P., et al. ''Power penalty due to decision-time jitter in optical communication systems,'' *Electron. Letters*, Vol. 22, No. 9, April 1986, pp. 45–451.

Shen, 1986] Shen, T. M. ''Power penalty due to decision-time jitter in receivers using avalanche photodiodes,'' *Electron. Letters*, Vol. 22, 1986, pp. 1043–1045.

Kazovsky, 1986] Kazovsky, L. G. ''Coherent optical receivers: performance analysis and laser linewidth requirements,'' *Optical Engineering*, Vol. 25, No. 4, 1986, pp. 575–579.

Tucker, 1986] Tucker, R. S., et al. *Electron. Letters*, Vol. 22, 1986, p. 917.

Iwashita, 1987] Iwashita, K., et al. ''Modulation and detection characteristics of optical continuous phase *FSK* transmission system,'' *IEEE J. Light. Technology*, Vol. LT-5, No. 4, April 1987, pp. 452–460.

Schumacker, 1987] Schumacker, K., et al. ''Power penalty due to jitter on optical communication systems,'' *Electron. Letters*, Vol. 23, No. 14, July 1987, pp. 718–719.

Linke, 1987] Linke, R. A., et al. ''Coherent optical detection: a thousand calls on one circuit,'' *IEEE Spectrum Magazine*, Feb. 1987, pp. 52–57.

Kimura, 1987] Kimura, T. ''Coherent optical fiber transmission,'' *IEEE J. Light. Technology*, Vol. LT-5, No. 4, 1987, pp. 414–428.

Pettit, 1987] Pettit, M. J., et al. Optical *FSK* transmission system using a phase-diversity receiver,'' *Electron. Letters*, Vol. 23, No. 20, Sept. 1987, pp. 1075–1076.

Davis, 1987] Davis, A. W., et al. ''Phase diversity techniques for coherent optical receivers,'' *IEEE J. Light. Technology*, Vol. LT-5, No. 4, April 1987, pp. 561–572.

Korotky, 1987] Korotky, S. K., et al. *Applied Phys. Letters*, Vol. 50, 1987, p. 1631.

Wakita, 1987] Wakita, K., et al. *IEEE Photon. Technology Letters*, Vol. 23, 1987, p. 303.

Kazovsky, 1987] Kazovsky, L. G. ''Multichannel coherent optical communication systems,'' *IEEE J. Light. Technology*, Vol. LT-5, 1987, p. 1095.

Okoshi, 1988] Okoshi, T., and K. Kikuchi. *Coherent Optical Fiber Communications*, Dordrecht (Holland): KTK Kluwer Academic Publishers, 1988.

Kasper, 1988] Kasper, B. L. ''Receiver design,'' in *Optical Fiber Telecommunications I,* Edited by S. Miller and A. G. Chynoweth, New York, NY: Academic Press, 1988, pp. 593–626.

Miller, 1988] Miller, S. E., and I. P. Kaminow, eds. *Optical Fiber Telecommunications-II*, New York, NY: Academic Press, Chapter 18, 1988, pp. 689–717.

Agrawal, 1988] Agrawal, G. P., et al. ''Dispersion penalty for 1.3 μm lightwave systems with multimode semiconductor lasers,'' *IEEE J. Light. Technology*, Vol. LT-6, No. 5, May 1988, pp. 620–625.

Imai, 1988] Imai, T., et al. ''Polarization fluctuations in a single-mode optical fiber,'' *IEEE J. Light. Technology*, Vol. LT-6, No. 9, 1988, pp. 1366–1375.

Welter, 1988] Welter, R., et al. ''150*Mb/s* phase diversity *ASK* homodyne receiver with a linewidth/bit rate of 0.5,'' *Electron. Letters*, Vol. 24, No. 4, Feb. 1988, pp. 199–201.

Calvani, 1989] Calvani, R. R., et al. ''Polarization measurements on single-mode fibers,'' *IEEE J. Light. Technology*, Vol. LT-7, No. 8, 1989, pp. 1187–1196.

Nicholson, 1989] Nicholson, G., et al. ''Polarization fluctuation measurements on installed single-mode optical fiber cables,'' *IEEE J. Light. Technology*, Vol. LT7, No. 8, 1989, pp. 1197–1200.

Kotoka, 1989] Kotoka, I., et al. *IEEE Photon. Technology Letters*, Vol. 1, 1989, p. 100.

Kazovsky, 1989] Kazovsky, L. G. ''Phase- and polarization diversity coherent optical techniques,'' *IEEE J. Light. Technology*, Vol. 7, No. 2, Feb. 1989, pp. 279–292.

Cimini, 1990] Cimini, L. J., et al. ''Optical equalization to combat the effects of laser chirp and fiber dispersion,'' *IEEE J. Light. Technology*, Vol. LT-8, No. 5, May 1990, pp. 649–659.

Mack, 1990] Mak, G., et al. *IEEE Photon. Technology Letters*, Vol. 2, 1990, p. 730.

[Goto, 1990] Goto, M., et al. *IEEE Photon. Technology Letters*, Vol. 2, 1990, p. 896.

[Kazovsky, 1990] Kazovsky, L. G., et al. "A 1320*nm* experimental optical phase-locked loop: performance investigation and homodyne *PSK* experiments at 140*Mb/s* and 2*Gb/s*," *IEEE J. Light. Technology*, Vol. LT-8, No. 9, 1990, pp. 1414–1425.

[Walker, 1990] Walker, N. G., et al. "Polarization control for coherent communications," *IEEE J. Light. Technology*, Vol. LT-8, No. 3, 1990, pp. 438–458.

[Toba, 1990] Toba, H., et al. "100-channel optical *FDM* transmission/distribution at 622*Mb/s* over 50*km*," *Proc. Opt. Fiber Communication Conf., OFC'90*, San Francisco, CA, PD1, Feb. 1990, see also *Electron. Letters*, Vol. 26, 1990, p. 376.

[Saleh, 1991] Saleh, B.E.A., and M. C. Teich. *Fundamentals of Photonics*, New York, NY: Wiley & Sons, 1991.

[Agrawal, 1992] Agrawal, G. P. *Fiber-Optic Communication Systems*, New York, NY: Wiley & Sons, 1992.

[Deri, 1992] Deri, R. J., et al. "High-speed heterodyne operation of monolithically integrated balanced polarization diversity photodetectors," *Electron. Letters*, Vol. 28, No. 25, Dec. 1992, pp. 2332–2334.

[Wenworth, 1992] Wenworth, R. H., et al. "Laser mode partition noise in lightwave systems using dispersive optical fibers," *IEEE J. Light. Technology*, Vol. LT-10, No. 1, Jan. 1992, pp. 84–89.

[Green, 1993] Green, P. E. *Fiber Optic Networks*, New York, NY: Prentice Hall, 1993.

[Powers, 1993] Powers, J. P. *An Introduction to Fiber Optic Systems*, Boston, MA: Irwin and Aken Associates, 1993.

[Kuebart, 1993] Kuebart, W., et al. "Monolithically integrated 10*Gb/s* *InP*-based receiver *OEIC*: design and realization," *Proc. European Conf. Opt. Commun., ECOC'93*, TuP6.4, Sept. 1993, pp. 305–308.

[Soole, 1993] Soole, J. B., et al. "Integrated grating demultiplexer and *pin* array for high-density wavelength division multiplexed detection at 1.5 μ*m*," *Electron. Letters*, Vol. 29, No. 6, March 1993, pp. 558–560.

[Zhou, 1993] Zhou, J., et al. "Polarization-insensitive coherent transmission at 622*Mbit/s* by synchronous half-bitrate polarization spreading," *Proc. European Conf. Opt. Commun., ECOC'93*, WeC10.4, Sept. 1993, pp. 537–540.

[Ichihashi, 1993] Ichihashi, Y., et al. "Four-channel *FDM* 300*km* repeaterless transmission using 2.5*Gb/s* *CPFSK*/heterodyne detection system," *Proc. European Conf. Opt. Commun., ECOC'93*, Sept. 1993, pp. 529–532.

[Van Bochoven, 1993] Van Bochoven, A. C., et al. Implementation of practical coherent trunk system," *Proc. Opt. Fiber Commun. Conf., OFC/IOOC'93*, ThH1, Feb. 1993, pp. 191–192.

Chapter 5
Optical Amplifiers

Probably one of the most important achievements in photonic device engineering during the last few years is the development of high-performance optical amplifiers, especially the Erbium-doped fiber amplifier.

Conventional submarine and long-haul terrestrial fiber-optic systems are optoelectronic hybrids that contain regenerative repeaters. These repeaters detect the light photoelectrically, amplify the resulting current electronically and use the amplified current to pump a laser diode, which reconverts the electric signal into an intense light signal that is sent on to the next repeater in the system.

Clearly, if the optical signal can be directly amplified with low noise and in a reliable way, these complex regenerative repeaters can be made simpler, smaller, and cheaper.

Transmission at gigabits per second over thousands of kilometers using optical amplifiers (OAs) as repeaters has been demonstrated. It is expected that more and more OAs will be employed in optical-fiber communication systems in the near future. For example, the first transatlantic transmission system employing OAs is targeted for service by 1995 [Zyskind, 1992].

Optical amplifiers are not only important devices in long-length point-to-point fiber-optic transmission systems, but also in many multiaccess networks, in which attenuation is caused by multiple tappings or by distribution of the same signal onto several receiver nodes.

There are essentially three distinct ways to use an optical amplifier, as shown in Figure 5.1:

1. The power amplifier, which boosts the laser output power to a higher level;
2. The line amplifier, which boosts the signal periodically along the transmission path to compensate for the losses due to attenuation or splitting;
3. The receiver preamplifier, which enhances receiver sensitivity.

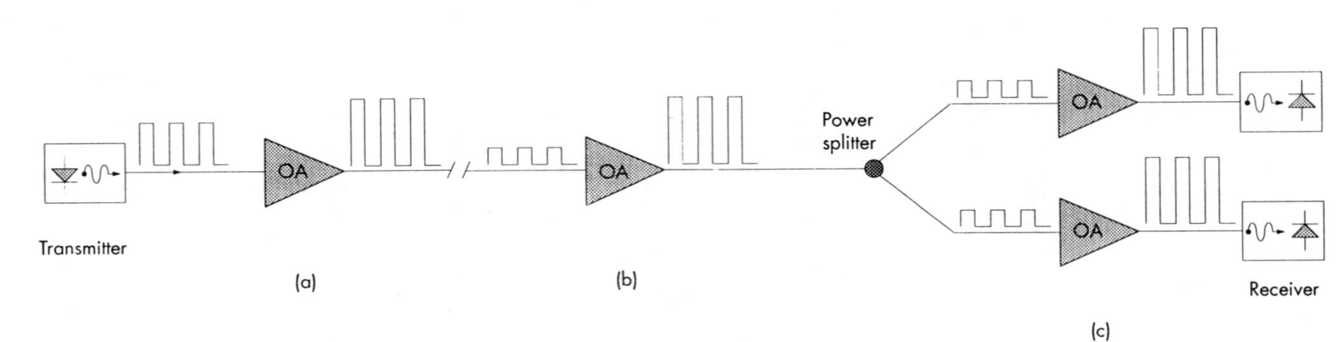

Figure 5.1 Application of optical amplifiers: (a) power amplifier, (b) line amplifier and (c) receiver preamplifier.

The optimum configuration depends upon technological tradeoffs and the envisaged applications, as well as system design constraints.

This chapter summarizes the current state of optical amplifiers with focus on the two most popular amplifier technologies; namely, semiconductor amplifiers and fiber-doped amplifiers. First, semiconductor amplifiers will be discussed in some detail with regard to the gain as a function of wavelength, the saturation output power, the crosstalk behavior, and the noise characteristics. We will then proceed to the discussion of doped-fiber amplifiers, which are at present considered to be more attractive than their competing semiconductor amplifiers because of their easier and better fiber coupling, their polarization independence, and their crosstalk-free behavior.

Today, however, the only practical doped-fiber amplifier operates in the 1.5-μm wavelength window, whereas most telecom systems operate in the 1.3-μm band. Semiconductor amplifiers can operate as easily at 1.3 μm as at 1.5 μm.

Furthermore, semiconductor amplifiers have the advantage of smaller power consumption and owing to their miniature size they can be integrated monolithically with other optical circuits, opening the way to high-performance optoelectronic integrated circuits (OEICs) and eventually to the development of optical data processing [Kalman 1992, Eiselt 1992].

Several other forms of optical amplifications, such as Raman and Brillouin amplifications, have also been investigated and experimentally demonstrated [Desurvire 1983, Atkins 1986, Mollenauer 1988]. Although these forms of optical amplification have resulted in impressive system performances, they have been superseded by erbium-doped fiber amplifiers and will not be considered here.

5.1 SEMICONDUCTOR AMPLIFIERS

A semiconductor amplifier is basically a laser diode that operates below the oscillation threshold.

Two basic types of semiconductor amplifiers can be distinguished, Fabry-Perot amplifiers (FPAs) and traveling wave amplifiers (TWAs). In FPAs, the two cleaved-crystal facets act as partially reflective end mirrors that form a Fabry-Perot cavity. The natural reflectivity of the facets is approximately 32%, but can be modified over a wide range by the addition of dielectric coatings. An incident light signal is coupled into the cavity, where it is amplified during successive passes between mirrors and emitted again at a higher intensity.

The structure of a TWA is identical to that of an FPA except that the facets of each end of the cavity are coated with a multilayer antireflection coating to prevent internal feedback (another way to avoid internal feedback is to cleave the facets so that they are no longer parallel). Thus, the light signal to be amplified travels through the device only once, emerging intensified at the other end. Since a TWA is essentially an FPA with zero end-facet mirror reflectivities, the performances of a TWA can be obtained from those of the FPA in the limit where reflectivities of both facets tend to zero.

5.1.1 The Fabry-Perot Amplifier

Consider Figure 5.2, which represents a schematic view of a Fabry-Perot amplifier. Let us assume, for the moment, that the single-pass power amplification factor is G_s.

An incident light of field strength E_i is directed towards the FPA and $t_1 E_i$ enters the cavity. After one single-pass in the amplifier of length L, the initial signal $t_1 E_i$ will be amplified by the gain medium and becomes equal to $t_1\sqrt{G_s}E_i e^{-jkL}$. Because the right-hand mirror has transparency t_2, the output signal amplitude after the first single-pass is $t_1 t_2\sqrt{G_s}E_i e^{-jkL}$. The reflected signal of amplitude $t_1 r_2\sqrt{G_s}E_i e^{-jkL}$ travels back to the left-hand mirror where its amplitude will be $t_1 r_2 G_s E_i e^{-2jkL}$. After reflection at this left-hand mirror, the reflected wave $t_1 r_2 r_1 G_s E_i e^{-2jkL}$ will be further amplified by the gain medium and at the output of the right-hand mirror the field will be given by $t_1 r_2 r_1 t_2 G_s\sqrt{G_s}E_i e^{-3jkL}$. Repeating this reasoning for all reflected waves, we find that the output field will be composed of an infinite series of transmitted components. If we suppose that the transit time into the cavity is much smaller than the duration of the incident signal field E_i [Kastler, 1974], the output field E_t will be the coherent sum of all these transmitted components. This leads to

$$E_t = E_i e^{-jkL} t_1 t_2 \sum_{m=0}^{\infty} (r_1 r_2 G_s)^m e^{-2mjkL} \tag{5.1}$$

which is a geometric progression with argument $(r_1 r_2 G_s)e^{-2jkL}$. For $|r_1 r_2 G_s| < 1$, (5.1) becomes

$$E_t = E_i \frac{\sqrt{G_s}t_1 t_2 e^{-jkL}}{1 - r_1 r_2 G_s e^{-2jkL}} \tag{5.2}$$

From (5.2), the power transfer function G of the FPA is given by

$$G = \left|\frac{E_t}{E_i}\right|^2 = \frac{G_s(1 - R_1)(1 - R_2)}{(1 - \sqrt{R_1 R_2}G_s)^2 + 4\sqrt{R_1 R_2}G_s \sin^2 kL} \tag{5.3}$$

where we have used (i = 1, 2):

$$T_i = t_i^2 \tag{5.4a}$$

$$R_i = r_i^2 \tag{5.4b}$$

$$\text{and } T_i + R_i = 1 \tag{5.4c}$$

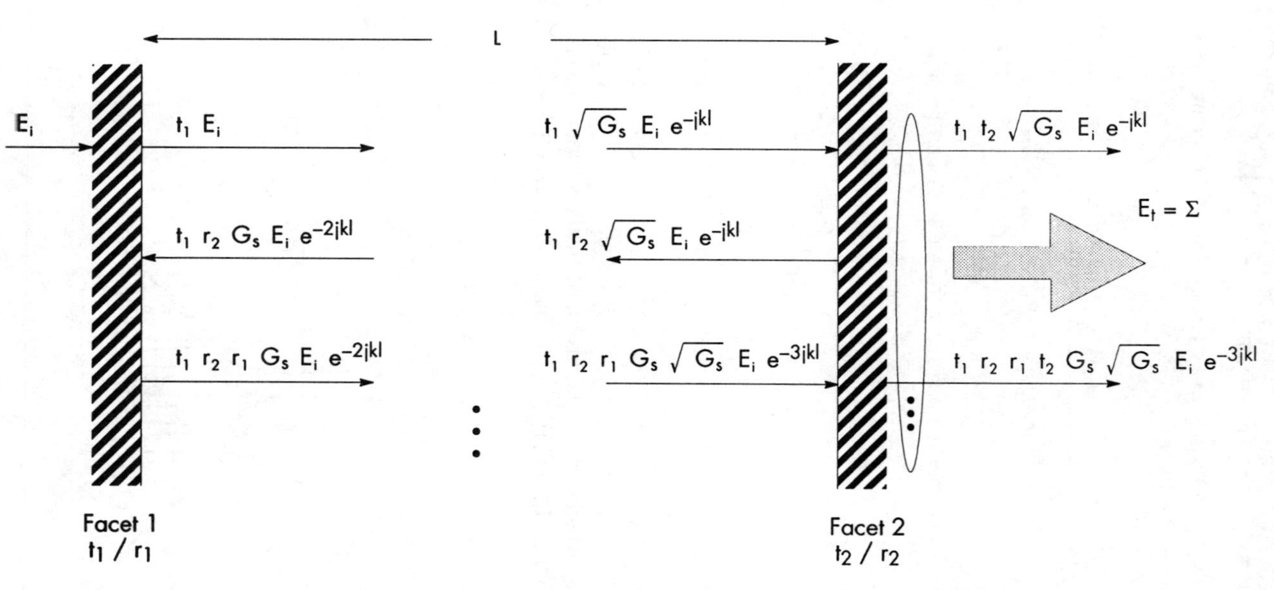

Figure 5.2 Schematic view of a Fabry-Perot amplifier. G_s is the power amplification factor of the amplifier medium. t_i and r_i ($i = 1, 2$) are the amplitude transmitivity and reflectivity of both facets, respectively.

Since $\sin^2(kL) = \sin^2\left(\frac{2\pi\nu}{c}L\right) = \sin^2\left(\frac{2\pi(\nu - \nu_0)}{c}L\right)$, where c is the light velocit in the amplifier medium and ν_0 the cavity resonant-mode frequency at which the gain i maximum, (5.3) can be rewritten as

$$G(\nu) = \frac{G_s(1 - R_1)(1 - R_2)}{(1 - \sqrt{R_1R_2}G_s)^2 + 4\sqrt{R_1R_2}G_s \sin^2\left(\frac{2\pi(\nu - \nu_0)L}{c}\right)} \quad (5.5$$

If the reflectivities of the two end facets of the FPA are the same, (5.5) becomes

$$G(\nu) = \frac{G_s(1 - R)^2}{(1 - RG_s)^2 + 4RG_s \sin^2\left(\frac{2\pi(\nu - \nu_0)L}{c}\right)} \quad (5.6$$

A typical gain curve $G(\nu)$ is depicted in Figure 5.3 for a Gaussian-shaped single pass gain G_s and for two values of $R = 0.32$ and $R = 0.02$.

For $R = 0.32$, which is typical for FPAs, $G(\nu)$ depicts large ripples due to th filtering function of the resonant cavity. For $R = 0.02$, $G(\nu)$ is very close to G_s but sti shows small ripples, which can be removed when the residual facet reflectivities ar further decreased so that the amplifier essentially becomes a TWA.

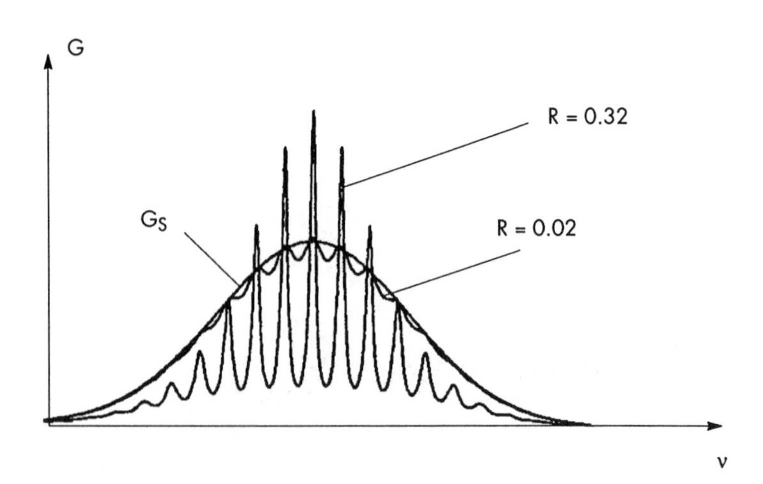

Figure 5.3 Typical gain curve $G(\nu)$ for a Gaussian-shaped single-pass gain G_s, Fabry-Perot amplifier ($R = 0.32$ and near-traveling wave amplifier ($R = 0.02$).

5.1.1.1 The 3-dB Bandwidth

The 3-dB bandwidth (full-width at half maximum, FWHM) is defined as the frequency detuning for which the power gain $G(\nu)$ is reduced by a factor 2. From (5.6), we immediately obtain

$$\Delta\nu_{3\mathrm{dB}} = \frac{c}{\pi L} \sin^{-1}\left[\frac{1 - RG_s}{2\sqrt{RG_s}}\right] \tag{5.7}$$

From (5.6) and (5.7), an important formula that relates the maximum gain to the achievable bandwidth can readily be obtained; for $RG_s \rightarrow 1$, we have $(1 - RG_s)/2\sqrt{RG_s} << 1$, and (5.7) can be approximated by

$$\Delta\nu_{3\mathrm{dB}} \cong \frac{c}{\pi L} \frac{1 - RG_s}{2\sqrt{RG_s}} \tag{5.8}$$

Combining (5.8) with (5.6), we obtain

$$\sqrt{G^{\max}} \cdot \Delta\nu_{3\mathrm{dB}} = \frac{c}{2\pi L} \frac{1 - R}{\sqrt{R}} = C^{\mathrm{ste}} \tag{5.9}$$

where $G^{\max} = G(\nu = \nu_0) = \dfrac{G_s(1 - R)^2}{(1 - RG_s)^2}$.

Equation (5.9) clearly expresses the tradeoff between bandwidth and gain for an FPA.

Note that the FPA is only useful if $G^{\max} > G_s$. Otherwise, it would be preferable to use an amplifier with gain G_s for a single-pass, that is a TWA. In order to realize $G^{\max} > G_s$, we see from (5.9) that the following condition must be satisfied: $1 < G_s < 1/R$. For $G_s = 1/R$, $RG_s = 1$ and from (5.9) $G^{\max} \rightarrow \infty$. This limit represents the oscillation threshold at which the amplifier behaves like a laser source (Chapter 2).

It can be shown that an asymmetrical cavity configuration with low input and high-output facet reflectivities leads to greater $G^{\max}$ than a symmetrical cavity configuration [Yamamoto, 1980]. As we will see shortly, this is closely related to the spontaneous emission noise.

Another important feature of semiconductor amplifiers (both FPA and TWA) is that gains for transverse electric (TE) and transverse magnetic (TM) polarized light modes are not the same [Jopson 1986, Simon 1987]. Distinct configurations that combine two laser-diode amplifiers have been proposed to obtain amplifier modules that are nearly polarization-insensitive [Grosskopf 1987, Magari 1990, Koga 1991, Eskildsen 1992].

5.1.1.2 Gain Saturation

The single-pass gain in the amplifier medium of length L obeys the exponential relation [Yamamoto 1980, Mukay 1981]

$$G_s = \exp\{(\Gamma g - \alpha)L\} \tag{5.10}$$

in which Γ is the confinement factor that represents the distribution of the optical field in the active layer, α is the loss coefficient (α = 14 cm^{-1} in bulk GaAs), and g is the gain coefficient of the amplifier per unit length.

The single-pass gain G_s therefore increases exponentially with increases in the device length.

However, the internal gain coefficient g is limited by gain saturation; that is, with a sufficiently large optical power, further increases in the input signal no longer result in appreciable change in output. Gain saturation occurs because when the signal power is too large, there are not sufficient excited carriers (electron-hole) to provide enough stimulated amplification. Gain saturation is expressed in terms of the amplifier output power by

$$g(\nu) = \frac{g_0(\nu)}{1 + \left(\frac{P_{out}}{P_{sat}}\right)} \tag{5.11}$$

Thus, with (5.10) and (5.11) we obtain

$$G_s = \exp\left[\left(\frac{\Gamma g_0}{1 + \left(\frac{P_{out}}{P_{sat}}\right)} - \alpha\right)L\right] \tag{5.12}$$

Once P_{out} exceeds the saturation level P_{sat} of the amplifier, the gain of the device will rapidly decrease from its maximal (small-signal) value $\exp[(\Gamma g_0 - \alpha)L]$.

5.1.2 The Traveling Wave Amplifier

An ideal TWA can be viewed as an FPA for which the end facet reflectivities are equal to zero. As shown in Figure 5.3, the 3-dB bandwidth of the TWA is determined by the full-gain width of the amplifier medium itself without being filtered by the Fabry-Perot cavity transfer function. The 3-dB bandwidth of TWA is about three orders of magnitude larger than that of the FPA. TWAs are therefore much more attractive candidates than FPAs, especially for network applications.

We have seen that the square-root-gain bandwidth product is determined only by the cavity length L and the facet reflectivity R. The lower the facet reflectivity is made, the larger the (square-root gain x bandwidth) product becomes. With present technology it is possible to reduce the residual facet reflectivities to as low as 10^{-4} to 10^{-5}, making it possible to obtain an amplification gain curve very close to the gain curve of the amplifier medium.

On the other hand, because the bandwidth of TWA is much larger than that of an FPA, the maximum gain of the TWA is expected to be very low. However, the very large bandwidth available from a TWA can be substantially reduced in order to increase its (single-pass) gain; that is, increasing the length L of the amplifier while keeping very small residual reflectivities results in an increase of the TWA gain.

Internal single-pass gains as high as 27.5 dB have been achieved. However, the fiber coupling loss per facet was 5.5 dB, resulting in a fiber-to-fiber gain of 16.5 dB. In general, it has been proved difficult to maintain a fiber-to-fiber gain of more than 20 dB or so [O'Mahony, 1985]. The 3-dB bandwidth of TWA is typically of the order of 35 nm, but new structures with multiple quantum wells (MQWs) have the capability to exhibit a 3-dB bandwidth as large as 240 nm [Tabuchi, 1990].

5.1.2.1 Gain Saturation

As with FPA, the maximum available signal gain of a TWA is limited by gain saturation. However, in the case of TWA, (5.12) no longer holds. The difference between TWA and FPA comes from the fact that in a TWA, all the gain occurs on one pass. The signal power near the input of the TWA may have reached saturation well before it arrives at the output of the TWA. The differential gain provided by the downward portion of the amplifier medium (near the output) will decrease until the input power is so strong that this portion no longer gives rise to significant amplification.

With the FPA, by contrast, the signal power is relatively uniform throughout the cavity since the light makes many passes back and forth between the end facets (typically of the order of 10).

According to [Ramaswami, 1990], the TWA gain G_s as a function of the input power $P_{\text{in}} = P_{\text{out}}/G_s$ is given by the transcendental equation

$$G_s = 1 + \frac{P_{\text{sat}}}{P_{\text{in}}} \ln\left[\frac{G_0}{G_s}\right] \tag{5.13}$$

where G_0 is the maximum amplifier gain, corresponding to the single-pass gain in the absence of input light.

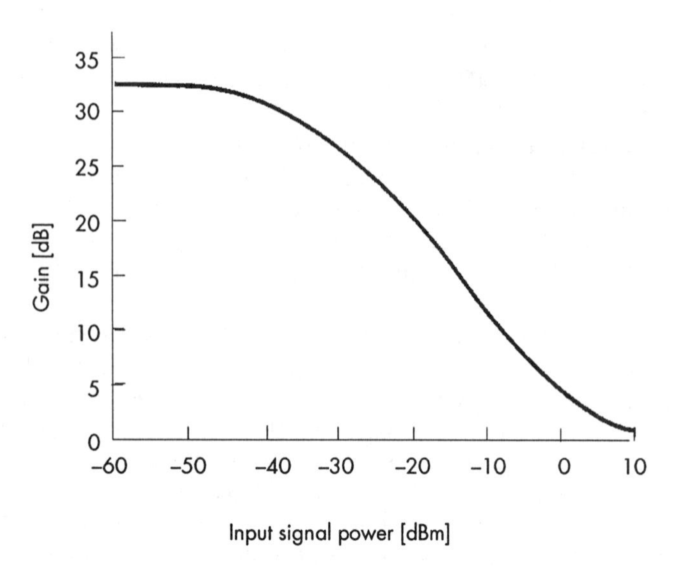

Figure 5.4 Gain saturation effect of TWA. P_{sat} = -6 dBm, and G_0 = 32.5 dB.

It can be shown that G_s is a monotonically decreasing function of P_{in}, and that th output power P_{out} is a monotonically increasing function of P_{in} [Ramaswami, 1990] Typical gain profile as a function of P_{in} calculated by (5.13) is depicted in Figure 5.4. A 1.3-μm laser-diode amplifier with MQW active layer has demonstrated a P_{sat} of +14 dBm (G_s = 18 dB) and a bandwidth of 110 nm [Sherlock, 1991].

5.1.2.2 Crosstalk

Crosstalk occurs when distinct optical signals, or channels, are allowed to superimpose i time and amplified simultaneously by the same amplifier. This results from gain satura tion, as shown in Figure 5.5.

In general, when several channels are simultaneously present inside the optica amplifier, the gain seen by a particular channel is affected by the intensity levels of th other channels.

The resulting effect depends on the carrier lifetime τ_e of the amplification medium With the help of the carrier-density rate equation [Yariv, 1991], it can be shown that th carrier density, and thus the amplifier gain, responds with a time-constant of the order o $\tau_e \cong 1$ ns.

Thus, when the input data rate is much smaller than the reciprocal of τ_e, th amplifier gain will vary according to the total instantaneous input power P_{in}. By contrast when the input data rate is much larger than $1/\tau_e$, P_{in} in (5.13) must be replaced by it time-averaged value.

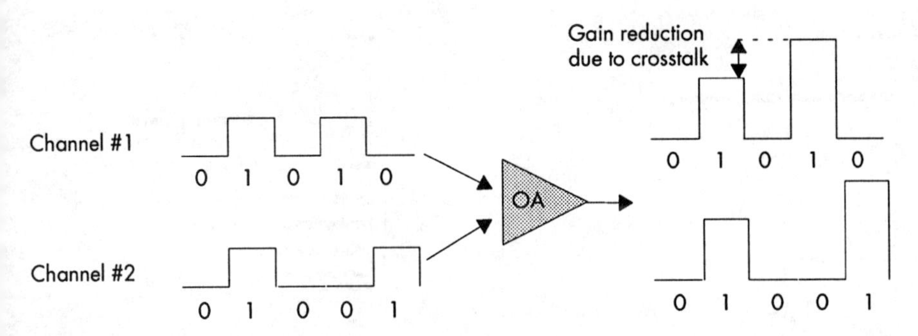

igure 5.5 Crosstalk effect in laser-diode amplifier for two-channel transmission.

Since the bit rates of most multiaccess optical-fiber networks range from few mega-its per second up to several gigabits per second, crosstalk in laser-diode amplifiers can ut limitations on network performance.

Analyses of crosstalk effects due to gain saturation in TWA can be found in Grosskopf 1986, Mukai 1987, Inoue 1989, Ramaswami 1990]. The treatment outlined ere follows the results of [Inoue 1989, Ramaswami 1990].

Consider N on-off keyed (OOK) channels, statistically independent but bit-synchro-ized, which are simultaneously amplified with a gain G_s given by (5.13). The channels ave the same average power, and bits "0" and "1" have the same probability of ccurrence.

Since the input signals are assumed to be bit-synchronized (worst-case assumption), he total input power P of the optical amplifier can take $(N + 1)$ values between 0 and NP_1, vhere P_1 is the power corresponding to a "1" bit (an extinction ratio $r = 0$ is assumed).

The probability that the input power takes the value kP_1 can be expressed by the inomial distribution

$$\text{Prob}\{P = kP_1\} = \frac{N!}{k!(N-k)!}\frac{1}{2^N} \tag{5.14}$$

The gain corresponding to input power kP_1 can be computed with (5.13).

The power of the amplified "1" bit of a particular channel will thus be affected by he statistics of the interfering channels according to (5.14) and (5.13), as shown quali-atively in Figure 5.6.

Assuming the receiver noise statistics are Gaussian, the bit error rate (BER) of a articular channel can be evaluated by

$$\text{BER} = \frac{1}{2}\sum_{k=0}^{N-1}\left\{\text{Prob}(kP_1)\text{erfc}\left(\frac{P_{k,1} - D}{\sqrt{2}\sigma_{k,1}}\right) + \frac{1}{2}\text{erfc}\left(\frac{D}{\sqrt{2}\sigma_0}\right)\right\} \tag{5.15}$$

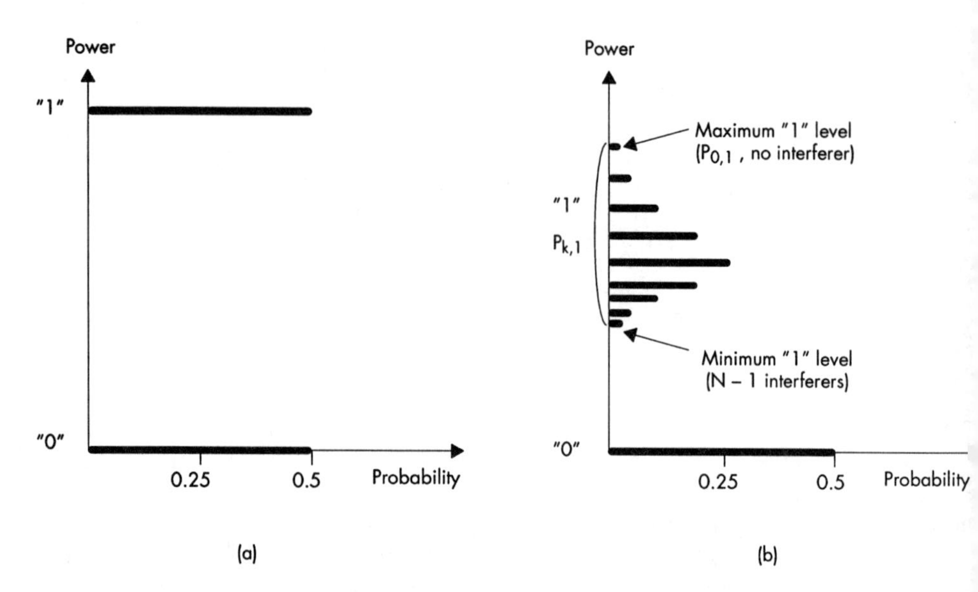

Figure 5.6 Amplified power levels of bits "1" and "0," and corresponding probability of occurrence: (a) without crosstalk, (b) with crosstalk ($N = 10$).

where $\mathrm{erfc}(x) = \frac{2}{\sqrt{\pi}}\int_x^{\infty} e^{-y^2}\,dy \cong \frac{1}{x\sqrt{\pi}}\exp(-x^2)$ is the complementary error function discussed in Section 4.2.1, D is the decision threshold, σ_0 and $\sigma_{k,1}$ are the standard deviations at the receiver for the "0" level and the kth "1" level, respectively.

Figure 5.7 shows the BER versus the average received signal power per channel for various numbers of channels. Also shown for comparison is the BER curve of the thermal-noise-limited receiver (without amplifier).

The system gain obtained when 10, 100, and 500 interfering channels are present is 12.2, 10.1, and 6.5 dB, respectively. Thus, it is seen that, even with crosstalk, the TWA provides a considerable net system gain.

As will be seen in Section 5.3, the carrier lifetime in erbium-doped fiber amplifier is of the order of 10 ms. Crosstalk effects are thus absent for channel data rates above few kilobits per second, leading to even more system performance improvements when such devices are used.

It must be noticed that another mechanism other than gain saturation may also introduce crosstalk between channels [Darcie, 1987]. Consider, for example, a multiwavelength coherent network in which the different channels are separated in frequency by $\Delta_{ch}\nu$. If these channels are amplified simultaneously by the same amplifier, the carrier density of the amplifier medium will be modulated at the beat frequencies of the incident signals. Since this crosstalk mechanism is strongly dependent on the product $\tau_e.\Delta_{ch}\nu$, it

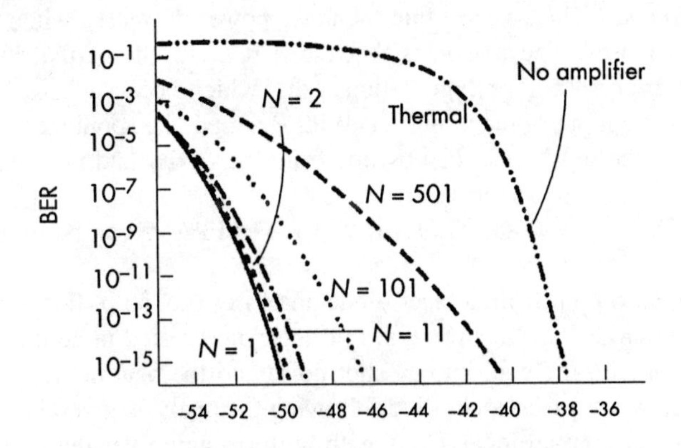

Figure 5.7 BER versus received signal power per channel. The channel bit rate is 200 Mbps and the thermal-noise current density is $3.8pA/\sqrt{Hz}$. [After Green, 1993.]

will set a limit on the minimum channel frequency separation. For $\tau_e = 1$ ns, this effect can be canceled out with $\Delta_{ch}\nu > 10$ GHz.

5.1.2.3 Amplifier Noise

Another limit to the effective gain is the fact that laser-diode amplifiers have noise, and the noise is amplified along with the signal.

The dominant noise generated by the amplifier is the amplified spontaneous emission (ASE). The amplification of spontaneous emission is triggered by the spontaneous recombination of electrons and holes in the amplifier medium. This spontaneous noise can be modeled as a random process in which the events are infinitely short pulses distributed all along the active medium. This random process is characterized by a flat-noise power spectrum and the ASE power spectral density is given by [Yariv, 1991]

$$N_{\text{ASE}} = \chi n_{\text{sp}}(G(\nu) - 1)\, h\nu \tag{5.16}$$

where n_{sp} is the spontaneous emission factor ($n_{\text{sp}} \geq 1$ with equality obtained when complete population inversion is achieved—typical values range from 1.4 to 4 depending both on the pumping rate and the operating wavelength), h is the Planck's constant, and χ is the excess noise factor given by [Mukay, 1982]

$$\chi = \frac{(1 + R_1 G_s)(G_s - 1)}{(1 - R_1)G_s} \tag{5.17}$$

From (5.16) and (5.17), we see that the noise power decreases when the input facet reflectivity R_1 decreases. The reason is that the part of the noise that has the greatest impact on amplifier performance degradation is that which occurs where the signal is the weakest; that is, at the input of the amplifier. With R_1 small, the spontaneous emission that is directed towards the input facet can escape from the cavity and no longer contributes to the ASE.

Since for TWA, $R_1 = 0$ and $G_s >> 1$, $\chi = 1$. For FPA, χ can be much larger since R_1 is not negligible.

When the total output from a laser-diode amplifier (i.e. amplified signal + ASE) is detected by a photodiode, two distinct types of noise are created in addition to electronic (thermal) noise; namely, the well-known shot noise and the beat noise.

These noises will now be quantified following the analysis given in [Olsson, 1989]. For simplicity, we assume an ideal TWA with uniform gain over the optical bandwidth $\Delta\nu_{opt}$ and the optical signal carrier centered in the middle of $\Delta\nu_{opt}$. Extension of the analysis to the FPAs can readily be obtained by modifying the optical bandwidths for the beat-noise components and by including the excess-noise factor χ.

Assume an ideal photodetector with unity quantum efficiency. The photocurrent generated by the optical signal power P_{in} is

$$i_{sig} = \frac{e}{h\nu} GP_{in} \tag{5.18}$$

The photocurrent generated by the ASE is

$$i_{ASE} = \frac{e}{h\nu} N_{ASE}\Delta\nu_{opt} = en_{sp}(G - 1)\Delta\nu_{opt} \tag{5.19}$$

From (4.23), the corresponding shot-noise power spectral densities are, respectively

$$N_{shot-sig} = 2ei_{sig} = 2\frac{e^2}{h\nu} GP_{in} \tag{5.20}$$

and

$$N_{shot-ASE} = 2ei_{ASE} = 2e^2 n_{sp}(G - 1)\Delta\nu_{opt} \tag{5.21}$$

The beat noises are the results of the square law-detection process of optical receivers. Signal-spontaneous beat noise is generated by mixing between the amplified signal and the ASE. Spontaneous-spontaneous beat noise is generated by mixing among the ASE components themselves.

The spontaneous-emission electric field can be expressed as a sum of cosine terms with random phase

$$E_{ASE}(t) = \sum_{k=-M}^{+M} \sqrt{2N_{ASE}\delta\nu}\cos\{(\omega_0 + 2\pi k\delta\nu)t + \phi_k\} \tag{5.22}$$

where $\delta\nu = \Delta\nu_{opt}/2M$.

The number M is introduced here for convenience of calculations. It does not appear in the final results and must therefore be seen as merely a mathematical parameter.

The photocurrent generated by the unity quantum efficiency photodetector is obtained by

$$i(t) = \frac{e}{h\nu}\langle(E_{sig}(t) + E_{ASE}(t))^2\rangle \tag{5.23}$$

where the brackets indicate time averaging over time constant of the order of the reciprocal of the receiver bandwidth.

Hence, with $E_{sig}(t) = \sqrt{2GP_{in}(t)}\cos(\omega_0 t)$ and with (5.22) and (5.23), we have

$$i(t) = \frac{e}{h\nu}GP_{in}(t) + \frac{4e}{h\nu}\left\langle\sum_{k=-M}^{+M}\sqrt{GP_{in}N_{ASE}\delta\nu}\cos(\omega_0 t)\cdot\cos\{(\omega_0 + 2\pi k\delta\nu)t + \phi_k\}\right\rangle$$

$$+\frac{2e}{h\nu}N_{ASE}\delta\nu\left\langle\left(\sum_{k=-M}^{+M}\cos\{(\omega_0 + 2\pi k\delta\nu)t + \phi_k\}\right)^2\right\rangle = i_{sig}(t) \tag{5.24}$$

$$+\, i_{sig-ASE}(t) + i_{ASE-ASE}(t)$$

The first term represents the signal photocurrent (5.18). The next two terms represent the beat photocurrent generated by the mixing between the signal-ASE and ASE-ASE, respectively. The dc term of the ASE-ASE is given by (5.19) and represents the photocurrent generated by the ASE alone; that is, i_{ASE}.

After time averaging, the second term in the right-hand side of (5.24) becomes

$$i_{sig-ASE}(t) = \frac{2e}{h\nu}\sqrt{GP_{in}N_{ASE}\delta\nu}\sum_{k=-M}^{+M}\cos(2\pi k\delta\nu t + \phi_k) \tag{5.25}$$

Examining the sum in (5.25), we see that for each frequency $k\delta\nu$, there are two components, but with random phase. The power spectrum of $i_{\text{sig-ASE}}(t)$ is therefore flat in the frequency interval 0-$\Delta\nu_{\text{opt}}/2$ and its density is given by

$$N_{\text{sig}-\text{ASE}} = \frac{4e^2}{(h\nu)^2} G P_{\text{in}} N_{\text{ASE}} \cdot 2 \cdot \frac{1}{2} \tag{5.26}$$

Using (5.16) within (5.26), we obtain

$$N_{\text{sig}-\text{ASE}} = \frac{4e^2}{h\nu} P_{\text{in}} n_{\text{sp}} (G - 1)G \tag{5.27}$$

Similarly, after time averaging, the last term in (5.24) becomes

$$i_{\text{ASE}-\text{ASE}}(t) = \frac{e}{h\nu} N_{\text{ASE}} \delta\nu \sum_{k=0}^{2M} \sum_{j=0}^{2M} \cos\{2\pi(k - j)\delta\nu t + \phi_k - \phi_j\} \tag{5.28}$$

Examining the double sum in (5.28), it is straightforward to see that the power spectrum of the spontaneous-spontaneous beat noise extends from 0 to $\Delta\nu_{\text{opt}}$ with a triangular shape. The dc term of the power density is obtained for $k = j$ in (5.28). Since there are $2M$ such terms, we have

$$i^{\text{dc}}_{\text{ASE}-\text{ASE}} = \frac{e}{h\nu} N_{\text{ASE}} \delta\nu \cdot 2M = e n_{\text{sp}} (G - 1)\Delta\nu_{\text{opt}} \tag{5.29}$$

in accordance with (5.19).

Since the terms in (5.28) having the same absolute frequency but of opposite sign add in phase, the power density around dc is given by

$$N_{\text{ASE}-\text{ASE}} = 2e^2 n_{\text{sp}}^2 (G - 1)^2 \Delta\nu_{\text{opt}} \tag{5.30}$$

Figure 5.8 shows the power spectral densities corresponding to the signal term, the signal-ASE beat term, and the ASE-ASE beat term.

The total noise power spectral density is given by

$$N_{\text{Tot}} = N_{\text{shot}-\text{sig}} + N_{\text{shot}-\text{ASE}} + N_{\text{sig}-\text{ASE}} + N_{\text{ASE}-\text{ASE}} + N_{\text{elect}} \tag{5.31}$$

in which $N_{\text{elect.}}$ is the electronic noise power spectral density.

With (5.18), (5.20), (5.21), (5.27), and (5.30), the signal-to-noise ratio (SNR) after detection of the amplified signal is given by

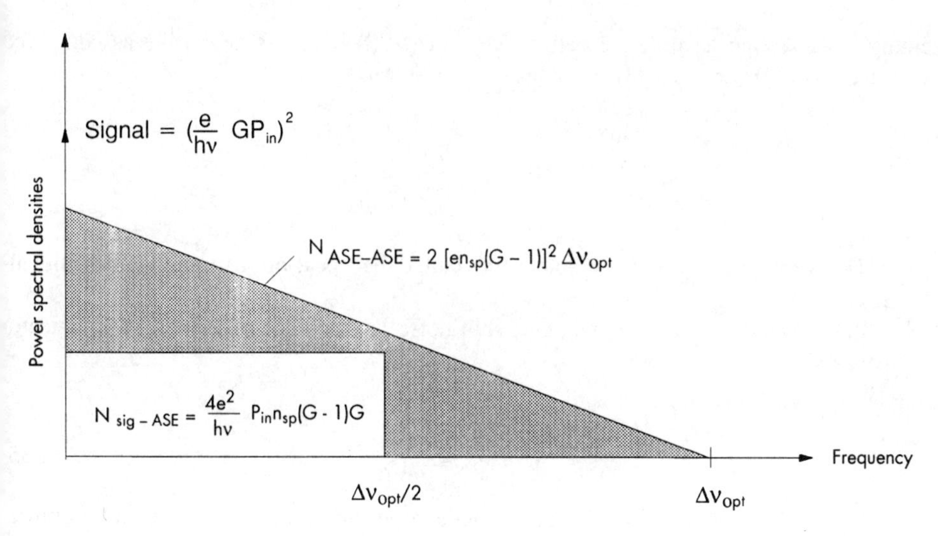

Figure 5.8 Power spectral densities of signal-spontaneous beat noise ($N_{sig\text{-}ASE}$) and spontaneous-spontaneous beat noise ($N_{ASE\text{-}ASE}$). The signal power is also shown (not scaled).

$$\mathrm{SNR} = \frac{(GP_{in})^2}{2h\nu B_e\{GP_{in} + n_{sp}h\nu(G-1)\Delta\nu_{opt} + 2n_{sp}P_{in}(G-1)G + n_{sp}^2 h\nu(G-1)^2\,\Delta\nu_{opt} + N_{elect}\}} \tag{5.32}$$

where B_e denotes the electrical bandwidth of the receiver.

When G is large enough, (5.32) reduces to

$$\mathrm{SNR} = \frac{P_{in}^2}{2h\nu B_e\{2n_{sp}P_{in} + n_{sp}^2 h\nu\Delta\nu_{opt}\}} \tag{5.33}$$

Therefore, for large gain, we see that beat-noise powers predominate over shot-noise powers. The total power level of the shot noise is about two orders of magnitude below that of the total beat noise.

Since the signal-ASE beat noise is proportional to the signal power while the ASE-ASE beat noise is not, the ASE-ASE beat noise will be the dominant noise for low-input signal levels, typically below -40 dBm.

On the other side, the signal-ASE beat noise will become dominant above about −40 dBm input signal levels.

The best way to suppress both the shot noise and the ASE-ASE beat noise is to insert a narrow bandpass filter at the output of the amplifier. If this filter bandwidth is small enough, the second term of the denominator of (5.33) becomes negligible as compared with the first term, and (5.33) becomes

$$[\mathrm{SNR}]_{\mathrm{BNL}} = \frac{P_{\mathrm{in}}}{4h\nu n_{\mathrm{sp}} B_e} \tag{5.34}$$

The SNR given by (5.34) is sometimes called the ''beat-noise-limited (BNL) signal-to-noise ratio.''

In Chapter 4, we have seen that the shot-noise-limited SNR of an ideal photodetector is given by

$$[\mathrm{SNR}]_{\mathrm{shot}} = \frac{P_{\mathrm{in}}}{2h\nu B_e} \tag{5.35}$$

Using (5.34) and (5.35), the SNR degradation introduced by the optical amplifier can be characterized by the so-called ''noise figure F'' defined by

$$F = \frac{[\mathrm{SNR}]_{\mathrm{shot}}}{[\mathrm{SNR}]_{\mathrm{BNL}}} = 2n_{\mathrm{sp}} \tag{5.36}$$

Therefore, even when n_{sp} approaches unity by enhancing the population inversion the noise figure of an ideal TWA is at least 3 dB. The best values of F obtained in research laboratories are around 5 dB [Mukai 1987, Tauber 1992].

5.2 DOPED-FIBER AMPLIFIERS

When optical fibers are doped with rare-earth ions, such as erbium (Er), neodymium (Nd), or praseodymium (Pr), the loss of the fiber can be drastically modified. For example, with Er^{3+} doping, the loss can be as high as about 1,500 dB/km at a wavelength $\lambda_p = 1.48$ μm while it would be less than 0.2 dB/km at the same wavelength in the absence of doping.

During the strong absorption process, the photons at λ_p are absorbed by the outer orbital electrons of the rare-earth ions and these electrons are raised to higher energy levels. The de-excitation of these high energy levels back to the ground state may arise in different ways, either nonradiatively or radiatively. If an intermediate energy level lying between the ground state and the initially excited state is filled in by de-excitation of the higher energy levels, further de-excitation from this particular intermediate energy level can be stimulated by signal photons provided that the bandgap between the intermediate excited state and the background state corresponds to the energy of the signal photons. The result would be an amplification of the optical signal at wavelength λ_s. This is effectively what happens in doped-fiber amplifiers, as will be discussed in the next section.

The main difference between doped-fiber amplifiers and semiconductor amplifiers is that the amplification gain of doped-fiber amplifiers is provided by means of optical pumping instead of electrical pumping.

Figure 5.9(a) shows a schematic representation of a rare-earth doped-fiber amplifier module. It consists of a pump laser, a wavelength selective coupler to combine the pump and the signal wavelengths (Chapter 6), a length of a suitable doped-fiber, an optical isolator, and (optionally) an optical filter to reject any unused pump light.

This latter component can be replaced by a second wavelength-selective coupler, in which case a second laser could be used to pump the doped fiber in a counter-propagating mode, as also shown in Figure 5.9(a).

Figure 5.9(b) shows the evolution of the pump and signal powers as they propagate through the amplifier module. It is clear that insufficient pump power or too-long doped fiber can result in the downward part of the amplifier providing loss rather than gain. Conversely, too-large pump power or too-short doped fiber leads to inefficient use of the pump power. Therefore, an optimum fiber length exists, which strongly depends on the available pump power.

At present, the only practical doped-fiber amplifiers are based on erbium doping. They are suitable for the 1.5-μm window signal wavelength. For the 1.3-μm window, research concentrates on two contenders: the praseodymium and the neodymium-doped fibers. Although encouraging results have been obtained very recently, many problems must still be solved before these amplifiers can be used in real-life systems. Nevertheless, a bright future can be anticipated for these devices as this active field of research unfolds over the next few years.

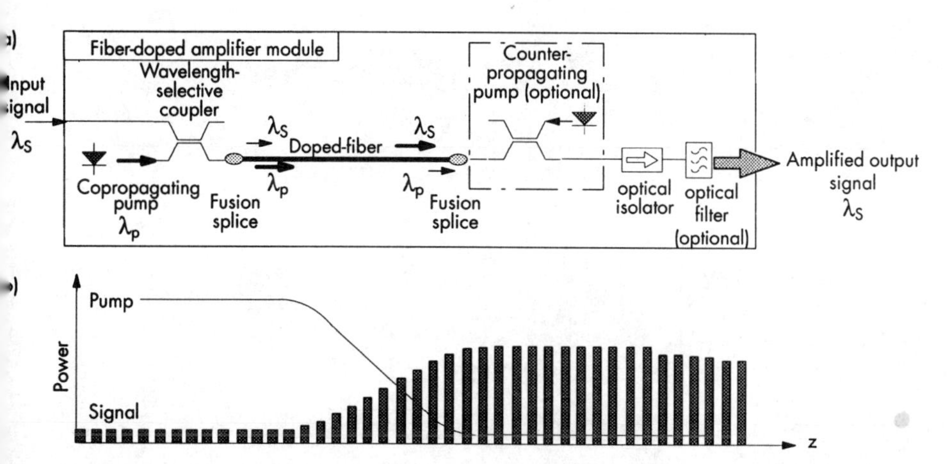

Figure 5.9 (a) Schematic of a fiber-doped amplifier module, and (b) pump and power evolution for copropagating pump.

5.2.1 The Erbium-Doped Fiber Amplifier

Figure 5.10 shows a schematic of the energy diagram of erbium ions [Mears, 1987; Miniscalco, 1991].

The useful transition, at 1,536 nm, occurs from the $^4I_{13/2}$ to the $^4I_{15/2}$ energy levels. The population inversion between these two energy levels can be achieved by optical pumping at appropriate wavelengths shorter than that at which stimulated emission is desired; that is, shorter than 1,536 nm. Currently, pump wavelengths of 980 nm or 1,480 nm have been revealed to be the most attractive.

When a 980-nm pump is used, the population inversion between the $^4I_{13/2}$ and the $^4I_{15/2}$ energy levels is achieved in two steps: the $^4I_{11/2}$ energy level is first excited and then, after a time τ_e = 1 μs, relaxed to the metastable energy level $^4I_{15/2}$ (τ_e = 14 ms).

For the 1,480-nm pump, the population inversion is achieved in one step. At present, 1,480-nm pump-laser diodes are more widely used, essentially because they are more readily available and they have a better reliability.

If the reliability of 980-nm pump-laser diodes improves in the future, they may become the pump of choice because they lead to better amplifier-noise figures. In addition, the design of the wavelength selective coupler is greatly simplified when the wavelength of the signal to be amplified (i.e., 1,525 to 1,560 nm) and the wavelength of the pump are well separated, as is the case for 980-nm pump.

Nevertheless, even if 980-nm pump-laser diodes will increasingly be used in the future, 1,480-nm laser diodes will remain attractive candidates in applications where pump transmission losses must be minimized, such as remote optically pumped amplifiers.

The spectral width of the 1,536-nm transition in Figure 5.10 is very limited. However, substantial broadening can be obtained by the addition of special codopants along

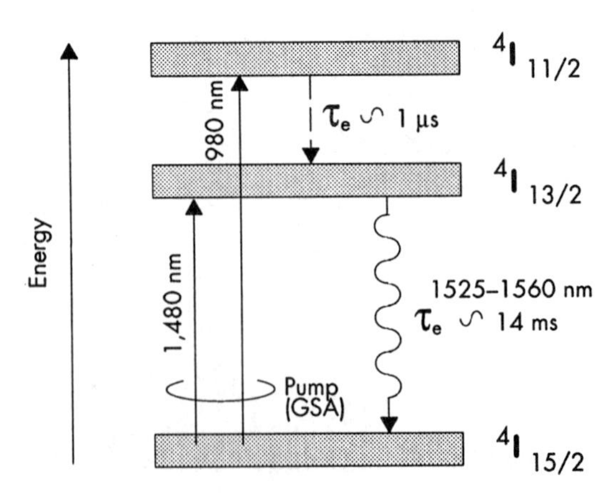

Figure 5.10 Schematic energy level diagram of Er^{3+} ions showing the 980- and 1,480-nm ground state absorption (GSA) lines.

with the erbium atoms. The inclusion of alumina (Al_2O_3) in the fiber core has been proven to be the most satisfactory solution. In addition to the resulting spectral broadening (and flattening), alumina codoping allows higher Er^{3+} concentration, resulting in a shorter length of doped fiber, which in turn allows a more compact erbium-doped fiber amplifier (EDFA) module to be realized. With alumina codoping, Er^{3+} concentrations of 1,000 ppm (part per million) are routinely achieved. Numbers as high as 8,900 ppm have been reported and have provided gain in excess of 23 dB over a bandwidth of 35 nm for a 1m-long fiber pumped at 980 nm with a pump power of 17.8 dBm [Kimura, 1992]. A maximum gain of 31.5 dB was obtained at 1,532-nm signal wavelength.

The gain of EDFA has been shown to be polarization-insensitive [Giles, 1989]. This is the result of random orientation of Er^{3+}-ions in the circularly symmetric fiber core. It has to be noticed that in order to have an EDFA module that is insensitive to the input signal polarization, polarization-insensitive wavelength selective couplers must also be used.

5.2.1.1 Gain Saturation

As with semiconductor amplifiers, the gain of an Er^{3+}-doped fiber amplifier also saturates. The gain saturation occurs when the amplified signal power grows large and approaches the pump power in the fiber, causing a substantial decrease of the population inversion.

The output saturation power is proportional to the pump power and is thus limited only by the availability of powerful pump sources. Saturation output power as high as 24.6 dBm has been demonstrated when pumped with 31.8 dBm (1.5W) at 1.047/1.053 μm [Grubb, 1992].

However, the key to practical EDFAs is the modest pump power needed to achieve high gain. Figure 5.11 shows the EDFA gain versus signal output power for three different pump powers, attainable with current semiconductor laser diodes [Spirit, 1991].

In order to minimize the pump power requirement while maximizing the pumping rate experienced by the Er^{3+} ions, the fiber-core diameter is made much smaller (typically 2.5 μm) than that of standard single-mode fibers. When small-core Er^{3+}-doped fibers are spliced with conventional single-mode fibers, each fusion splice produces a loss of 0.1 to 0.5 dB.

In Section 5.1.2.2, we have seen that another consequence of gain saturation is crosstalk. However, with EDFAs crosstalk effects are only significant for channel bit rates below 100 Kbps or so, thanks to the long fluorescence lifetime (τ_e = 14 ms) of the $^4I_{13/2}$ upper state of the amplifying transition [Pettit 1989, Meli 1992].

5.2.1.2 Amplifier Noise

Noise contributions of EDFAs arise primarily from ASE, similarly to semiconductor amplifiers. Equation (5.16) with $\chi = 1$ (since fiber amplifiers are essentially traveling wave amplifiers) holds also for Er^{3+}-doped fiber amplifiers.

Figure 5.12 shows typical curves of gain and noise figures as a function of Er^{3+}-doped fiber length. Clearly, there is a tradeoff between gain and noise figure. It is also

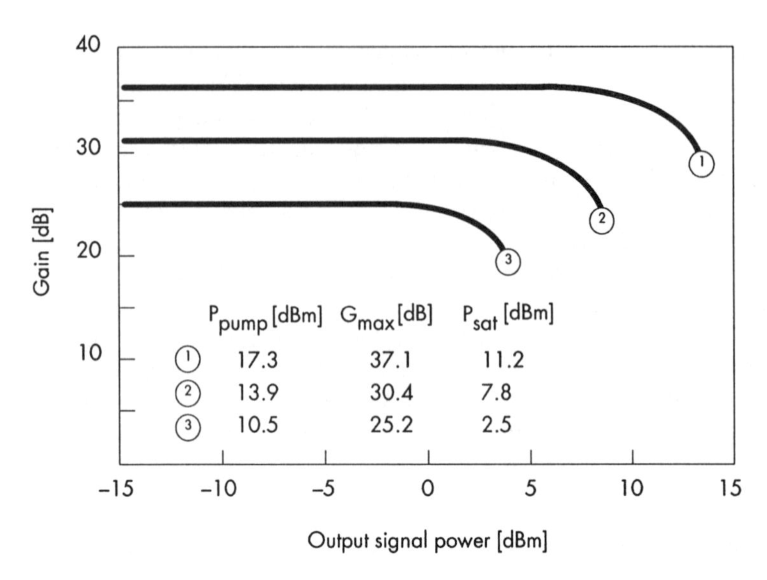

Figure 5.11 Variation of EDFA gain as a function of output signal power for three different pump powers as indicated.

apparent from the figure that noise performance is worse for a counterpropagating pump. This is because, in a counterpropagating pump configuration, the population inversion at the input of the amplifier is relatively low leading to a higher n_{sp} and thus worse SNR.

In a copropagating pump configuration, a quantum-limit noise figure of 3.2 dB has been achieved at 980-nm pump [Zyskind, 1992].

At 1,480-nm pump, the noise figure is worse because the pump and signal wavelengths are very close to each other, and emission stimulated by pump light restricts the population inversion. Noise figures of 5.2 dB have been reported for a copropagating pump at 1,490 nm [Giles, 1992].

5.2.2 1.3-μm Doped-Fiber Amplifiers

In the previous section we saw that EDFAs offer high gain, high efficiency, low noise, low fiber-coupling loss, low crosstalk and polarization-insensitivity. However, the EDFA operates only in the 1.5-μm window.

With the same principle as EDFA, one can expect that by doping the fiber with materials other than erbium, a doped-fiber amplifier can be fabricated that has similar performance characteristics as EDFA but in the 1.3-μm band.

Two dopant candidates that have the requested 1.3-μm transition have been found: the neodymium (Nd) and the praseodymium (Pr).

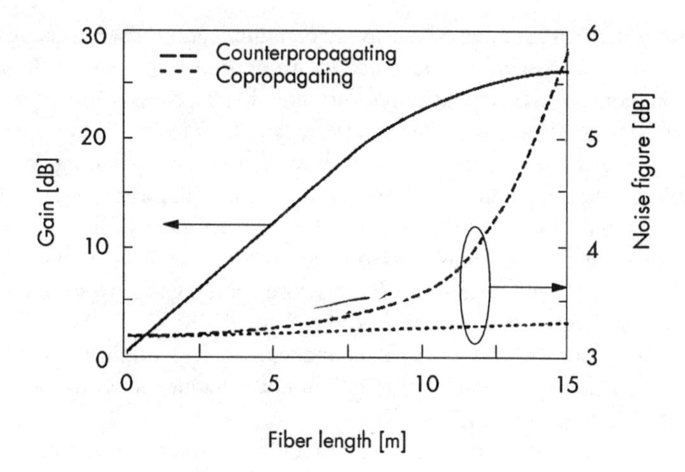

Figure 5.12 Typical variation of gain and noise figure as a function of erbium-doped fiber length. Pump and input signal powers are 20 and −30 dBm, respectively.

Unfortunately, for several reasons explained in [Miyajima, 1992], neither of these dopants can be used in conventional silica fibers. Fluoro-zirconate fibers (ZrF_4-BaF_2-LaF_3-AlF_3-NaF or "ZBLAN" fibers) have been successfully doped with these two rare-earth elements and have shown optical gain in the 1.3-μm window.

For Nd^{3+}-doped fluoride-fiber amplifiers (NDFAs), the best results published to date report 10-dB gain at 1.343 μm with 17-dBm pump power supplied by a commercially available AlGaAs laser diode operating at 0.82 μm [Miyajima, 1991]. A major problem, which has yet to be solved, is that NDFA operates at a wavelength band that is slightly longer (1.32 to 1.38 μm) than the 1.3-μm fiber-window.

For the Pr^{3+}-doped fluoride-fiber amplifiers (PDFAs), a gain of 38.2 dB at 1.31 μm was obtained for a launched pump power of 24.8 dBm at 1.05 μm [Miyajima, 1991]. The 3-dB bandwidth is 35 nm. Also, 11-dB gain has been obtained with a laser-diode pump source [Shimizu, 1992].

It is expected that improvements in pump efficiency, allowing higher gain or less required pump power, will be obtained in the near future. However, before 1.3-μm doped-fiber amplifiers become practical several issues remain to be addressed, such as mechanical doped-fiber properties, splicing techniques, and reliable high-power 1.017-μm laser diodes.

5.3 SUMMARY

Since the beginning of the 1980s, several kinds of optical amplifiers have been studied and developed. Two distinct technologies, semiconductor-based and fiber-based amplifiers, have matured to such an extent that fiber-optic communication systems using those optical amplifiers are likely to become widespread as we approach the end of the century.

Semiconductor-based optical amplifiers are based on laser-diode technology and therefore have similar attractive features, such as a miniature size, a low power consumption and a potential for low-cost large-scale production. By providing low-facet reflectivity, suitable electrical pump power and high internal differential gain, semiconductor optical amplifiers have the ability to provide fiber-to-fiber gains of up to about 20 dB over a bandwidth of (typically) 35 nm. They operate as easily at 1.3 μm as at 1.5-μm signal wavelengths. With multiple quantum-well structures, large gains over the desired 240-nm bandwidth have recently been achieved.

One of the major drawbacks of semiconductor optical amplifiers is that they introduce a large amount of interchannel crosstalk in multichannel networks, a peculiarity that is absent with fiber-based amplifiers.

Fiber-based optical amplifiers have attracted considerable attention since (a few years ago) the erbium dopant was proved effective to provide amplification characteristics similar to those of semiconductor-based amplifiers. At present, only erbium-doped fiber amplifiers have shown satisfactory properties for use in real-life environments. However, they only operate for signal wavelengths in the 1.5-μm window.

Although encouraging results have recently been obtained for 1.3-μm window fiber amplifiers (mainly based on neodymium and praseodymium dopants), several issues remain to be addressed before they become practical.

In common with both kinds of optical amplifiers, whether semiconductor or doped-fiber amplifiers, gain saturation and noise degrade the performance that could be expected with ideal devices. These effects must be carefully considered when designing the physical topology of multiaccess networks, as will be seen in Chapter 6.

REFERENCES

[Kastler, 1974] Kastler, A. "Transmission d'une impulsion lumineuse par un interferometre Fabry-Perot," *Nouv. Rev. Optique*, Tome 5, No. 3, April 1974, pp. 133–139.

[Yamamoto, 1980] Yamamoto, Y. "Characteristics of *AlGaAs* Fabry-Perot cavity type laser amplifiers," *IEEE J. Quant. Electron.*, Vol. QE-16, No. 10, Oct. 1980, pp. 1047–1052.

[Mukay, 1981] Mukay, T., et al. "Gain, frequency bandwidth, and saturation output power of *AlGaAs DH* laser amplifiers," *IEEE J. Quant. Electron.*, Vol. QE-17, No. 6, June 1981, pp. 1028–1034.

[Mukay, 1982] Mukay, T., et al. "*S/N* and error rate performance in *AlGaAs* semiconductor laser preamplifier and linear repeater systems," *IEEE J. Quant. Electron.*, Vol. QE-18, No. 10, 1982, pp. 1560–156.

[Desurvire, 1983] Desurvire, E., et al. "High-gain optical amplification of laser diode signal by Raman scattering in single-mode fibers," *Electron. Letters*, Vol. 19, No. 19, Sept. 1983, pp. 751–753.

[Kobayashi, 1984] Kobayashi, S., et al. "Semiconductor optical amplifier," *IEEE Spectrum*, May 1984, pp. 26–33.

[O'Mahony, 1985] O'Mahony, M. J., et al. "Low-reflectivity semiconductor laser amplifier with 20*dB* fiber-to-fiber gain at 1500*nm*," *Electron. Letters*, Vol. 21, No. 11, May 1985, pp. 501–502.

[Atkins, 1986] Atkins, C. G., et al. "Application of Brillouin amplification in coherent optical transmission," *Electron. Letters*, Vol. 22, No. 10, May 1986, pp. 556–558.

[Grosskopf, 1986] Grosskopf, G., et al. "Crosstalk in optical amplifiers for two-channel transmission," *Electron. Letters*, Vol. 22, No. 17, Aug. 1986, pp. 900–902.

[Jopson, 1986] Jopson, R. M., et al. "Polarisation-dependent gain spectrum of a 1.5 μm traveling-wave optical amplifier," Electron Letters, Vol. 22, No. 21, Oct. 1986, pp. 1105–1107.

[Mukai, 1987] Mukai, T., et al. "5.2*dB* noise figure in a 1.5 μm *InGaAsP* traveling-wave laser amplifier," *Electron. Letters*, Vol. 23, No. 5, Feb. 1987, pp. 216–217.

[Simon, 1987] Simon, J. C., et al. "Gain polarisation sensitivity and saturation power of 1.5 μm near-traveling-wave semiconductor laser amplifier," Electron. Letters, Vol. 23, No. 7, March 1987, pp. 332–334.

[Mukai, 1987] Mukai, T., et al. "Signal gain saturation in two-channel common amplification using a 1.5 μm *InGaAsP* traveling-wave laser amplifier," *Electron. Letters*, Vol. 23, No. 8, April 1987, pp. 396–397.

[Wagner, 1987] Wagner, S. S. "Optical amplifier applications in fiber optic local networks," *IEEE Trans. on Communications*, Vol. COM-35, No. 4, May 1987, pp. 419–426.

[Mears, 1987] Mears, R. J., et al. "Low-noise Erbium-doped fiber amplifier operating at 1.54μm," *Electron. Letters*, Vol. 23, No. 19, Sept. 1987, pp. 1026–1028.

[Grosskopf, 1987] Grosskopf, G., et al. "Optical amplifier configurations with low polarisation sensitivity," *Electron. Letters*, Vol. 23, No. 25, Dec. 1987, pp. 1387–1388.

[Darcie, 1987] Darcie, T. E. "Intermodulation distortion in optical amplifiers from carrier-density modulation," *Electron. Letters*, Vol. 23, No. 25, Dec. 1987, pp. 1392–1394.

[Mollenauer, 1988] Mollenauer. L. F., et al. "Demonstration of solitons transmission over more than 4000*km* in fiber loss periodically compensated by Raman gain," *Optics Letters*, Vol. 13, July 1988, pp. 675–677.

[O'Mahony, 1988] O'Mahony, M. J. "Semiconductor laser optical amplifiers for use in future fiber systems," *IEEE J. Light. Technology*, Vol. 6, No. 4, April 1988, pp. 531–544.

[Giles, 1989] Giles, C. R., et al. *Proc. IOOC'89* , Paper 20A4–3, 1989.

[Pettit, 1989] Pettitt, M. J., et al. "Crosstalk in Erbium doped fiber amplifiers," *Electron. Letters*, Vol. 25, No. 6, March 1989, pp. 416–417.

[Olsson, 1989] Olsson, N. A. "Lightwave systems with optical amplifiers," *IEEE J. Light. Technology*, Vol. 7, No. 7, July 1989, pp. 1071–1082.

[Inoue, 1989] Inoue, K. "Crosstalk and its power penalty in multichannel transmission due to gain saturation in a semiconductor laser amplifier," *IEEE J. Light. Technology*, Vol. LT-7, No. 7, July 1989, pp. 1118–1124.

[Tabuchi, 1990] Tabuchi, H., et al. "External grating tunable *MQW* laser with wide tuning of 240*nm*," *Electron. Letters*, Vol. 26, No. 11, May 1990, pp. 742–746.

[Magari, 1990] Magari, K., et al. "Polarisation insensitive traveling wave type amplifier using strained multiple quantum well structure," *IEEE Photon. Technology Letters*, Vol. 2, No. 8, Aug. 1990, pp. 556–558.

[Ramaswami, 1990] Ramaswami, R., et al. "Amplifier induced crosstalk in multichannel optical networks," *IEEE J. Light. Technology*, Vol. LT-8, No. 12, Dec. 1990, pp. 1882–1896.

[Spirit, 1991] Spirit, D. M. "Silica fiber amplifiers and systems," in *Optical Fiber lasers and amplifiers*, Edited by P. W. France, Blackie & Son Ltd, 1991.

[Sherlock, 1991] Sherlock, G., et al. "1.3 μm *MQW* semiconductor optical amplifiers with high gain and output power," *Electron. Letters*, Vol. 27, Jan. 1991, pp. 165–166.

[Yariv, 1991] Yariv, A. *Optical Electronics*, published by Saunders College Publishing, Holt, Rinehart and Winston, 1991.

[Koga, 1991] Koga, M., et al. "High-gain polarisation-insensitive optical amplifier consisting of two serial semiconductor laser amplifiers," *IEEE J. Light. Technology*, Vol. LT-9, No. 2, Feb. 1991, pp. 284–290.

[Miniscalco, 1991] Miniscalco, W. J. "Erbium-doped glasses for fiber amplifiers at 1500 nm," *IEEE J. Light. Technology*, Vol. LT-9, No. 2, Feb. 1991, pp. 234–250.

[Miyajima, 1991] Miyajima, Y., et al. "Nd^{3+}-doped fluoride fiber amplifier module with 10*dB* gain and high pump efficiency," in *Optical Amplifiers and their Applications*, IEEE LEOS, and OSA Photonic Reports, Snowmass, CO., July 24–26, 1991, Vol. 3, pp. 16–19.

[Zyskind, 1992] Zyskind, J. L., et al. "Erbium-doped fiber amplifiers and the next generation of lightwave systems," *AT&T Technology J.* , Jan./Feb. 1992, pp. 53–62.

[Tauber, 1992] Tauber, D., et al. ''Low-noise figure 1.5 *μm MQW* optical amplifier,'' *OFC'92*, San Jose, CA, TuM2, Feb. 1992, p. 71.

[Kalman, 1992] Kalman, R. F., et al. ''Space-division optical switches based on semiconductor optical amplifiers,'' *OFC'92*, San Jose, CA, WH4, Feb. 1992, p. 131.

[Eiselt, 1992] Eiselt, M., et al. ''Photonic packet switching using cascaded semiconductor optical amplifier gates,'' *OFC'92*, San Jose, CA, WH3, Feb. 1992, p. 130.

[Grubb, 1992] Grubb, S. G., et al. ''+24.6*dBm* output power *Er/Yb* codoped optical amplifier pumped by diode-pumped *Nd:YLF* laser,'' *Electron. Letters*, Vol. 28, No. 13, June 1992, pp. 1275–1276.

[Kimura, 1992] Kimura, Y., et al. ''Gain characteristics of erbium-doped fibre amplifiers with high erbium concentration,'' *Electron. Letters*, Vol. 28, No. 15, July 1992, pp. 1420–1422.

[Meli, 1992] Meli, F., et al. ''Gain crosstalk in saturated EDFA for *WDM* applications,'' *Electron. Letters*, Vol. 28, No. 20, Sept. 1992, pp. 1896–1897.

[Miyajima, 1992] Miyajima, Y. ''Progress towards a practical 1.3 *μm* optical fiber amplifier,'' *ECOC'92*, Berlin, Germany, A2.1, Sept. 1992, pp. 687–694.

[Eskildsen, 1992] Eskildsen, L., et al. ''Polarisation-insensitive semiconductoroptical amplifier at 1.55 *μm*,'' *Electron. Letters*, Vol. 28, No. 21, Oct. 1992, pp. 2019–2021.

[Shimizu, 1992] Shimizu, M., et al. ''Optical Amplifiers and their Applications,'' IEEE, LEOS, and OSA Photonic Reports, Santa Fe, CA, PD3, July 1992.

Part II
Multiaccess Networks

Chapter 6

Passive Devices and Network Topologies

The implementation of multiaccess optical-fiber networks requires several new components, commonly referred to as passive optical devices. These devices can be categorized according to the functions they perform, as shown in Figure 6.1.

One can distinguish: power splitters or taps, which split the optical power from the single-input fiber into several ($\geq$ 2) output fibers, as shown in Figure 6.1(a); power combiners, which perform the opposite function of the former by mixing the output of several transmitters and launching it into a single output fiber, as shown in Figure 6.1(b); star couplers, which mix the optical signals injected at input fibers and divide each of them equally among the output fibers, as shown in Figure 6.1(c); wavelength multiplexers, which combine channels at distinct wavelengths and launch them into a single optical fiber, as shown in Figure 6.1(d); wavelength demultiplexers, which split the multichannel input into different output fibers according to their wavelength, as shown in Figure 6.1(e); and optical isolators, which protect the transmitter or any other sensitive device from undesired reflections, as shown in Figure 6.1(f).

In addition to these functions, several multiaccess optical networks require tunable components, such as tunable transmitters, whose wavelength can be tuned, and tunable optical filters, which dynamically filter out one channel at a specific wavelength. Tunable transmitters have been discussed in Chapter 2. The description of tunable filters will be given in Chapter 7 because, strictly speaking, they are active components and they are related specifically to the WDMA multiaccess scheme.

This chapter deals first with the basic properties of passive devices shown in Figure 6.1 and is intended to provide an understanding of the basic concepts behind them and of their characteristics that are important from a practical point of view. Design aspects are not considered here, but the reader can consult [Tomlinson, 1988], which covers the subject in some detail. Section 6.1 considers the simple case of a 2×2 coupler and will serve to illustrate several useful properties common to many passive optical devices. Section 6.2 is devoted to the important case of star couplers, and various implementations

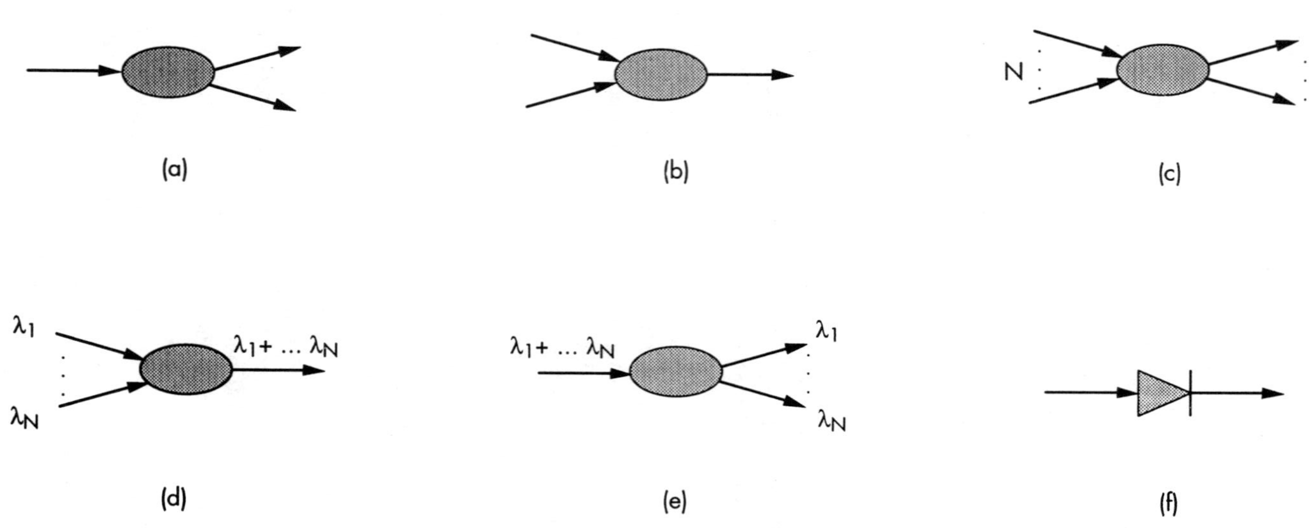

Figure 6.1 Passive optical device functions: (a) power splitter on tap, (b) power combiner, (c) star coupler, (d) wavelength multiplexer, (e) wavelength demultiplexer, and (f) optical isolator.

will be discussed. Section 6.3 reviews the distinct options for WDM devices and Section 6.4 presents the working principle of optical isolators.

The second part of the chapter is devoted to alternative network topologies that can be constructed using the passive devices described in the first part of the chapter. In Section 6.5, distinct topologies are discussed and special attention is given to the losses of the star and bus networks. These two network topologies are further compared on the basis of maximum achievable capacity, which is defined as the product of the number of nodes times the bit rate per node. Section 6.6 presents some proposals to overcome the network power budget limitations and focuses on the use of optical amplifiers for this purpose.

An important aspect for the evaluation of the network capacity is the means for signaling between the network nodes. This is required to set up and remove any connection within the network. Signaling can be supported by a common control channel that transports only the necessary signaling information. Since this control channel may have a limited capacity, it can put some restrictions on the reconfiguration speed of the network, which can lead to a limitation of the overall network capacity. This will be discussed in Section 6.7. Finally, Section 6.8 summarizes the main results obtained in this chapter.

6.1 THE 2×2 COUPLER

Several useful properties of many passive optical devices can be obtained by considering the simple case of a 2×2 coupler. As shown in Figure 6.2, a 2×2 coupler is a device that couples two input beams and provides two output beams with a predetermined power distribution. As an example, a 3-dB coupler distributes half the light power at each input port to each output port.

Technologies used to fabricate these devices are classified as micro-optic, all-fiber-optic, and integrated-optic. Early developments were based on micro-optic technology [Kobayashi 1980, Winzer 1984, Ishio 1984, Senior 1989]. 2×2 couplers have been fabricated using GRIN-rod lenses and beam-splitter films to obtain both power splitting and wavelength multiplexing/demultiplexing functions [Tomlinson, 1980].

Although micro-optic technology can perform most of the functions of other

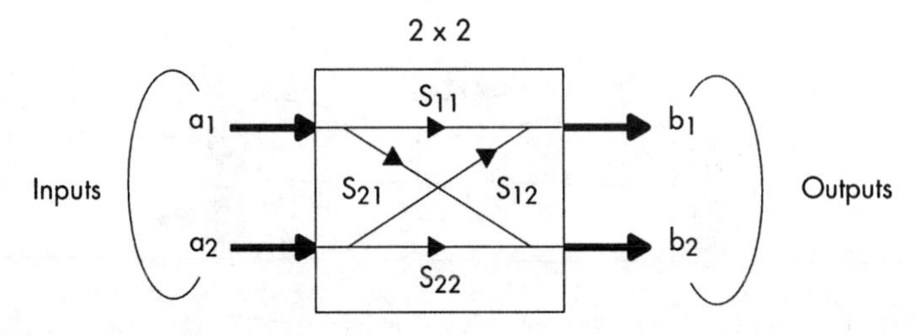

Figure 6.2 A 2×2 coupler as a black box. The relevant S-parameters of the scattering matrix are also shown.

passive-component technologies, and for some devices is the only known practical solution (e.g., Fabry-Perot filters), designers concentrate their efforts on all-fiber-optic and integrated-optic technologies. The major reasons are that fiber attachment is much easier with all-fiber-optic devices while integrated-optic technology using lithographic processes (also referred to as planar-waveguide technology) has the potential of dramatic cost reduction when large-scale production is envisaged. In what follows, only all-fiber and integrated-optic devices will be considered.

For both technologies, the physical mechanisms that act within 2×2 couplers are basically the same except that the waveguide cross-sections are circular in all-fiber devices while usually rectangular in integrated-optic devices.

Consider the fused-fiber 2×2 coupler obtained by melting and pulling two single-mode fibers together over a uniform section of length L, as shown in Figure 6.3 [Oshima, 1985]. Each input and output fiber has a long tapered section since the transversal dimensions are gradually reduced down to that of the coupling region by pulling the fiber during the fusion process. As the light propagates along the taper and into the coupling region, there is a decrease in V-number (the ratio a/λ decreases, see (3.24)), which means that much of the input field propagates outside the original fiber core. By careful dimensioning of the coupling region, this decoupled field can be recoupled into the other fiber. The fraction of the power launched at one input that appears at one of the two outputs can be varied between zero and unity. The same physical mechanism occurs for the coupling between integrated-optic waveguides.

Guided-wave 2×2 couplers, whether all-fiber or integrated-optic, can be analyzed in terms of the partial scattering matrix S, which defines the relationship between the two input field strengths a_1 and a_2, and the two output field strengths b_1 and b_2 (Figure 6.2) [Abbas 1985, Pietzsch 1989, Weissman 1993].

By definition, we have:

$$\begin{pmatrix} b_1 \\ b_2 \end{pmatrix} = \begin{pmatrix} S_{11} & S_{12} \\ S_{21} & S_{22} \end{pmatrix} \cdot \begin{pmatrix} a_1 \\ a_2 \end{pmatrix} \quad \text{or} \quad b = S \cdot a \tag{6.1}$$

where S_{ij} represents the coupling coefficient from input port j to output port i.

Two fundamental physical restrictions apply to the partial scattering matrix S. They

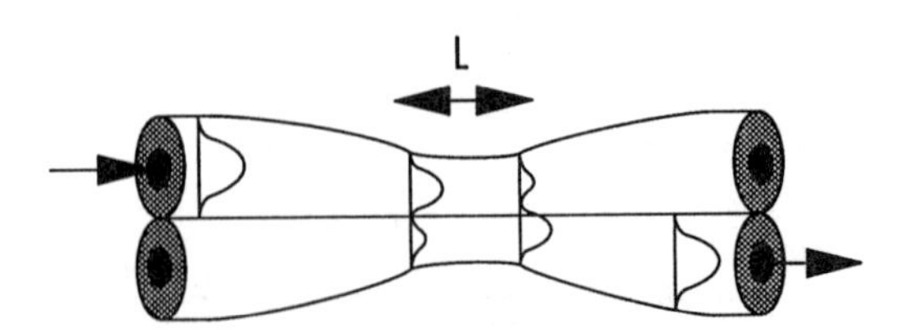

Figure 6.3 A 2×2 fused-fiber coupler obtained by melting and pulling two single-mode fibers together over a uniform section of length L.

result from the reciprocity condition assuming single-mode operation and the energy conservation assuming a lossless device (no excess loss).

The reciprocity condition is a consequence of the fact that solutions of Maxwell's equations are invariant for time inversion. It follows that [Abbas, 1985]

$$S_{12} = S_{21} \tag{6.2}$$

Energy conservation implies that the sum of the output intensities I_0 is equal to the sum of the input intensities I_i:

$$I_0 = |b_1|^2 + |b_2|^2 = I_i = |a_1|^2 + |a_2|^2 \tag{6.3}$$

Writing out this last equation using (6.1) and (6.2) gives the following set of three equations:

$$|S_{11}|^2 + |S_{12}|^2 = 1 \tag{6.4}$$

$$|S_{22}|^2 + |S_{12}|^2 = 1 \tag{6.5}$$

$$S_{11}S_{12}^* + S_{12}S_{22}^* = 0 \tag{6.6}$$

Now, assume that the coupler has been designed so that $S_{11} = \sqrt{1 - \alpha}$, α being a real number between 0 and 1; that is, a fraction $(1-\alpha)$ of the optical power at input port 1 appears at output port 1, while a fraction α of the same input power appears at output port 2.

Inserting $S_{11} = \sqrt{1 - \alpha}$ into (6.4), and subsequently using (6.5), we immediately obtain

$$S_{12} = S_{21} = \sqrt{\alpha} \cdot e^{j\phi_{12}} \tag{6.7}$$

$$S_{22} = \sqrt{1 - \alpha} \cdot e^{j\phi_{22}} \tag{6.8}$$

Using (6.7) and (6.8) with (6.6), we have

$$e^{j(2\phi_{12} - \phi_{22})} = -1 \tag{6.9}$$

which gives

$$\phi_{12} = \frac{\phi_{22}}{2} + (2n + 1)\frac{\pi}{2} \qquad (n = 0, 1, 2. \; . \; .) \tag{6.10}$$

Without loss of generality, we can impose $\phi_{22} = 0$, and finally obtain

$$S = \begin{pmatrix} \sqrt{1 - \alpha} & j\sqrt{\alpha} \\ j\sqrt{\alpha} & \sqrt{1 - \alpha} \end{pmatrix} \tag{6.11}$$

Thus, independently of α, $S_{i \neq j}$ has a $\pi/2$ phase shift relative to S_{ii} $(i,j = 1,2)$. This property is used in balanced coherent detectors, as we saw in Chapter 4.

It is also clear that (at one wavelength) the 2×2 coupler cannot couple the full power from both inputs into one output; if the coupler is designed to deliver a large amount of power from input port 1 to output port 1 (small α, in this case the coupler is generally referred to as a tap), then the amount of power reaching output port 1 from input port 2 is small.

Because of the reciprocity condition, it is also clear that if the role of input and output ports is inverted, the same matrix S still applies. For example, if 1 mW is injected at input port 1 of a 3-dB coupler, 0.5 mW will appear at each output port. Inverting the role of input and output ports, it appears from the reciprocity condition that if 1 mW is injected at each output port, then only 1 mW will appear at input port 1, with the remaining power (also 1 mW) appearing at input port 2.

Equation (6.11) applies to lossless devices since it has been obtained assuming energy conservation. In practice, passive devices always have excess losses resulting from insertion losses, intrinsic losses in the constitutive coupler medium, and output coupling losses. The excess loss β is defined as the fraction of the total input power that appears as total output power. Typical excess losses, expressed in decibels, are $\beta \approx 0.3$ to 0.5 dB $(= -10 log_{10}[\Sigma P_{out}/\Sigma P_{in}])$, with some suppliers offering selected units with ≤ 0.1 dB excess loss.

Inherent advantages of all-fiber technology include simple fiber attachment, very good directivity (i.e., the part of the launched power that is reflected back towards the input ports, typically 50 dB), low reflections, long-term stability, and low material costs. Disadvantages include difficulties for automated assembly, some packaging problems, and little potential for integration.

Integrated-optic technology, by contrast, has great economic potential when the demand will justify large scale manufacturing. Passive planar-waveguide components have been fabricated with lithium niobate ($LiNbO_3$) and along with semiconductors (e.g., GaAs and InGaAsP) are attractive candidate materials for active optical circuits integrating passive components with lasers, LEDs, PINs, and so forth [Horimatsu, 1989]. Another approach consists of using some low-loss transparent substrate materials such as certain polymers or silica. Up to now, the most promising technology for passive-integrated optical components is SiO_2 waveguides on Si (silica-on-silicon) [Adar, 1992].

6.2 STAR COUPLERS

Star couplers can be classified into two broad categories: $N \times M$ full-star couplers and $1 \times M$ splitters/combiners (which are a special case of the former with $N = 1$), as depicted in Figures 6.1(a–c).

The common role of all star couplers is to combine the optical signal(s) injected at input port(s) and divide each of them equally among the output ports. A particular example of a star coupler is the 3-dB coupler discussed in the preceding section.

The fiber-fusion technique can be applied to form $N \times M$ or $1 \times M$ star couplers. $N \times N$ full-star couplers with N up to 7 and splitters/combiners with M up to 19 have been constructed using this technique with excess losses as low as 0.4 dB and 0.85 dB, respectively, at 1.3-μm wavelengths [Arkwright 1989, 1991]. However, large-port-count couplers have proven to be difficult to fabricate because the coupling behavior between the large number of fibers must be controlled during the melting and pulling processes. Moreover, since the coupling between the different fibers is wavelength-dependent (see Section 6.3), it is also technologically difficult to fabricate so-called "achromatic" or "broadband" star couplers that perform as well at 1.3 μm as at 1.55 μm windows.

A more attractive alternative consists of cascading 3-dB couplers, as shown in Figure 6.4 for an 8×8 star [Marhic, 1984]. It is seen that the required number of 3-dB couplers to form a $N \times N$ full-star coupler is

$$N_c = \frac{N}{2} \log_2 N \tag{6.12}$$

and that a fraction $1/N$ of the launched power at each input port effectively appears at all output ports.

If β ($0 \leq \beta \leq 1$) is the excess loss of each 3-dB coupler, then the loss incurred by a signal passing through the $N \times N$ star is given (in dB) by

$$\text{Loss[dB]} = -10\log_{10}\left\{\frac{\beta^{\log_2 N}}{N}\right\} \cong 10(1 - 3.3\log_{10}\beta) \cdot \log_{10} N \tag{6.13}$$

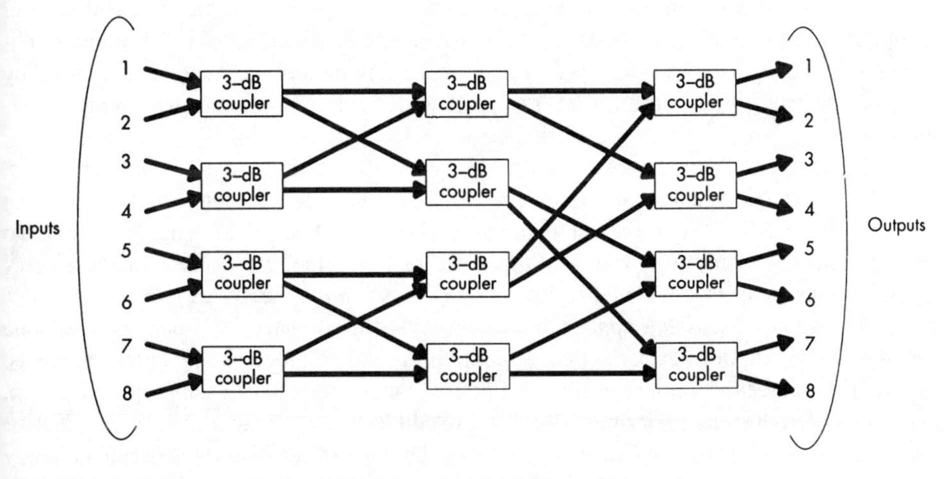

Figure 6.4 An 8×8 star coupler formed by interconnecting 3-dB couplers.

Thus, the loss in dB builds up logarithmically with N. The importance of this fact will become more apparent later in this chapter when we will compare the star and bus network topologies.

Another important fact resulting from the cascading of 3-dB couplers is that N is a multiple of 2 (i.e., $N = 2^n$ with n = 1, 2, . . .). This affects the network "modularity" because when just one additional node must be connected to the already fully connected *N*×*N* star network, the *N*×*N* star coupler must be replaced by a 2*N*×2*N* coupler, leaving 2(N—1) ports unused (except if a 3-dB additional loss is allowed for this extra node).

32×32 single-mode full-star couplers made from a cascade of 3-dB fused-fiber couplers have been fabricated and are now commercially available. Such devices have only about 1.0-dB excess loss and port-to-port uniformity of ±0.5 dB (i.e., ±11%).

The cascading technique can also be applied to integrated-optic devices and this is the current trend for several suppliers because of the potential for cost reduction with high-volume production. 32×32 single-mode full-star couplers lithographed in silica-on-silicon are commercially available with about 4.0 dB excess loss and ±0.5 dB port-to-port uniformity.

Another possibility with integrated-optic technology is the so-called planar-lens star coupler. It consists of a planar-input waveguide array that radiates into a "free-space" planar region (also referred to as the planar lens), which distributes the input powers uniformly among the output waveguides. 19×19 planar-lens full-star couplers have been reported with excess loss kept to within 3.5 dB and port-to-port uniformity of ±1.0 dB [Dragone, 1989]. In principle, the excess loss is independent of N and the coupling is wavelength-independent [Okamoto, 1992]. Since arrays of 100×100 seem feasible, this technology would represent a significant advance for multiaccess optical-fiber networks in the future.

An important factor that distinguishes integrated-optic technology from fiber-fusion is that Y-shaped waveguides can be implemented with integrated-optics as easily as 2×2 couplers. This is exploited for 1×*M* splitters/combiners, which are then inherently achromatic because, in contrast to a 2×2 coupler, the splitting/combining ratio experienced by a signal passing through a uniform Y-shaped waveguide is by nature wavelength-independent.

Recently, an integrated-optic 1×128 power splitter composed of a slab waveguide, funnel-shaped waveguides, and output waveguides has been demonstrated with a low excess loss of 2.3 dB and a small standard deviation of 0.63 dB [Takahashi, 1993]. To achieve this uniform power splitting, the width of the funnel waveguides increases progressively away from the center following the Gaussian diffraction pattern.

The above discussion applies to transmissive star couplers for which (at least) one input and one output fiber per node are required to send and receive the optical channels. In order to reduce the amount of fiber, reflective star couplers have been proposed [Saleh, 1988] and demonstrated [Hermes, 1991]. An example of construction of an 8×8 reflective star is shown in Figure 6.5. The power entering any one of the N ports is divided equally and reflected back out of all ports. Each link may operate in single-wavelength full duplex,

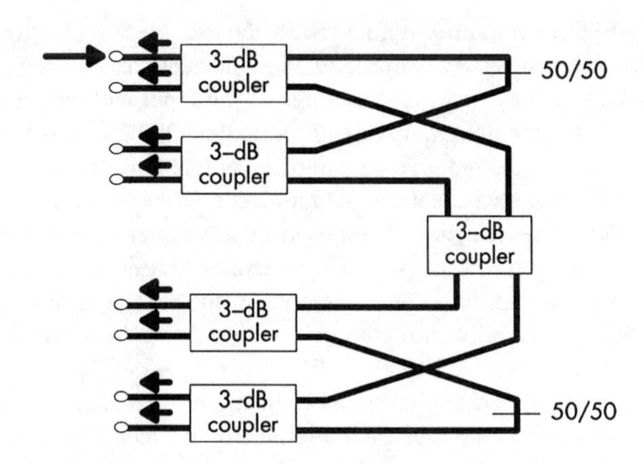

Figure 6.5 A reflective 8×8 star coupler formed by interconnecting 3-dB couplers and using half-reflecting mirrors within two fibers.

in which case each node must incorporate a 3-dB coupler to separate the transmitted and received signals.

6.3 WAVELENGTH MULTIPLEXERS/DEMULTIPLEXERS (WDMs)

Most of the discussion above applies to single-mode operation at a unique wavelength. Solving Maxwell's equations for actual 2×2 coupler construction parameters (length of interaction region, difference in transversal waveguide dimensions, etc.) reveals that the power coupling coefficient α from one waveguide to the other depends on the operating wavelength λ as [Tagaki, 1992]

$$\alpha^2(\lambda) = C(\lambda) \cdot \sin^2\{K(\lambda) \cdot L\} \tag{6.14}$$

where $C(\lambda)$ and $K(\lambda)$ are two functions that depend on the wavelength and other construction parameters, and L is the length of the coupling region.

Thus, the power transferred from one waveguide to the other is a periodic function of L. By proper dimensioning of the coupled region, the coupler can be made achromatic so that the coupling coefficient α is constant over a large range of wavelengths covering both the 1.3- and 1.55-μm windows (acromatic or broadband couplers mentioned in the preceding section).

It is also clear from (6.14) that by proper design the coupler can be made to deliver almost all the power at one wavelength λ_1 to one output and almost all the power at another wavelength λ_2 to the other output. Such devices are referred to as wavelength demultiplexers (WDM demux). By reciprocity, these devices can also serve as wavelength multiplexers (WDM mux) by inverting the roles of the input and output ports.

Thus, in contrast to the unique wavelength situation, 2×2 couplers used with distinct wavelengths can couple all the input powers into one output port. It will be seen in Section 6.6 that this has important consequences for multiaccess fiber-optic networks.

Waveguide WDM devices have been used in wavelength division multiplexed bidirectional links in which a 1.3-μm signal wavelength is used for transmission in one direction and a 1.5-μm signal wavelength is used in the reverse direction. Typical performances of such commercially available devices are 1.5- to 4.0-dB loss and demultiplexing crosstalk in the range −25 to −50 dB.

In order to multiplex or demultiplex more than two wavelengths, such devices can be placed in cascade. An example is given in Figure 6.6, in which three fused-fiber WDMs are cascaded to form a multiplexer at 1,200, 1,240, 1,280, and 1,320 nm [Fussgaenger, 1986].

When many wavelengths are considered, alternative techniques have been revealed to be more practical. These alternatives utilize the angularly dispersive feature of some elements.

Prisms were employed in the early development of angularly dispersive WDM devices [Ishio, 1984]. However, prism-based WDM devices have been proven to be difficult and expensive to produce, and typically exhibit relatively low angles of angular dispersion, leading to quite bulky devices. These inherent drawbacks have resulted in the use of the diffraction grating as the major angularly dispersive element within WDM devices.

A diffraction grating reflects light in particular directions according to the grating constant (the number of ruled lines per unit distance, usually between 1,000 and 5,000 lines per millimeter), the angle at which the light is incident on the grating, and the light wavelength.

The main structural type of grating-based WDM devices is the so-called Littrow configuration, as illustrated in Figure 6.7 [Bousquet, 1968].

In this configuration, a single lens and a separate plane grating are employed. The

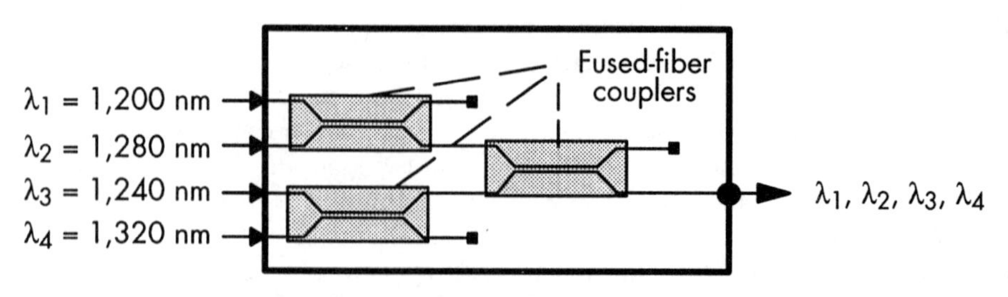

Figure 6.6 Cascade of fused-fiber couplers to form a WDM for four wavelengths.

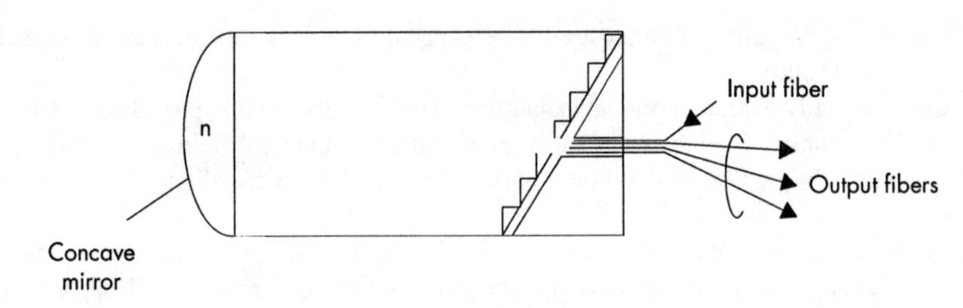

Figure 6.7 Schematic representation of a Littrow-configured WDM device for single-mode operation.

blaze angle of the grating is such that the incident and reflected light beams follow virtually the same path. Several lens types have been implemented within such devices, but the GRIN-rod lens has proven to be the most attractive due to its compactness and easy and stable alignment.

The Littrow configuration has been exploited very cleverly in the so-called STI-MAX device [Laude, 1983, 1992]: up to 20 channels with 1.6-nm wavelength spacing in the 1.5-μm window have been multiplexed and demultiplexed with adjacent-channel crosstalk of −30 to −40 dB. In the 1.3-μm window, 1×16 WDMs are available with 2- or 4-nm channel spacing. Overall attenuation is typically 5.0 dB or less for such commercially available devices.

Recently, arrayed waveguide gratings with 28 WDM channels spaced by 1 nm in the 1.3-μm wavelength band have been fabricated in research laboratories [Takahashi, 1991]. This technology may represent a significant advance for integrated WDM devices of the future.

WDM devices that have just been discussed are static devices and can serve as building blocks to build specific network topologies. An example of such a construction is discussed in Sub-section 6.6.3. In some multiaccess networks (e.g., the broadcast-and-select network discussed in Sections 6.6.1 and 7.1.1) a dynamic WDM device is required to select a specific channel in the optical domain. Such devices are referred to as tunable optical filters and are described in Chapter 7.

6.4 OPTICAL ISOLATOR

The performance of certain optical-fiber communication systems can be significantly degraded if part of the launched power is reflected back towards the transmitter. This is notably the case for analog-fiber transmission systems (e.g., cable television, CATV), very-high-bit-rate digital systems using single-longitudinal mode (SLM) lasers, or systems using optical amplifiers. Reflections may arise from splices, connectors, couplers, filters, or even from the fiber itself (backscattering).

In order to prevent such undesired reflections, optical isolators are placed immediately after the sensitive device.

Commercially available optical isolators make use of the nonreciprocal nature of the Faraday effect, which changes the state of polarization of incident fields in the presence of a magnetic field parallel to the direction of propagation. Consider Figure 6.8, showing the structure of an optical isolator.

Suppose a plane-polarized wave is impinging on a polarizer placed at the input of the device. This polarizer passes only the vertically polarized component. When passing through the Faraday rotator, the field will experience a rotation α of its polarization with

$$\alpha = \varrho \cdot H \cdot L \tag{6.15}$$

where H is the magnetic field strength, L the interaction length, and ϱ the Verdet's constant (for diamagnetic materials, such as silica fibers, ϱ is independent of temperature).

Assume that α has been adjusted so as to produce a 45-deg rotation and that a second polarizer (analyzer) rotated 45 deg relative to the input polarizer is placed at the output of the device. The incident light will pass through the device with typically 1 to 2 dB of loss provided that the incident light is linearly polarized with polarization parallel to the transmission direction of the input polarizer. The backward light coming from undesired reflections will be rotated by another 45 deg when passing once again through the Faraday rotator, but in the opposite direction with respect to the direction of propagation of the signal. This is the result of the nonreciprocity of the Faraday effect. The reflected light facing the input polarizer will be orthogonal to the input polarizer's orientation and will thus be blocked by it. Typical isolation coefficients for commercially available devices are of the order of 30 dB.

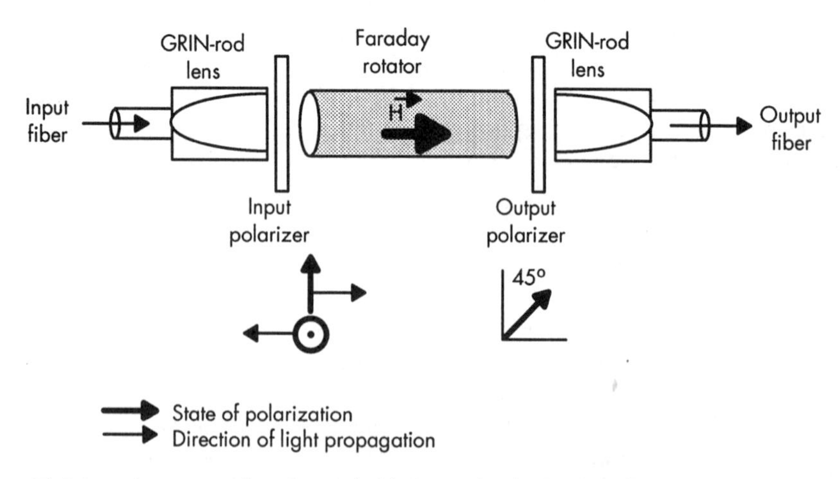

Figure 6.8 Schematic representation of an optical isolator using the Faraday effect.

6.5 NETWORK TOPOLOGIES

The placement of the actual nodes and the way in which they are physically connected to the network define the physical-network topology. This must be distinguished from the logical-path topology, which defines how the traffic flows between the different end nodes. For example, a star, bus, or tree physical-network topology can serve to broadcast the traffic to each node from a head-end. This one-to-all traffic flow characterizes the logical-path topology on top of the physical-network topologies. This section considers only the physical-network topologies, while various logical-path topologies will be considered in the next chapter.

There are three basic physical-network topologies for multiaccess optical-fiber networks: the star, the bus, and the ring, as shown in Figure 6.9(a–c) [Limb 1984, Brackett 1991, Wu 1992]. With these three basic topologies, a multitude of other network topologies can be constructed, with a few examples illustrated in Figure 6.9(d–f) [Nassehi 1985, Gerla 1988, Albanese 1988]. Each of these has advantages and disadvantages compared with the others, and none is universally favored for multiaccess networks. However, the star or multilevel star have proven to be the most practical (up to now), mainly because they relieve some of the technological constraints.

One of the most important constraints in multiaccess optical-fiber networks is the power budget limitation. In contrast to point-to-point links that are generally bandwidth-limited, multiaccess networks suffer from power losses. This network power-budget shortage will now be analyzed on the basis of two distinct topologies, namely the single star (Figure 6.9(a)) and the double bus (Figure 6.9(b)). The analysis can easily be extended to other types of topologies.

In the $N \times N$ single-star network, the output of all transmitters is combined by the passive star coupler and equally distributed to all receivers. All channels are available at each receiver, and a particular channel can be selected by using a tuning scheme that depends on how the channels are multiplexed together (time, wavelength, subcarrier or code division multiplexing).

In the double-bus network, each node sends to nodes farther to the right on the upper bus and to nodes farther to the left on the lower bus. Here also, channel selection at each receiver is achieved by using the adequate tuning scheme.

The double-bus network has the disadvantage that it requires twice as many transmitters and receivers per node as the single-star network. This disadvantage can be avoided with the folded-bus (Figure 6.9.(e)), but the folded-bus cannot support as many nodes as the double-bus because the average connection passes through twice as many taps. This drawback can be overcome by placing a regenerator in the folding part of the network, as shown in the figure, and it is straightforward to see that as long as the power budget is the only consideration, the situation in this case is the same as for the double-bus.

Neglecting fiber and coupling losses, the attenuation (in dB) of a signal passing through the $N \times N$ star coupler of Figure 6.9(a) is given by (6.13). It has been seen that the loss in dB of the $N \times N$ star network builds up logarithmically with N.

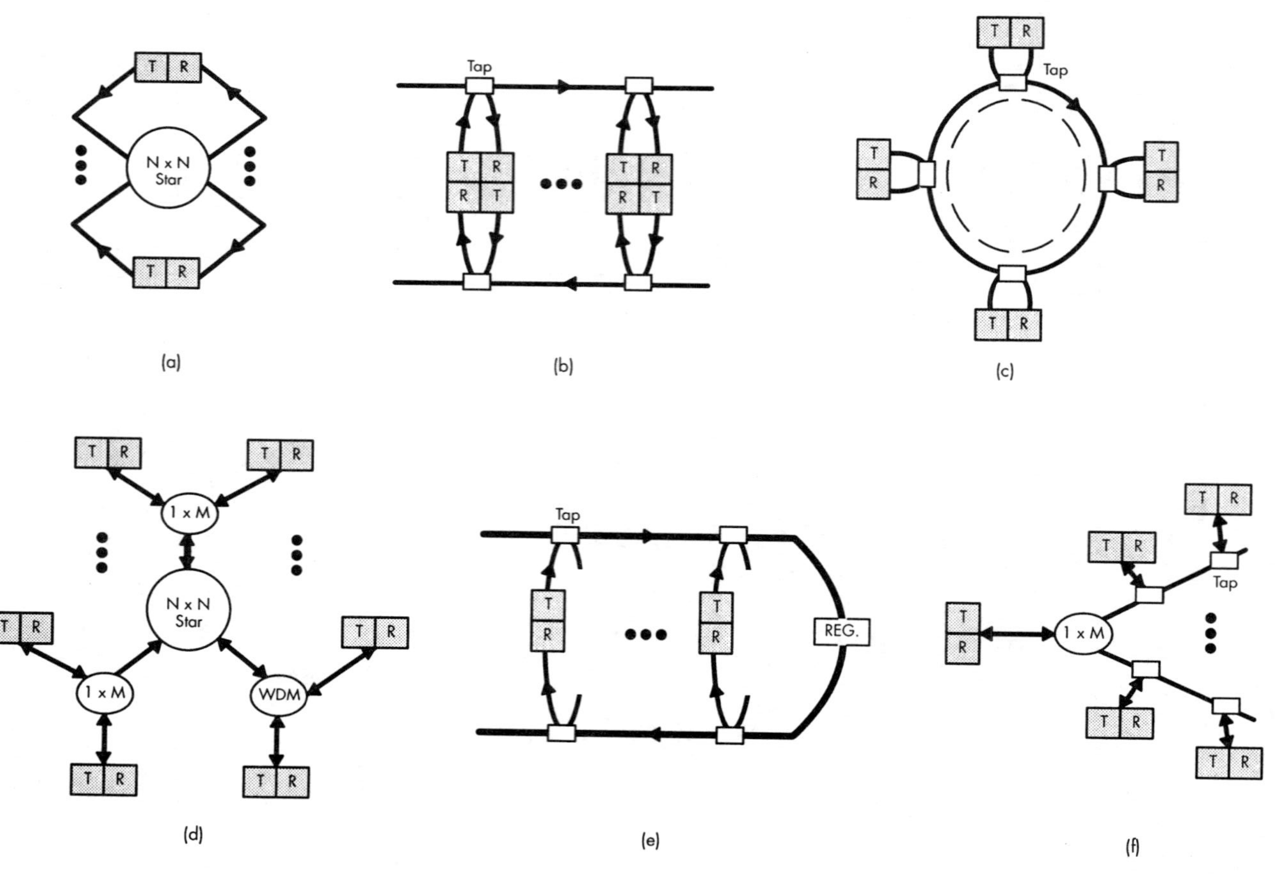

Figure 6.9 Some examples of physical-network topologies of multiaccess optical-fiber networks: (a) $N \times N$ single-star, (b) double-bus, (c) ring, (d)

For the double-bus network, the worst-case loss occurs for the link between the two end nodes; that is, the leftmost and the rightmost nodes. For a given number N of nodes, the worst-case loss can be minimized if the coupling coefficient is individually selected for each tap on the fiber cable [Limb, 1984]. In practice, however, this is unrealistic because it would require a large variety of taps that have to be changed whenever nodes are added to or removed from the network. Moreover, in many cases the optimal tap coefficients are so small that they are comparable to the excess loss and hence the taps themselves will be difficult to manufacture. In the following analysis, it is assumed that the taps are the same for all nodes.

Let α be the tap coefficient (that is, $|S_{12}|^2 = |S_{21}|^2 = \alpha$, which means that $|S_{11}|^2 = |S_{22}|^2 = 1 - \alpha$) and β the excess loss of the tap. Neglecting fiber and coupling losses, the worst-case ratio A of output (P_{rec}) to input (P_0) power is readily obtained and is given by

$$A = P_{rec}/P_0 = \alpha^2(1-\alpha)^{N-2}\beta^N \tag{6.16}$$

which is maximized when $\alpha_{opt} = 2/N$.

Inserting the optimum value $\alpha = 2/N$ into (6.16), we obtain

$$A = \frac{4}{N^2}\left(1 - \frac{2}{N}\right)^{N-2}\beta^N \xrightarrow{\text{Large}N} \frac{4}{e^2}\frac{\beta^N}{N^2} \tag{6.17}$$

or expressed in dB for large N

$$\text{Loss}_{\text{Bus}}[\text{dB}] = -10\log_{10}A \cong 2.7 + 10\log_{10}N^2 - 10N\cdot\log_{10}\beta \tag{6.18}$$

Thus, in contrast to the star topology, the loss in dB for the double-bus topology builds up logarithmically with N^2. Moreover, as seen from (6.18), the effect of the tap excess loss is much more serious than with the star since the resulting loss (in dB) grows linearly with N instead of logarithmically. This also means that the required receiver dynamic range for the nodes connected to the bus must be higher than for the star [Ota, 1992].

Figure 6.10 shows the difference of network losses for the star and the double-bus topologies calculated from (6.13) and (6.18), respectively.

If we suppose, for example, that the optical channels in the network can support a maximum link loss of 40 dB, the maximum number of nodes that can be connected to the bus network is limited to 22, assuming $\beta = 0.5$ dB. By comparison, the star network with the same link loss of 40 dB (and $\beta = 0.5$ dB) can support over 3,000 nodes.

The advantage of the star network can further be highlighted by considering the network capacity, defined as the product of the node's bit rate times the number of nodes ($C = B\cdot N$). The network capacity must not be confused with the network throughput, which takes into account many factors relating to protocol, traffic statistics, and

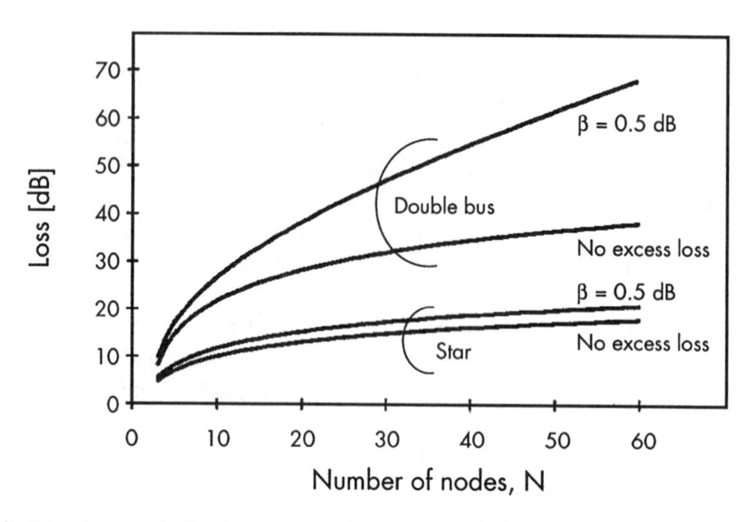

Figure 6.10 Network losses for the single-star and double-bus topologies as a function of the number of nodes. For both topologies, the losses are shown for the ideal case (no excess loss) and for realistic couplers (β = 0.5 dB).

network-control overhead (which are not considered in this chapter). The network capacity can thus be viewed as the interconnection capacity of the network, assuming that there is no overhead, contention, or blocking.

For a given bit rate B, the mean optical power required at the receiver to achieve a specified bit error rate (BER) can be expressed as (Chapter 4)

$$\overline{P}_s = \overline{n}_p \cdot h\nu \cdot B \tag{6.19}$$

where $\overline{n}_p$ is the mean number of photons per bit required at the receiver.

If $\overline{P}_T$ is the average transmit power, which is assumed to be the same for all nodes, then the received power for the star and the double-bus topologies is, using (6.13) and (6.17), respectively,

$$\overline{P}_{\text{rec}}^{\text{Star}} = \overline{P}_T \cdot \frac{\beta^{\log_2 N}}{N} \tag{6.20}$$

and

$$\overline{P}_{\text{rec}}^{\text{Bus}} = \overline{P}_T \cdot \frac{4}{N^2}\left(1 - \frac{2}{N}\right)^{N-2} \beta^N \xrightarrow{\text{Large}\,N} \frac{4\overline{P}_T}{e^2}\frac{\beta^N}{N^2} \tag{6.21}$$

The product B·N is maximized when $\overline{P}_{\text{rec}} = \overline{P}_S$ since this corresponds to the maximum allowable network loss and thus the maximum number of nodes N for a given bit rate B. Putting $\overline{P}_{\text{rec}} = \overline{P}_S$ in (6.20) and (6.21), the capacity of the star and the double-bus networks is given by

$$C_{\text{Star}} = (B \cdot N)_{\text{Star}} = \left(\frac{\overline{P}_T}{\overline{n}_p \cdot h\nu}\right) \cdot \beta^{\log_2 N} \tag{6.22}$$

and

$$C_{\text{Bus}} = (B \cdot N)_{\text{Bus}} \xrightarrow{\text{Large}N} = \left(\frac{\overline{P}_T}{\overline{n}_p \cdot h\nu}\right) \cdot \frac{4}{e^2} \cdot \frac{\beta^N}{N} \tag{6.23}$$

Figure 6.11(a, b) represents the B·N product for the star and the double-bus calculated from (6.22) and (6.23) for the lossless case and for β = 0.2 dB and 0.5 dB. In the figure $\overline{n}_p$ = 1,000, which is representative for direct detection (DD) receivers, the transmitted power $\overline{P}_T$ = 1 mW and the systems are assumed to operate in the 1.55-μm window (i.e., $h\nu$ = 0.8 eV).

Consider first the star network. For the lossless star network the maximum capacity is C = 7.8 Tbps, independent of N. This value increases to 78 Tbps if coherent detection with sensitivity $\overline{n}_p$ = 100 is used instead of direct detection. For lossy couplers, C decreases slowly as N increases, as shown in Figure 6.11(a) for β = 0.2 dB. It is noteworthy that such extremely high capacities can only be approached with large channel numbers because the bit rate per channel is limited by the end-node electronics. The dashed lines correspond to constant bit rates and the network capacity follows the path indicated by the arrows in the figure: for small values of N, the capacity increases linearly with N until the power-budget limit is encountered. For larger values of N, the bit rate per node must be reduced, resulting in an overall degradation of C. As an example, for B = 5 Gbps, C approaches its power budget limit of 5 Tbps for $N \approx$ 1,000 (β = 0.2 dB) when direct detection is used. Theoretically, this value increases by a factor of 10 when coherent detection is used.

In practice, the achievable capacity is considerably lower unless optical amplifiers are used. This is mainly because the receiver sensitivity degrades significantly for increasing bit rates. The spot lines of Figure 6.11(a) show representative values of C for realistic receivers [Brackett, 1990]. The maximum achievable capacity approaches 1 Tbps for direct detection at $B \approx$ 1 Gbps and 3 Tbps for coherent detection.

From Figure 6.11(b), it is seen that the situation is much more dramatic for the double-bus network. Even for lossless taps (no excess loss) and lossless transmission, the power budget limit is encountered at $N \approx 30$ with $B = 5$ Gbps, leading to $C \approx 0.15$ Tbps. For realistic values of the tap excess loss and transmission losses (including connector and splice losses), the maximum achievable capacity is further reduced to

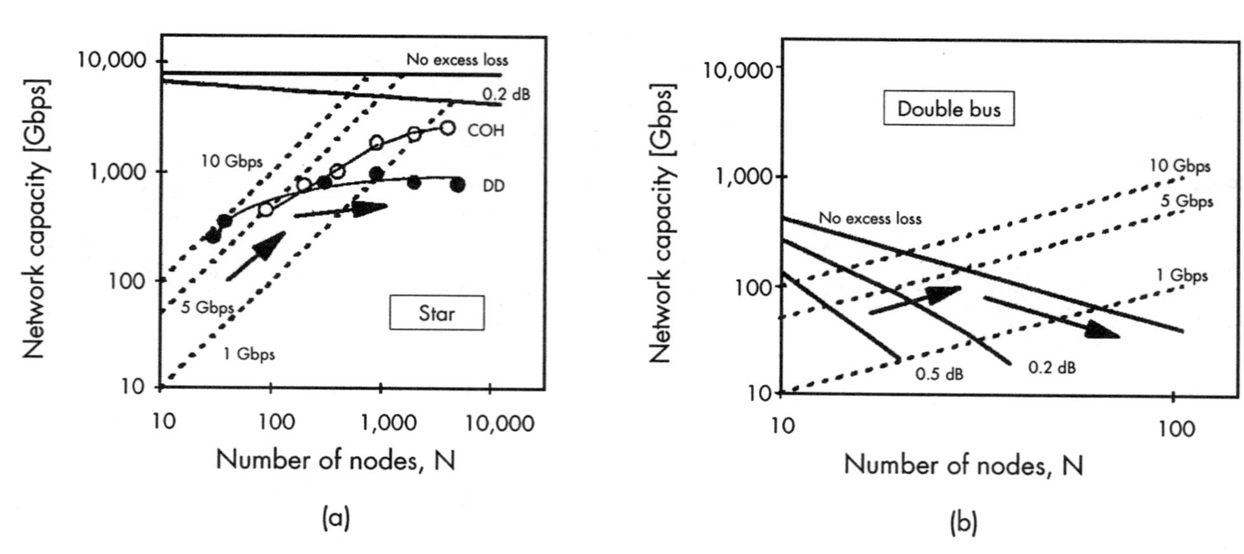

Figure 6.11 Network capacity C as a function of the number of nodes N for (a) the star and (b) the double-bus physical-network topologies. The curves are calculated assuming a transmit power $P_T = 1$ mW and a receiver sensitivity of 1,000 photons/bit at a wavelength $\lambda = 1.55$ μm. The dashed lines correspond to constant bit rates. The curves corresponding to 0.2 and 0.5 dB illustrate the effect of the excess loss per 2×2 coupler used as tap in the double-bus or as basic element to construct the $N \times N$ star coupler. DD and COH indicate direct detection and coherent detection, respectively.

rather small values. With $\beta = 0.2$ dB and B = 5 Gbps, the maximum number of nodes that can be connected to the double-bus is only $N \approx 20$, resulting in $C \leq 0.1$ Tbps.

Increasing the capacity would require either an increase in transmit power, an improvement in the receiver sensitivity, or a lossless interconnection network using optical amplifiers. This subject is discussed in Section 6.6.

6.6 NETWORK POWER BUDGET ENHANCEMENT

As seen in the previous section, one of the fundamental capacity limitations is due to the network losses or, in other words, to the restricted-power budget. Improving the power budget would make it possible to approach to the fundamental capacity limit as closely as possible, or to relax the requirements on optical powers at network terminations with overall cost reduction as a benefit.

A number of options can be envisaged to overcome this power-budget limitation. First, increasing the transmitter power of each node would improve the end-to-end power budget and would make it possible to further increase the number N of nodes connected to the network. However, several system parameters could make it impossible to do this for many reasons. Crosstalk induced by nonlinear effects will prohibit reliable communication if the launched power exceeds 1–10 mW (Chapter 7) [Chraplyvy, 1990]. The optical power within the network might be limited for eye-safety reasons. Finally, since the cost of laser transmitters increases with increasing output power, the resulting total network cost may become prohibitive.

A second option might consist of increasing the receiver sensitivity; that is, decreasing the minimum allowed value of P_{rec}. This may prove to be effective, especially for the star-type networks, and will be discussed in the next paragraph. For bus-type networks, a worthwhile observation of (6.17) shows that N_{max} is approximately inversely proportional to $\sqrt{P_{rec}}$. Hence, decreasing P_{rec} will not be an effective means of improving N in bus-type topologies.

Clearly, to improve the power budget appreciably we must deal with the network losses more directly. To this end, we will now investigate the use of optical amplifiers for both the star and the bus-type networks.

6.6.1 The Amplified Star Network

Consider Figure 6.9(a) and assume for the moment that DD receivers are used. There are essentially two distinct options for the placement of optical amplifiers within such networks (1) within the star coupler, which is the network-equivalent of the in-line amplifier in point-to-point links; or (2) immediately preceding the receiver, in which case the amplifier is used as a preamplifier.

The first option (1) can be implemented in different ways, either by placing the amplifiers at the inputs of the star coupler, inside the star coupler (thus forming an active-star coupler), or immediately at the outputs of the device. Of these three possibilities, the most attractive is revealed to be the active-star coupler because it allows a significant saving on the number of amplifier pump diodes that would otherwise be required for the other two possibilities.

Various constructions of active-star couplers have been proposed and demonstrated [Willner 1990, 1991, Presby 1991, Irshid 1992 a,b, Chen 1992]. Consider Figure 6.12, illustrating the basic concept behind active-star couplers. An active $N \times N$ star coupler is obtained by interconnecting smaller star couplers with N fiber amplifiers. In this configuration, the K($M \times M$) input star couplers serve not only to split the N input signals equally among the N outputs, but also to distribute the pump power to the N fiber amplifiers inside the device.

It is seen that only K = N/M pump lasers are required, whose number can be considerably reduced if large values of M are possible. Planar-star couplers, discussed in Section 6.2, may prove to be particularly attractive here because of their low excess loss independent of M. In addition, and in contrast to fused-fiber couplers, their insertion loss is essentially wavelength-flat, which would allow the active-star coupler to be pumped either at 1,480 nm (remote pumping) or at the more efficient 980-nm wavelength.

However, M may not be too large since only 1/M of the pump laser power is injected in each fiber amplifier and that power must be high enough to provide sufficient gain to overcome end-to-end losses of the $N \times N$ active star. Therefore, a tradeoff exists between M and the pump laser power, and the optimum solution will depend on several factors such as the availability of star couplers and pump lasers, and of course the related cost.

The placement of an optical amplifier in front of each receiver (2) would make it possible to reduce the required received power P_{rec}, resulting in an improved end-to-end power budget [Inoue, 1991]. The resulting improvement will depend on the detection mechanism used. Let us first consider direct detection, and assume zero-pigtailing coupling losses at the amplifier, ideal lossless coupling between the amplifier and the photodiode, and unity photodiode quantum efficiency. The signal-to-noise ratio (SNR) after detection of the amplified signal has been determined in Chapter 5 and is given by (5.32), or by (5.33) for large amplifier gain. The receiver BER can be obtained following the analysis of Section 4.2.1. For instance, it is given by (4.44) or by

$$\mathrm{BER} = \frac{1}{2}\,\mathrm{erfc}\left(\sqrt{\frac{\mathrm{SNR}}{2}}\right) \tag{6.24}$$

The receiver sensitivity is obtained by substituting (5.33) in (6.24), and solving for P_{in}. The result (with $Q = \sqrt{\mathrm{SNR}}$) is

$$P_{in} = h\nu \cdot F \cdot B_e \cdot \left\{ Q^2 + Q\sqrt{Q^2 + \left(\frac{\Delta\nu_{opt}}{2B_e}\right)} \right\} \tag{6.25}$$

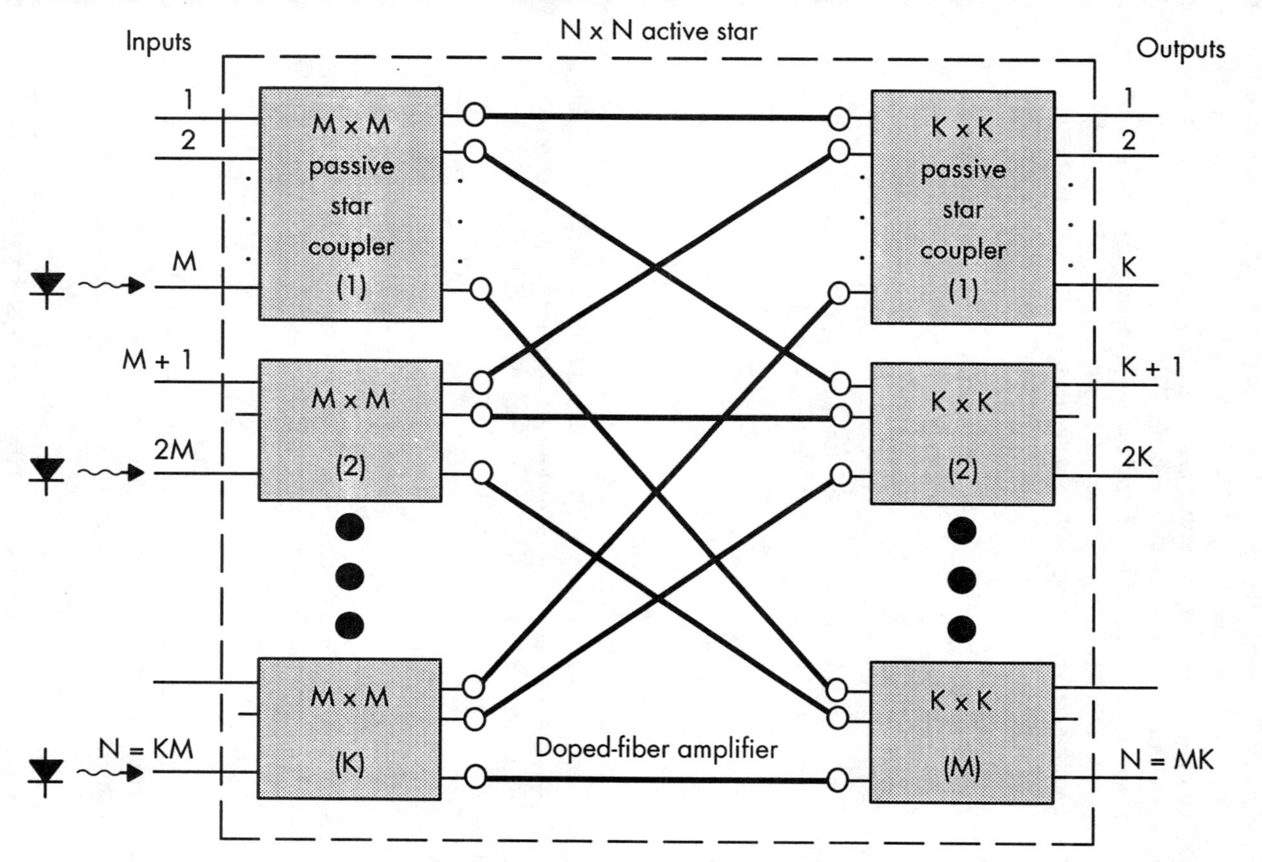

Figure 6.12 Lossless $N \times N$ star coupler implemented using K($M \times M$) input star couplers, N lengths of doped-fiber amplifiers pumped with K pump lasers, and M($K \times K$) output star couplers. N = KM. (© 1992 IEEE. [After Irshid, 1992.])

Equation (6.25) clearly shows why optical amplifiers with a small noise figure F must be used, as already pointed out in Chapter 5. It also shows how optical filters can improve the receiver sensitivity by decreasing $\Delta\nu_{opt}$. The minimum optical bandwidth is equal to the electrical bandwidth (optical-matched filter, $\Delta\nu_{opt} = B_e$).

The receiver sensitivity can be written in terms of the average number of photons/bit $\overline{n}_p$, using $P_{in}/2 = \overline{n}_p \cdot h\nu \cdot B$ where B is the bit rate. By taking $\Delta\nu_{opt} = B_e = B/2$ as a typical value of the receiver bandwidth, $\overline{n}_p$ is given by

$$\overline{n}_p = \frac{F}{2} \cdot Q^2 \tag{6.26}$$

The minimum value of F is 2 for an ideal amplifier. Thus, using Q = 6 for a BER of 10^{-9}, the absolute minimum value of (6.26) is $\overline{n}_p = 36$. This is exactly the same sensitivity as that of coherent (synchronous) heterodyne ASK receivers. A value of $\overline{n}_p = 46$ has been achieved experimentally using an optically preamplified DD receiver [Gabla, 1992].

The combination of an optical preamplifier and DD receiver thus provides a simple means to overcome the thermal noise of the DD receiver and, hence, to approach the network power-budget capacity limit.

The preceding analysis can be easily extended to the case of coherent receivers using (5.33), provided we add to the receiver noise the other contributions resulting from local-oscillator shot noise and local-oscillator spontaneous beat noise [Okoshi, 1988]. As an example, it can be shown that the receiver sensitivity of an asynchronous FSK heterodyne receiver is given by [Agrawal, 1992]

$$\overline{n}_p = 40 \cdot [F + h\nu \cdot G \cdot F^2 \cdot \Delta\nu_{opt}/(2P_{LO})] \tag{6.27}$$

Since $\overline{n}_p = 40$ in the absence of an optical amplifier, it is seen that the receiver sensitivity degrades by the factor in the square brackets of (6.27). The second term of (6.27) can be made negligible by increasing the local-oscillator power, thereby avoiding the use of an optical filter that would otherwise be necessary. In the limit of large P_{LO}, the sensitivity degrades by F. The same conclusion holds for other modulation formats as well. However, in a heterodyne scheme, the shot-noise-limited receiver sensitivity can be achieved, at least theoretically, without using a preamplifier. Therefore, the advantage of the combination of an optical preamplifier and a heterodyne coherent receiver is rather small.

In the above discussion we have assumed the utilization of fiber-doped amplifiers, which do not suffer from crosstalk for the bit rates of interest. For semiconductor amplifiers we can refer to the analysis carried out in Section 5.1.2 [O'Mahony 1988, Olsson 1989, Inoue 1989, Ramaswami 1990, Willner 1992].

6.6.2 The Amplified Bus Network

In the case of multiaccess bus networks, optical amplifiers can be positioned periodically along the bus, with each amplifier having sufficient gain to compensate for attenuation experienced by the signal. Unfortunately, the number of amplifiers that can be placed along the bus is limited because of the accumulation of optical-amplifier noise and because of the optical-gain-saturation phenomenon.

At each amplification stage, amplified spontaneous emission (ASE) noise is added to the signal, and the SNR degrades accordingly. After a certain number of amplifications, the ASE level eventually overwhelms the SNR on the downstream portion of the bus, prohibiting reliable communication.

Furthermore, since the total power entering a given amplifier consists of both signal power and ASE power from previous amplifiers, the total power grows as more amplifiers are added to the bus. At some point, the aggregate power at the input of the amplifier will be so high that it will cause gain saturation, with the amplifier no longer being able to supply the required gain.

The optimum amplifier placement and the maximum number allowed along the bus has been analyzed at length [Wagner 1987, Liu 1989, Ramaswami 1991 & 1993]. Although both of the above-mentioned effects act simultaneously, in most circumstances the major limitation on the number of amplifiers stems from the effect of gain saturation.

Gain-saturation limitations are discussed here for two distinct situations. First, the single-channel situation in which a time division multiple access (TDMA) scheme is used (Chapter 9); that is, the optical channels do not overlap along the bus and any amplifier sees only one channel at any one time. Second, the multichannel situation, in which all nodes are allowed to be active everywhere on the bus at all instants of time. This corresponds to the cases where WDMA, SCMA, or CDMA schemes are used (Chapters 7, 8, and 10).

Consider Figure 6.13 showing the overall structure of the network, which consists of a number of subbuses, each accommodating M nodes connected together by (traveling wave) optical amplifiers.

Assuming identical taps with tap coefficient α ($|S_{12}|^2 = \alpha$) and excess loss β, the loss experienced by a signal propagating the entire length of a subbus between amplifiers is given by

$$\Lambda_0 = (1 - \alpha)^M \cdot \beta^M \tag{6.28}$$

We assume that each amplifier has sufficient gain, G_s, to offset the loss incurred since the previous amplifier; that is, $G_s \cdot \Lambda_0 = 1$. We also assume ideal amplifiers with an excess noise factor $\chi = 1$, a spontaneous emission factor $n_{sp} = 1$ and zero-pigtailing coupling losses. The single-channel and the multiple-channel situations will now be discussed separately.

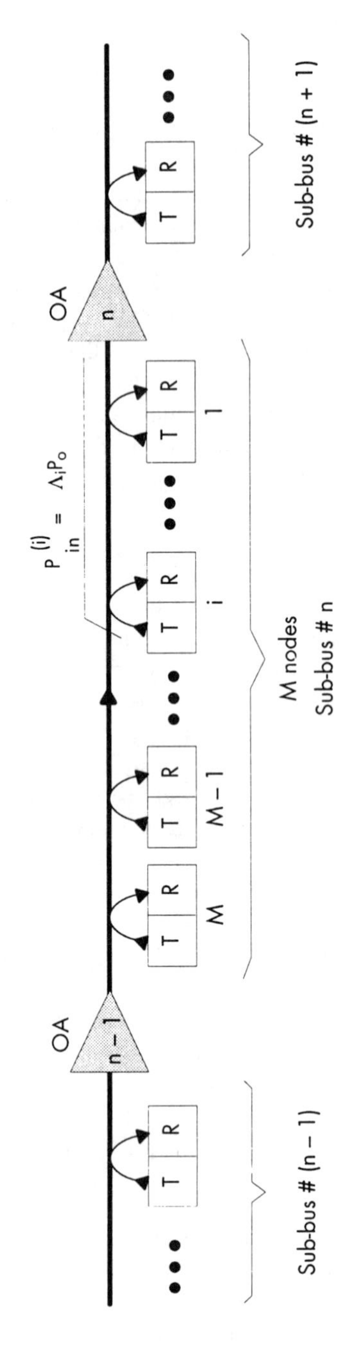

Figure 6.13 Bus network employing optical amplifiers, with M identical taps between each amplifier.

6.6.2.1 The Single-Channel Situation

Since we consider the limitation due to amplifier-gain saturation, we must look at the worst-case situation, which corresponds to the maximum amplifier-input power. For the single-channel bus, the worst-case results from the signal power launched from the rightmost node on the subbus (the closest to the amplifier) plus the accumulated ASE power from the previous amplifiers. The power from the rightmost node arriving at the input of the amplifier is $\alpha\beta P_0$ where P_0 is the launched power, supposed to be identical for all nodes. From (5.16), the ASE power at the output of each amplifier is given by

$$P_{\mathrm{ASE}} = (G_s - 1) \cdot h\nu \cdot \Delta\nu_{\mathrm{opt}} \tag{6.29}$$

where $\Delta\nu_{\mathrm{opt}}$ is the optical-amplifier bandwidth.

The ASE power generated by each amplifier is attenuated by a factor Λ_0 before being incident at the input of the following amplifier. Since the amplifier gain is $G_s = \Lambda_0^{-1}$, each amplifier adds a noise power of P_{ASE} given by (6.29). The (worst-case or maximum) output power of amplifier n is thus given by

$$\begin{aligned} P_{\mathrm{out}}^{(n)} &= G_s \cdot \alpha\beta \cdot P_0 + n \cdot P_{\mathrm{ASE}} \\ &\cong G_s \cdot [\alpha\beta \cdot P_0 + n \cdot h\nu \cdot \Delta\nu_{\mathrm{opt}}] \end{aligned} \tag{6.30}$$

The upper bound on n is obtained by putting the condition $P_{\mathrm{out}}^{(n)} \leq P_{\mathrm{sat}}$, where P_{sat} is the saturation output power of the amplifier. It follows that

$$n \leq (P_{\mathrm{sat}} - \alpha\beta G_s P_0) \cdot (G_s \cdot h\nu \cdot \Delta\nu_{\mathrm{opt}})^{-1} \tag{6.31}$$

As an example, consider the following representative values: $\alpha = 0.2$, $\beta = 0.89$ (= 0.5 dB), $G_s = 30$ (15 dB), $h\nu = 0.8$ eV ($\lambda = 1.55$ μm), $\Delta\nu_{\mathrm{opt}} = 3.7$ THz ($\Delta\lambda = 30$ nm), $P_{\mathrm{sat}} = +5$ dBm (3.16 mW) and $P_0 = -10$ dBm (0.1 mW). Equation (6.31) provides $n \leq 75$, resulting in a maximum number of nodes along the bus $N = [1 + int(\mathrm{n})]M \cong 750$ where *int*(n) indicates the integral part of n. $M = 2/\alpha$ has been assumed, which corresponds to the optimum number of nodes on each subbus, as can be seen from (6.16).

This example reveals that a significant power-budget improvement of the single-channel TDMA bus network can be achieved with the help of optical amplifiers. The maximum number of nodes can be increased by one to two orders of magnitude over the case of a completely passive bus.

Further improvements of the gain-saturation limit may be achieved in a number of ways. For instance, with M smaller than that of the above example (larger α), the number of nodes can still be increased. It has been shown that over 5,000 nodes can be supported with 10-dB amplifiers between every two nodes [Ramaswami, 1991]. Although this configuration significantly increases N, it has the real disadvantage that it requires as many amplifiers as the number of nodes (in the double-bus configuration), which may result in a prohibitive cost.

Another way to increase N can be achieved by limiting the optical bandwidth of the amplifiers. This can be seen from (6.31), which shows that the upper bound of n is inversely proportional to $\Delta\nu_{opt}$. By placing optical filters in front of each amplifier, we can decrease $\Delta\nu_{opt}$, thus allowing for a greater value of N. For example, limiting $\Delta\nu_{opt}$ to 1 THz ($\Delta\lambda = 8$ nm) and using the same values as those of the above example results in $N \approx 2{,}800$. This approach does reduce the optical bandwidth available for information transmission, but this is not a real drawback since the single-channel TDMA bus allows each node to transmit at the same wavelength. The same approach may not be effective in the multichannel situation, which is discussed in the following paragraph.

6.6.2.2 The Multichannel Situation

This situation is complicated by the fact that all nodes are allowed to transmit simultaneously on the bus. The total power in front of the first amplifier, $P_{in}^{(1)}$, is the sum of the launched power P_0, from each node balanced by the corresponding attenuation factor. The attenuation factor for the power launched by node i, as shown in Figure 6.13, is readily obtained by

$$\Lambda_i = \alpha \cdot (1 - \alpha)^{i-1} \cdot \beta^i \tag{6.32}$$

Thus, we have

$$P_{in}^{(1)} = P_0 \cdot \sum_{i=1}^{M} \Lambda_i = P_0 \cdot \Lambda \text{ with } \Lambda = \alpha\beta \cdot \frac{1 - (1 - \alpha)^M \beta^M}{(1 - \beta + \alpha\beta)} \tag{6.33}$$

Similar to (6.30), the output power of the first amplifier is given by

$$P_{out}^{(1)} \cong G_s \cdot [\Lambda P_0 + h\nu \cdot \Delta\nu_{opt}] \tag{6.34}$$

Since each subbus adds the same amount of power to the total power, the output power of amplifier n is now given by

$$P_{out}^{(n)} \cong n \cdot G_s \cdot [\Lambda P_0 + h\nu \cdot \Delta\nu_{opt}] \tag{6.35}$$

Using the same values as those used for the example of (6.31) reveals that $h\nu \cdot \Delta\nu_{opt} \ll \Lambda P_0$, such that the effect of ASE on the output power can be neglected compared to the effect of the aggregate power launched by all nodes. Imposing $P_{\text{out}}^{(n)} \leq P_{sat}$, we obtain from (6.35)

$$n \leq \frac{P_{\text{sat}}}{G_s \cdot \Lambda \cdot P_0} \tag{6.36}$$

With the same representative values as those previously used, (6.36) reduces to $n \leq 565 \cdot P_{\text{sat}}$. It is clear from this last result that in order to obtain acceptable values of n (say $n \geq 10$), the saturation output power of the amplifier applicable in the multichannel bus must be much higher than in the single-channel bus. In the above example, P_{sat} must be as high as +15 dBm ($N \approx 180$) or +20 dBm ($N \approx 570$) to increase N by an order of magnitude, at best.

The product N = nM is optimized when M = 1; that is, an amplifier at each tap. Even in this case, which relaxes the stringent requirements on P_{sat}, N can be increased to only a few hundreds [Ramaswami, 1991; Chen, 1993].

6.6.3 Wavelength Routing

A number of physical-network topologies have been suggested in which the nodes are interconnected by WDM devices, such as those discussed in Section 6.3 [Hill 1988, Irshid 1992 a&b].

One advantage of WDMs over the conventional wavelength-insensitive couplers is that they avoid the splitting loss. An input optical signal at a given wavelength is directed only to a single output port rather than being split among all the output ports. This leads to a reduction in the power loss imposed on the individual signals (the excess loss remains) and may be exploited to improve the overall network power budget.

Figure 6.14 illustrates a $N \times N$ wavelength-routing network implemented using WDM devices. A connection between node i and node j is set up by tuning the transmitter T_i and the receiver R_j to λ_j. Meanwhile, some other pairs of nodes could also be tuned to λ_j. Thus, this approach offers considerable wavelength reuse, reducing significantly the number of different wavelengths required. The price paid involves the fact that the transmitters and/or the receivers must be tunable and that the wavelengths of the lasers must correspond to those of the WDMs.

In order to avoid the required tunability features of the transmitters and receivers, the same structure as that depicted in Figure 6.14 has been proposed for multihop lightwave networks [Irshid, 1992c]. We shall discuss such types of networks in Section 7.5.

The wavelength-routing concept has been demonstrated in the field to increase the capacity and routing flexibility of installed fiber-based systems interconnecting central offices [Westlake, 1991].

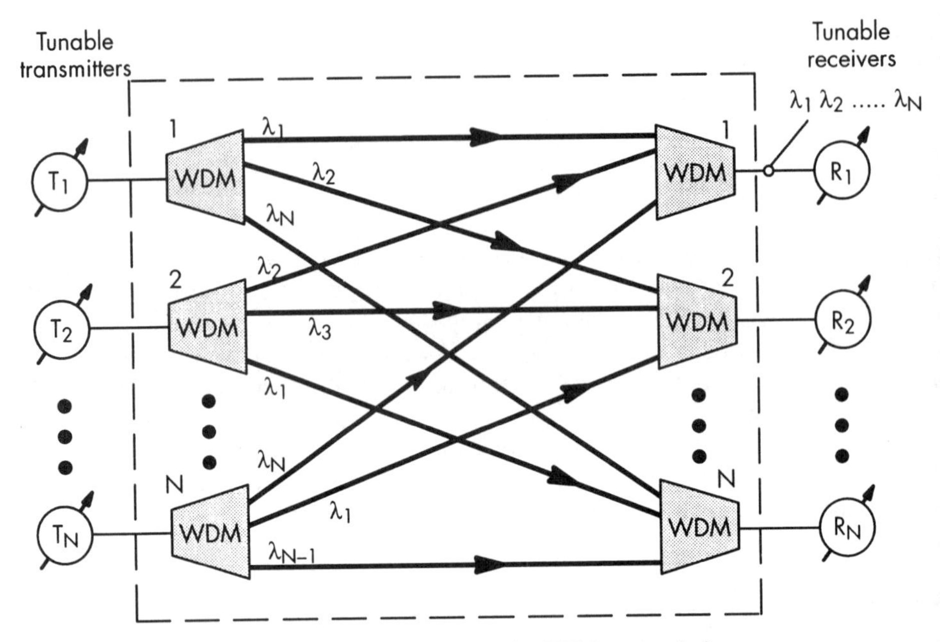

Figure 6.14 A $(N \times N)$ wavelength-routing network employing WDM passive devices.

6.7 CONTROL-CHANNEL LIMITATION

So far, we have been concerned with the limitations due to the power budget and we saw how these limitations can be overcome with the help of optical amplifiers. In fact, other factors may also further limit the network capacity.

For example, one of these factors is related to the accessibility of the available optical spectrum within WDMA networks.

Another factor is related to the medium access control (MAC) and is applicable to most of the multiaccess schemes [Henry, 1988]. More specifically, consider, as an example, the broadcast-and-select multiwavelength star network (that will be discussed in more detail in Section 7.1.1) in which N lasers tuned to a fixed and unique wavelength are interconnected with N tunable receivers through a $N \times N$ star coupler (Figure 6.15). In such a network, a connection is set up by tuning one of the receivers to one of the laser wavelengths. All nodes can communicate with each other provided that the tunable receivers are capable of tuning over the entire range of transmitter wavelengths. A control channel is required to communicate the tuning information to the receivers. Even if the processing time required to resolve the contention of the network is negligibly small, the limited bandwidth of the control channel can restrict the network capacity. To see that quantitatively, consider the architecture of Figure 6.15 [Brackett, 1991].

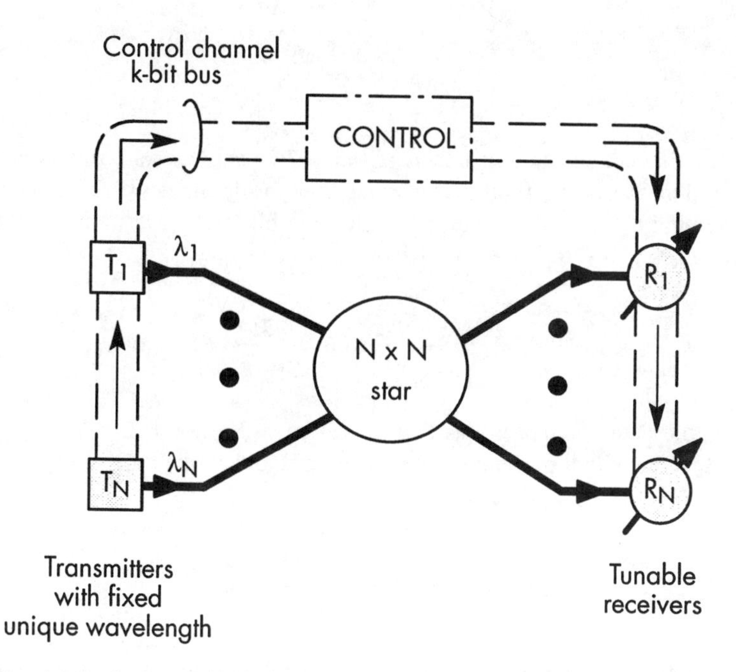

Figure 6.15 Broadcast-and-select multiwavelength star network showing the control channel that provides signaling information between the nodes.

Suppose that the transmitters are synchronized and send packets of the same length of P bits. A connection is established when one transmitter/receiver pair is tuned to the same wavelength. The receiver is informed as to which input wavelength to tune via the control channel. This requires an address specification of at least $\log_2 N$ bits. Since the network is composed of N input/output pairs, the control channel must transport at least $N\log_2 N$ signaling bits to set up all the connections within the network. The contention resolution can be performed centrally, as shown in Figure 6.15, or individually at each output port following a specific distributed protocol [Habbab 1987, Goodman 1989, Mehravari 1990, Willner 1992, Jeon 1992].

The network configuration (i.e., the logical-path topology) may change every cycle, which is defined as the holding time of a session. Long cycles characterize circuit switched traffic (in the present analysis, this corresponds to extremely long packets), while short cycles characterize bursty traffic where packet connection must be set up and taken down far more rapidly.

In both cases, the control channel must transport $N\log_2 N$ signaling bits per cycle.

Let T_{cycle} be the duration of a cycle and C_c the required capacity of the control channel. Then

$$C_c = \frac{N\log_2 N}{T_{\text{cycle}}} \tag{6.37}$$

In practice, we can assume that $T_{\text{cycle}} = m \cdot T_P$ where m is a positive integer and T_P is the time duration of the P bit-packets that flow inside the network.

Assuming that each transmitter operates at a bit rate B, we immediately obtain $T_P = P/B$ and $T_{\text{cycle}} = mP/B$, and (6.37) becomes

$$C_c = \frac{N \cdot \log_2 N}{m \cdot P} \cdot B = \frac{\log_2 N}{m \cdot P} \cdot C \tag{6.38}$$

where C is the network capacity as defined earlier in Section 6.5.

Let k be the width of the control-channel bus and B_c the bit rate of each of the bus links. Thus $C_c = k \cdot B_c$, and (6.38) gives

$$C = B \cdot N = k \cdot \frac{mP}{\log_2 N} \cdot B_c \tag{6.39}$$

Equation (6.39) expresses the network capacity as a function of the control-channel parameters. Two special cases of (6.39) will be examined. The first is $k = 1$, the serial control channel for which $N\log_2 N$ signaling bits are transported serially over one control link. The second is $k = \log_2 N$, the parallel control channel for which the $N\log_2 N$ signaling bits are transported over $\log_2 N$ control links.

With (6.39), we immediately have

$$k = 1: \qquad C = \frac{m}{\log_2 N} \cdot P \cdot B_c \tag{6.40}$$

$$k = \log_2 N: \qquad C = m \cdot P \cdot B_c \tag{6.41}$$

It is seen that two critical parameters in both cases are the session holding time, characterized by m, and the product of the packet length times the control channel bit rate, P · Bc.

The session holding time is determined by the envisaged network functionality. For example, if the network is intended for crossconnect functions, the session holding time is generally much longer than the packet duration T_P, resulting in very large values of m. In this case, the control channel has little to do and the limited network power-budget discussed in Sections 6.5 and 6.6 is the major factor determining the maximum achievable network capacity.

In the opposite situation in which the network is intended for packet switching or for computer local-area networking, the session holding time is of the order of T_P (small m) with the limit $T_{\text{cycle}} = T_P$ (m = 1). It is clear from (6.40) and (6.41) that in such a situation the network capacity limit set by the restricted control-channel capacity is fully determined by the product $P \cdot B_c$. This product must be made as large as possible by increasing either the packet length or the control-channel bit rate in order to push away the limit imposed by the control-channel capacity.

However, the packet length is usually dictated by issues such as the desired service needs (e.g., minimum packetization delay), the network efficiency and latency, as well as standards issues.

With respect to this last point, two distinct cases are of special interest: asynchronous transfer mode (ATM) packets defined by the ITU standardization committee as the solution for broadband integrated digital services networks (B-IDSN) [de Prycker, 1991], and computer data transfer packets (DATA).

For ATM, the packet (officially named the cell) length is P = 424 bits (53 bytes). For DATA transfer, the packet length varies from the standard but is generally much longer than for ATM. We assume here that P = 16,000 bits (2,000 bytes).

The control-channel bit rate is dictated by practical considerations such as the choice of technology and the implementation complexity, as well as the cost. We assume here that B_c = 1 Gbps. Therefore, we have for the two cases:

$$\begin{aligned} \text{ATM:} \quad & P \cdot B_c = 424 \text{ Gb/s} \\ \text{DATA:} \quad & P \cdot B_c = 16 \text{ Tb/s} \end{aligned}$$

Figure 6.16 shows the network capacity limit for ATM and DATA packets using either the one-bit bus or the $\log_2 N$-bit bus control channel. The two oblique lines correspond to $B = 5 \cdot B_c = 5$ Gbps and $B = B_c = 1$ Gbps.

Let us first discuss the ATM case. For relatively small values of N, the network capacity increases linearly with N following C = B · N. For the one-bit bus, the limit due to the control-channel capacity is encountered at $N \approx 20$ (B = 5 Gbps) and $N \approx 70$ (B = 1 Gbps). For higher number of nodes, C decreases logarithmically with N. The maximum network capacity is about 100 Gbps for B = 5 Gbps and 70 Gbps for B = 1 Gbps.

For the $\log_2 N$-bit bus, the control-channel limit is encountered at $N \approx 85$ (B = 5 Gbps) and $N = 424$ (B = 1 Gbps), above which the network capacity stays constant at a value of 424 Gbps (= P · Bc).

Thus, in both of these ATM cases, the network capacity is restricted by the control-channel limit rather than by the power-budget limit. In particular, there is not much reason to use coherent detection or optically preamplified direct detection to improve the power budget. In addition, since the full benefit is achieved for $N \approx 100 - 400$ there is no real advantage in opting for large network size if maximum capacity is the goal. In this case, we have seen that a significant improvement in network capacity may be achieved by using a parallel-bus control channel.

Let us now turn to the discussion of the DATA case. For the one-bit bus, the

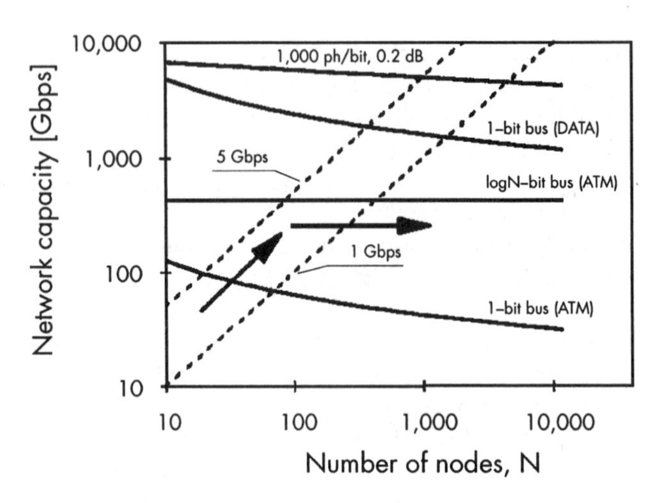

Figure 6.16 Network capacity limits due to the control-channel capacity limitation for both ATM (424 bits) and DATA (16,000 bits) packets examples.

control-channel limit is encountered at N ≈ 375 (B = 5 Gbps) and N ≈ 1,515 (B = 1 Gbps), above which the network capacity decreases logarithmically with N. The maximum value of C is about 1.9 Tbps (B = 5 Gbps) and 1.5 Tbps (B = 1 Gbps). Referring to the discussion of Section 6.5 (Figure 6.11), it is seen that for N beyond about 100, the capacity is limited by the power budget to ≈ 1 Tbps if direct detection is used, so that coherent detection (or direct detection with optical amplifier) is advisable to maximize the capacity. Even in this case, the network size must not be larger than about $N \approx 1{,}000$ if the one-bit bus control-channel limitation is to be avoided.

If larger network size is the objective, the implementation of the control channel must be directed towards the parallel bus.

The above discussion is not restricted to the broadcast-and-select multiwavelength star network used here as an illustrative example. It also applies to other multiaccess schemes such as SCMA and CDMA, which will be considered in Chapters 8 and 10.

It is also important to note that the means by which the contention resolution is resolved has not been included in the above control-channel limit. This might represent a further limitation on network capacity [Habbab 1987, Goodman 1989, Mehravari 1990].

6.8 SUMMARY

Most passive optical devices that have been developed so far can be classified into two main categories: wavelength-insensitive couplers such as the 3-dB and the $N \times M$ star couplers, and wavelength-selective devices such as the WDM multiplexers and demultiplexers.

With the two technologies currently used to design these devices, namely fused-fiber and planar technologies, the maximum number of optical channels (i.e., the maximum number of device's ports) is limited to 10–100.

We have seen that these devices serve as building blocks to passively interconnect the network nodes and to form the physical topology of the multiaccess networks. The passive nature of the interconnection pattern results in the possibility to transport distinct signals without altering their information content.

The primary parameters of passive optical devices are the splitting loss (for couplers), the directivity (for WDMs), and the excess loss (for both).

Despite the low excess losses that have been achieved for commercially available devices (and even lower excess losses for devices still in the research stage), multiaccess networks suffer from a power-budget shortage, which restricts the maximum achievable network capacity.

We have seen that optical amplifiers are helpful to improve the network capacity, especially for multichannel star networks. However, another limiting factor imposed by the limited control-channel capacity may further restrict the overall network capacity. This limiting factor becomes increasingly important as the packet length shortens, but it becomes almost completely negligible as the packet length enlarges, eventually up to (virtually) infinity as is the case for circuit-switched networks.

REFERENCES

[Bousquet, 1968] Bousquet, P. Spectroscopie Instrumentale, Dunod Universite, 1968.

[Tomlinson, 1980] Tomlinson, W. J. "Applications of GRIN-rod lenses in optical fiber communication systems," Applied Optics, Vol. 19, 1980, pp. 1127–1138.

[Kobayashi, 1980] Kobayashi, K., et al. "Microoptic grating multiplexers and optical isolators for fiber optic communications," IEEE J. Quant. Electron., Vol. QE-16, No. 1, Jan. 1980, pp. 11–22.

[Laude, 1983] Laude, J. P. "STIMAX, a grating multiplexer for monomode and multimode fibers," Proc. European Conf. Optical Communication, ECOC'83, 1983, pp. 417–420.

[Limb, 1984] Limb, J. O. "On fiber optic taps for local area networks," Links for the future, IEEE/Elsevier Science Publishers B. V. (North-Holland) , 1984, pp. 1130–1136.

[Winzer, 1984] Winzer, G. "Wavelength multiplexing components—A review of single-mode devices and their applications," IEEE J. Light. Technology, Vol. 2, No. 4, Aug. 1984, pp. 369–378.

[Ishio, 1984] Ishio, H., et al. "Review and status of wavelength-division multiplexing technology and its application," IEEE J. Light. Technology, Vol. 2, No. 4, Aug. 1984, pp. 448–463.

[Marhic, 1984] Marhic, M. E. "Combinatorial star couplers for single-mode optical fibers," Proc. FOC/LAN 84, Boston, 1984, pp. 175–179.

[Oshima, 1985] Oshima, S., et al. "Small loss-deviation tapered fiber star coupler for LAN," IEEE J. Light. Technology, Vol. 3, No. 3, June 1985, pp. 556–560.

[Abbas, 1985] Abbas, G. L., et al. "A dual-detector optical heterodyne receiver for local oscillator noise suppression," IEEE J. Light. Technology, Vol. 3, No. 5, Oct. 1985, pp. 1110–1122.

[Nassehi, 1985] Nassehi, M. M., et al. "Fiber optic configurations for local area networks," IEEE J. Selected Areas of Communication, Vol. 3, No. 6, Nov. 1985, pp. 941–949.

[Fussgaenger, 1986] Fussgaenger, K., et al. "4×560Mbit/s WDM system using 3 wavelength selective fused single mode fiber coupler as demultiplexer," Proc. European Conf. Optical Communication, ECOC'86, 1986, pp. 447–450.

[Habbab, 1987] Habbab, I. M., et al. "Protocols for very high speed optical fiber local area networks using a passive star topology," IEEE J. Light. Technology, Vol. 5, No. 12, 1987., pp. 1782–1784

[Wagner, 1987] Wagner, S. S. "Optical amplifier applications in fiber optic local networks," IEEE Trans. Communication, Vol. 35, No. 4, April 1987, pp. 419–426.

[Okoshi, 1988] Okoshi, T., and K. Kikuchi. *Coherent Optical Fiber Communications*, Kluwer Academic Pub., 1988.

[Tomlinson, 1988] Tomlison, W. J., et al. "Integrated optics: basic concepts and techniques" and "Passive and low-speed active optical components for fiber systems," in Optical Fiber Telecommunications II, Ch. 9 & 10 Edited by Miller & Kaminow, New York, NY: Academic Press, 1988.

[Saleh, 1988] Saleh, A.A.M., et al. "Reflective single-mode fiber-optic passive star couplers," IEEE J. Light Technology, Vol. 6, No. 3, March 1988, pp. 392–398.

[O'Mahony, 1988] O'Mahony, M. J. "Semiconductor laser optical amplifiers for use in future fiber systems," IEEE J. Light. Technology, Vol. 6, No. 4, April 1988, pp. 531–544.

[Henry, 1988] Henry, P. S. "Very-high-capacity lightwave networks," Proc. IEEE International Communication Conference, June 1988, pp. 1206–1209.

[Gerla, 1988] Gerla, M., et al. "Tree structured fiber optics MAN's," IEEE J. Selected Areas of Communication Vol. 6, No. 6, July 1988, pp. 934–943.

[Albanese, 1988] Albanese, A., et al. "Loop distribution using coherent detection," IEEE J. Selected Areas of Communication, Vol. 6, No. 6, July 1988, pp. 959–973.

[Hill, 1988] Hill, G. R. "A wavelength routing approach to optical communication networks," British Telecom Technology J., Vol. 6, No. 3, July 1988, pp. 24–31.

[Senior, 1989] Senior, J. M., et al. "Devices for wavelength multiplexing and demultiplexing," IRE Proc., Vol. 136, No. 3, June 1989, pp. 183–202.

[Arkwright, 1989] Arkwright, J. W., et al. Electron. Letters, Vol. 26, 1989, pp. 1534–1535.

[Goodman, 1989] Goodman, M. S. "Multiwavelength networks and new approaches to packet switching," IEEE Communication Magazine, Vol. 27, Oct. 1989, pp. 27–35.

[Dragone, 1989] Dragone, C. "Efficient N×N star coupler based on Fourier optics," IEEE J. Light. Technology Vol. 7, No. 3, 1989, pp. 479–489.

[Pietzsch, 1989] Pietzsch, J. "Scattering matrix analysis of 3×3 fiber couplers," IEEE J. Light. Technology, Vol. 7, No. 2, Feb. 1989, pp. 303–307.

[Olsson, 1989] Olsson, N. A. "Lightwave systems with optical amplifiers," IEEE J. Light. Technology, Vol. 7 No. 7, July 1989, pp. 1071–1081.

[Inoue, 1989] Inoue, K. "Crosstalk and its power penalty in multichannel transmission due to gain saturation in a semiconductor laser amplifier," IEEE J. Light. Technology, Vol. 7, No. 7, July 1989, pp. 1118–1124.

[Liu, 1989] Liu, K., et al. "Analysis of optical bus networks using doped-fiber amplifiers," Conf. Digest IEEE/LEOS Summer Topical on Optical Multiple Access Networks, Monterey, CA, July 25–27, 1989, pp 41–42.

[Horimatsu, 1989] Horimatsu, T., et al. "OEIC technology and its applications to subscriber loops," IEEE J. Light. Technology, Vol. 7, No. 11, Nov. 1989, pp. 1612–1622.

[Keck, 1989] Keck, D. B., et al. "Passive components in the subscriber loop," IEEE J. Light. Technology, Vol. 7, No. 11, Nov. 1989, pp. 1623–1633.

[Willner, 1990] Willner, A. E., et al. "1.2Gb/s closely-spaced FDMA-FSK direct detection star network," IEEE Photon. Technology Letters, Vol. 2, No. 3, 1990, pp. 223–226.

[Kirby, 1990] Kirby, P. A. "Multichannel wavelength-switched transmitters and receivers—New component concept for broadband networks and distributed switching systems," IEEE J. Light. Technology, Vol. 8, No. 2, Feb. 1990, pp. 202–211.

[Brackett, 1990] Brackett, C. A. "Dense wavelength division multiplexing networks: principles and applications," IEEE J. Selected Areas of Communication, Vol. 8, Aug. 1990, pp. 948–964.

Mehravari, 1990] Mehravari, N. ''Performance and protocol improvements for very high speed optical fiber local area networks using a passive star topology,'' IEEE J. Light Technology, Vol. 8, No. 4, 1990, pp. 520–530.

Chraplyvy, 1990] Chraplyvy, A. R. ''Limitations on lightwave communications imposed by optical fiber nonlinearities,'' IEEE J. Light. Technology, Vol. 8, No. 10, 1990, pp. 1548–1557.

Ramaswami, 1990] Ramaswami, R., et al. ''Amplifier induced crosstalk in multichannel optical networks,'' IEEE J. Light. Technology, Vol. 8, No. 12, Dec. 1990, pp. 1882–1896.

de Prycker, 1991] de Prycker, M. *Asynchronous Transfer Mode: solution for BroadBand-ISDN*, Ellis Horwood series in computer and networking, 1991.

Ramaswami, 1991] Ramaswami, R., et al. ''Analysis of multiple-channel optical bus networks using doped fiber amplifiers,'' Proc. Opt. Fiber Communication Conf., OFC91, FE5, Feb. 18–22, 1991, p. 219.

Wisely, 1991] Wisely, D. R. ''32 channels WDM multiplexer with 1nm channel spacing and 0.7nm bandwidth,'' Electron. Letters, Vol. 27, No. 6, March 1991, pp. 520–521.

Karol, 1991] Karol, M. J. ''Exploiting the attenuation of fiber-optic passive taps to create a large high-capacity LAN's and MAN's,'' IEEE J. Light. Technology, Vol. 9, No. 3, March 1991, pp. 400–408.

noue, 1991] Inoue, K., et al. ''Multichannel amplification utilizing an Er3+-doped fiber amplifier,'' IEEE J. Light. Technology, Vol. 9, No. 3, March 1991, pp. 368–374.

Arkwright, 1991] Arkwright, J. W., et al. ''Monolithic 1×19 single-mode fused fiber couplers,'' Electron. Letters, Vol. 27, No. 9, April 1991, pp. 737–738.

Takahashi, 1991] Takahashi, H., et al. ''Multi-demultiplexer for nanometer spacing WDM using arrayed-waveguide gratings,'' Integrated Photonic Research, Monterey, CA, April 1991, pp. 5–7.

Brackett, 1991] Brackett, C. A. ''On the capacity of multiwavelength optical-star packet switches,'' IEEE Light. Trans. Systems, May 1991, pp. 33–37.

Willner, 1991] Willner, A. E., et al. ''Star couplers with gain using multiple erbium-doped fibers pumped with a single laser,'' IEEE Photon. Technology Letters, Vol. 3, No. 3, 1991, pp. 250–252.

Presby, 1991] Presby, H. M., et al. ''Amplified integrated star couplers with zero loss,'' IEEE J. Light. Technology, Vol. 3, No. 8, 1991, pp. 170–173.

Hermes, 1991] Hermes, T., et al. ''Coherent bidirectional broadband communication experiment using a reflective star coupler,'' Proc. European Conf. Opt. Communication, ECOC'91, Paris, Sept. 9–12, 1991, pp. WeB9–6.

Westlake, 1991] Westlake, H. J., et al. ''Reconfigurable wavelength routed optical networks: a field demonstration,'' Proc. European Conf. Optical Communication, Sept. 1991, pp. 753–756.

Laude, 1992] Laude, J. P. *Le Multiplexage de Longueur D'Onde*, Masson, Paris, 1992.

Agrawal, 1992] Agrawal, G. P. *Fiber-Optic Communication Systems*, New York, NY: Wiley & Sons, 1992.

Adar, 1992] Adar, R., et al. ''Adiabatic 3dB couplers, filters, and multiplexers made with silica waveguides on silicon,'' IEEE J. Light. Technology, Vol. 10, No. 1, Jan. 1992, pp. 46–50.

Irshid, 1992.a] Irshid, M. I., et al. ''Star couplers with gain using fiber amplifiers,'' IEEE Photon. Letters, Vol. 4, No. 1, Jan. 1992, pp. 58–60.

Irshid, 1992.b] Irshid, M. I., et al. ''A fully transparent fiber-optic ring architecture for WDM networks,'' IEEE J. Light. Technology, Vol. 10, No. 1, Jan. 1992, pp. 101–108.

Okamoto, 1992] Okamoto, K., et al. ''Fabrication of wavelength-insensitive 8×8 star coupler,'' IEEE Photon. Technology Letters, Vol. 4, No. 1, Jan. 1992, pp. 61–63.

Willner, 1992] Willner, A. E. ''SNR analysis of crosstalk and filtering effects in an amplified multichannel direct-detection dense-WDM system,'' IEEE Photon. Letters, Vol. 4, No. 2, Feb. 1992, pp. 186–189.

Ota, 1992] Ota, Y., et al. ''DC-1Gb/s burst-mode compatible receiver for optical bus applications,'' IEEE J. Light. Technology, Vol. 10, No. 2, Feb. 1992, pp. 244–249.

Gabla, 1992] Gabla, P. M., et al. ''92 photons/bit sensitivity using an optically preamplified direct detection receiver,'' Proc. Opt. Fiber Communication Conf., OFC92, San Jose, CA, 1992, p. 245.

[Irshid, 1992.c] Irshid, M. I., et al. ''A WDM cross-connected star topology for multihop lightwave networks,'' IEEE J. Light. Technology, Vol. .10, No. 6, June 1992, pp. 828–835.

[Tagaki, 1992] Tagaki, A., et al. ''Wavelength characteristics of (2×2) optical channel type directional couplers with symmetric or nonsymmetric coupling structures,'' IEEE J. Light. Technology, Vol. 10, No. 6, June 1992, pp. 735–746.

[Chen, 1992] Chen, Y. K. ''Amplified distributed reflective optical star couplers,'' IEEE Photon. Letters, Vol. 4, No. 6, June 1992, pp. 570–573.

[Wu, 1992] Wu, T-H., et al. ''A novel passive protected SONET bidirectional self-healing ring architecture,'' IEEE J. Light. Technology, Vol. 10, No. 9, Sept. 1992, pp. 1314–1322.

[Irshid, 1992] Irshid, M. I., et al. ''Star couplers with gain using fiber amplifiers,'' IEEE Photon. Technology Letters, Vol. 4, No. 1, Jan. 1992, pp. 58–60.

[Willner, 1992] Willner, A. E., et al. ''Comparison of central and distributed control in WDMA star network,'' Proc. Int, Conf. Communication, ICC92, 1992, 330.2, pp. 824–828.

[Jeon, 1992] Jeon, H. B., et al. ''Contention-based reservation protocols in multiwavelength optical networks with a passive star topology,'' Proc. Int. Conf. Communication, ICC92, 349.5, 1992, pp. 1473–1477.

[Weissman, 1993]Weissman, Y. *Optical Network Theory*, Norwood, MA: Artech House, 1993.

[Takahashi, 1993] Takahashi, H., et al. ''Integrated-optic 1×128 power splitter with multifunnel waveguide,'' IEEE Photon. Technology Letters, Vol. 5, No. 1, Jan. 1993, pp. 58–60.

[Goldstein, 1993] Goldstein, E. L., et al. ''Multiwavelength fiber-amplifier cascades in unidirectional interoffice ring networks,'' Proc. Opt. Fiber Communication Conf., OFC93, TuJ3, Feb. 21–26, 1993, pp. 44–46.

[Chen, 1993] Chen, D. N., et al. ''Transparent optical bus using a chain of remotely pumped erbium-doped-fiber amplifiers,'' Proc. Opt. Fiber Communication Conf., OFC93, WI10, Feb. 21–26, 1993, pp. 144–145.

[Ramaswami, 1993] Ramaswami, R., et al. ''Analysis of effective power budget in optical bus and star networks using erbium-doped fiber amplifiers,'' submitted to IEEE J. Light. Technology, 1993.

Chapter 7
Wavelength Division Multiaccess Networks

Unlike other communication technologies, optical technology offers a new dimension, the *color of light*, to perform such network functions as multiplexing, routing, and switching. In essence, it is this new dimension that distinguishes optical networks, in general, from other network technologies.

Wavelength division multiaccess networks (WDMA networks) effectively use the wavelength as an additional degree of freedom by concurrently operating distinct portions (i.e., distinct *colors* or *wavebands*) of the 1.3- to 1.6-μm wavelength spectrum accessible within the fiber network. Each waveband supports a network channel that can operate at peak electronic processing speed of, say, a few gigabits per second. With several hundreds to 1,000 wavebands, such networks have the potential to realize Tbps capacity.

The implementation of WDMA networks generally requires wavelength-tunable optical components such as tunable lasers and/or tunable optical filters. These tunable optical components form the tunable transceiver, which is incorporated into each network node. Tunable transceivers are used differently depending on the type of WDMA network architecture chosen [Mukkerjee, 1992].

In *single-hop* WDMA networks (sometimes also called *all-optical* WDMA networks), the data stream, once transmitted as light, reaches its final destination directly without being converted to electronic form in between. For a packet transmission to occur, one of the laser transmitters of the sending node and one of the optical receivers of the destination node must be tuned to the same wavelength for the duration of the packet transmission.

In circuit-switched networks, the required speed of transceiver tunability is usually slow. By contrast, in packet-switched networks the node transceivers must be able to tune to different wavelengths rapidly in order to send and receive packets in quick succession. Besides this technological problem of fast wavelength tuning, the other key challenge in single-hop WDMA networks is to develop protocols for efficiently coordinating the connections at different wavelengths within the network.

Multihop WDMA networks get around these problems by avoiding transceiver tunability. Each node transceiver is provided with a small number (two, for example) of fixed-tuned optical transmitters and fixed-tuned optical receivers. Each transmitter in the network is tuned to a different wavelength. Direct connection (i.e., single-hop connection) between two nodes is only possible if the destination node has one of its receivers tuned to one of the transmitter-node wavelengths. Connectivity between any other pair of nodes is achieved by routing through intermediate nodes, where the optical channel is converted to electronic form, the packet's destination address is decoded, and the packet is then switched electronically and retransmitted on the appropriate wavelength to reach its final destination or another intermediate node where the process is repeated. Therefore, in general, a packet will experience multihopping through a number of intermediate nodes before it reaches its final destination.

A number of different multihop WDMA network architectures are possible, each having distinct performance characteristics such as the average packet delay and the statistics of the number of hops that must be traversed.

From a performance point of view, single-hop and multihop WDMA networks are equally attractive. Both approaches apply two technologies where they can best be exploited: optics for high-speed transport and routing, and electronics for memory and logic functions.

This chapter is intended to provide a comprehensive account of WDMA networks that have been proposed, and in some cases also prototyped. It should be noted, however, that the field of research in WDMA networks is rapidly moving, with new innovations occurring every few months. Hence, within the restricted space of this book, we will focus on the fundamentals and highlight some of the most promising network architectures that have been investigated to date.

In the optical communication community, it is common to categorize WDMA networks by indicating the wavelength scales of interest and the degree of complexity required in the selection and control of wavelengths [Brackett, 1990]:

1. In the so-called *dense* WDMA networks, the spacing between different wavelengths is on the order of 1 nm. Intensity modulation/direct detection (IM/DD) transceivers are used in such networks.
2. In the so-called *optical frequency division multiaccess* networks (FDMA networks), the wavelength spacing is on the order of the signal bandwidth or bit rate (i.e., a few gigahertz in frequency; note that 1-nm wavelength spacing around 1.5-μm center wavelength corresponds to a frequency spacing of 130 GHz—see appendix B). Coherent transceivers are generally used in such networks.

This distinction is useful when considering the required implementation technology but makes no fundamental architectural difference. We will therefore refer to the common term of WDMA networks for both dense WDMA and FDMA networks and indicate which technology is the most suitable only when appropriate.

The chapter is organized as follows. In Section 7.1 we introduce the basic concept behind single-hop WDMA networks by classifying them within two distinct categories; namely, the *broadcast-and-select* and the *wavelength-routing* networks. Section 7.2 is devoted to tunable optical filters that play a major role in WDMA networks. Design and performance issues of single-hop WDMA networks are discussed in Section 7.3, although most of these issues are applicable to multihop WDMA networks (to be discussed in Section 7.5) as well. Some relevant examples of single-hop WDMA networks are described in Section 7.4, along with a discussion of the corresponding network protocols that control the data flow within the networks. Section 7.5 focuses on multihop WDMA networks with emphasis on the so-called Shufflenet. The basic concepts are first introduced in Section 7.5.1, while the performance analysis is carried out in Section 7.5.2. Two examples of multihop WDMA networks that have been experimentally demonstrated are discussed in Section 7.6. Finally, Section 7.7 summarizes the main results obtained in the present chapter.

7.1 SINGLE-HOP WDMA NETWORKS

Single-hop WDMA networks can be classified into two broad categories: the broadcast-and-select networks and the wavelength-routing networks. An example of each of them is illustrated in Figures 7.1 and 7.2, respectively. The following two sections discuss each of them separately.

7.1.1 Broadcast-and-Select WDMA Networks

Figure 7.1 represents a broadcast-and-select star network (other topologies such as a bus, a ring, or a tree may be implemented as well): all transmitted channels, each operating at a distinct wavelength, are combined in an $N \times N$ star coupler and broadcast to all node receivers. Several functional possibilities exist, depending on whether the transmitters, the receivers, or both are made tunable. Notice that in general each network node can be equipped with a number of transmitters and receivers, some of which are dynamically tunable while others are fixed-tuned to some specific wavelengths. For simplicity of explanation, we will, for instance, assume that each node contains only one transmitter and one receiver.

1. When the transmitters are tunable, while the receivers are tuned to fixed but distinct wavelengths, a connection is established by tuning the transmitter to the wavelength accessible by the addressed receiver. The network is basically an input-queued space-division switch in function. Collisions may occur in such a network since two or more packets with the same destination may arrive simultaneously at different node inputs (i.e., at different inputs to a centralized switch or at different distributed nodes of a LAN or MAN). This contention problem is resolved by means of the network protocol that coordinates the connections within the network.

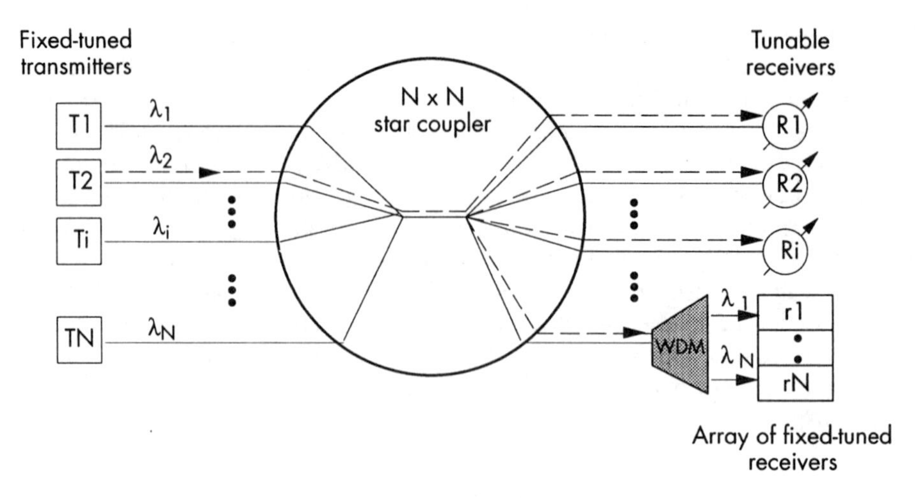

Figure 7.1 The broadcast-and-select single-hop WDMA star network.

Single-hop WDMA networks with a single tunable transmitter and a single fixed-tuned receiver (later referred to as TT-FR networks) are restricted to point-to-point connections. Many-to-one connections are possible provided that network nodes are equipped with more than one (say, m) fixed-tuned receiver (TT-FR^m networks). Similarly, multicast capability (i.e., one-to-many) is made possible by equipping each network node with more than one tunable transmitter (TT^m-FR networks). The price to be paid for such additional functionalities is increased system cost brought about by the required additional hardware and a higher complexity of the network protocol.

2. A more attractive WDMA network can be built by using fixed-tuned transmitters and tunable receivers (later referred to as FT-TR networks). In addition to point-to-point capability, such a network inherently provides multicast capability. This is achieved by simultaneously tuning more than one node receiver to the same wavelength. Similarly to the TT-FR networks, many-to-one connections are also possible provided that network nodes are equipped with more than one tunable receiver (FT-TR^m networks). Notice that collisions are automatically avoided in FT-TR networks since each channel uses a different wavelength. Packet loss (or packet blocking) may, however, occur if the receivers can tune to only one wavelength at a time. In this case, the performance of FT-TR networks can be severely limited by the means to inform the receivers what wavelength to tune to at what times. Several proposed network protocols will be discussed in Section 7.4.
3. The third possibility for broadcast-and-select WDMA networks is obtained when both the transmitters and the receivers are made tunable (later referred to as TT-TR networks). TT-TR networks can support point-to-point and many-to-one as well as

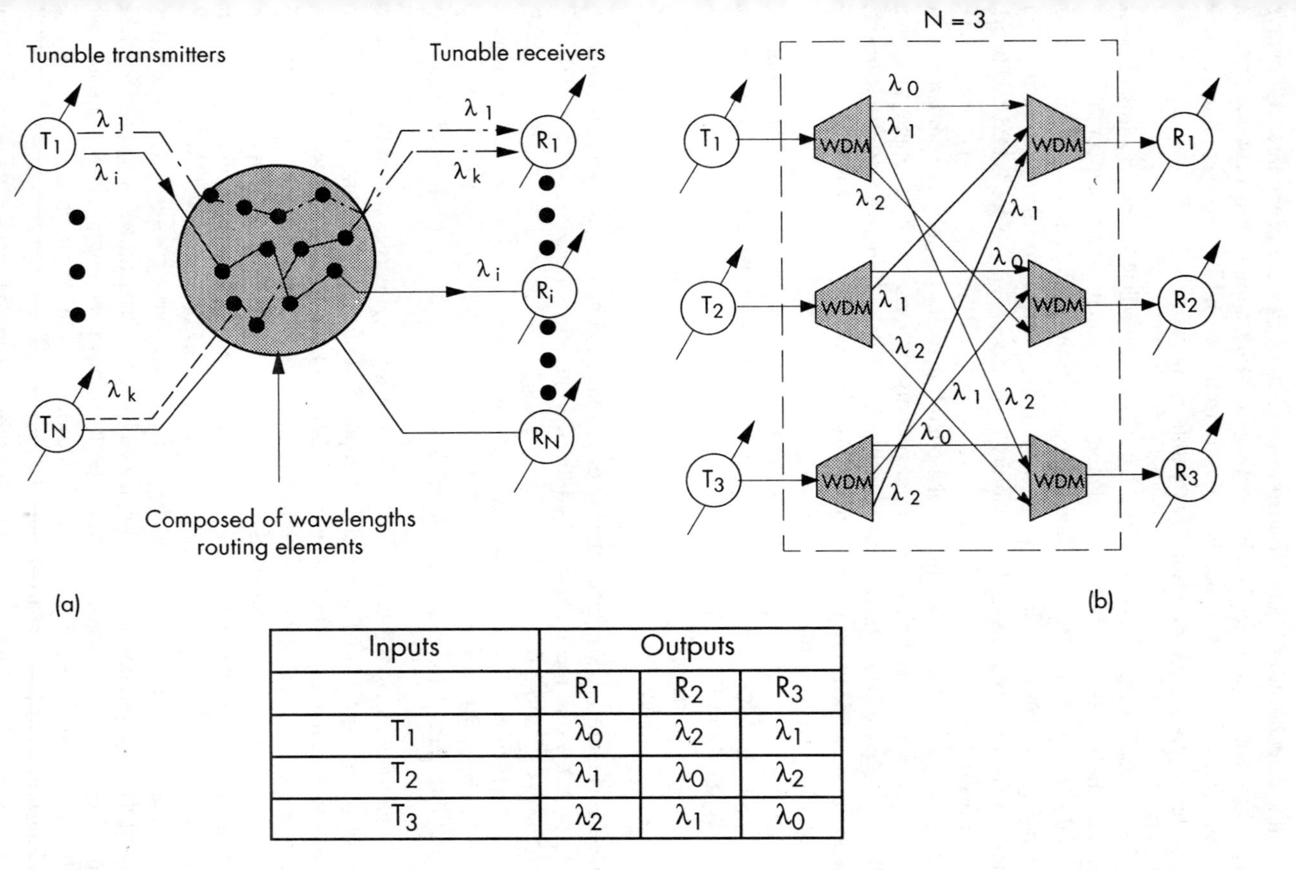

Inputs	Outputs		
	R_1	R_2	R_3
T_1	λ_0	λ_2	λ_1
T_2	λ_1	λ_0	λ_2
T_3	λ_2	λ_1	λ_0

Figure 7.2 (a) Principle of wavelength-routing single-hop WDMA networks, (b) example of implementation for $N = 3$ nodes using WDM passive devices. (© 1990 IEEE. [After Brackett, 1990.])

multicast connections, and are therefore the most flexible of the three alternative for broadcast-and-select WDMA networks. More complicated network protocol are, however, required since both the transmitters and the receivers must be con trolled to coordinate the data flow within the network.

The preceding description of broadcast-and-select WDMA networks implicitly as sumes that the number of usable wavelengths W is equal to the number of nodes N connected to the network. However, the number of usable wavelengths may be restricte for technological reasons such that, practically, W is generally much smaller than N. It i therefore interesting to analyze the capability of broadcast-and-select WDMA network under the restriction $W < N$.

At first glance, this would reduce the overall network capacity as compared to th case where $W = N$. Indeed, there are no longer enough wavelengths available to suppor simultaneous $N \times N$ connections. It turns out that when $W < N$, there exists a fundamenta difference in network capacity when tunability is provided at the transmitter or the re ceiver, or when it is provided at both sides.

Prior to discussing the general case, let us first consider the following example i which four nodes are connected to each other by a 4×4 wavelength-independent sta coupler, as shown in Figure 7.3.

Each node is equipped with a single transmitter and a single receiver. Three distinc cases will be considered:

- *Case a* (Figure 7.3(a)): each transmitter can tune to one of two distinct wavelength (λ_1 or λ_2), while receivers are fixed-tuned to either λ_1 (receivers #1 and #2) or λ (receivers #3 and #4).
- *Case b* (Figure 7.3(b)): the transmitters are fixed-tuned to either λ_1 (transmitters # and #2) or λ_2 (transmitters #3 and #4), while receivers can tune to λ_1 or λ_2.
- *Case c* (Figure 7.3(c)): all transmitters and all receivers can tune to either λ_1 or λ_2

In the three cases, the number of available wavelengths is less than the number o nodes ($W = 2$ while $N = 4$). Each node is composed of one input port and one output port

Suppose that the destination port of data packets arriving at the input ports i indicated by the packet's label, as shown in Figure 7.3. The packet at input port #1 i destined for output port #1. The packet at input port #2 is destined for output port #2. Th packets at input ports #3 and #4 are empty packets and can thus be discarded for th present analysis since they do not contribute to the network capacity.

In Case a (tunable transmitters/fixed-tuned receivers), packet transmission fror input port #1 to output port #1 is accomplished by tuning transmitter #1 to λ_1. The packe transmission from input port #2 to output port #2 is, however, not possible simultaneousl (i.e., it is blocked) since transmitter #2 should be tuned to λ_1, which is already busy fo the connection from input #1 to output #1.

In Case b (fixed-tuned transmitters/tunable receivers), packet transmission fror input port #1 to output port #1 is accomplished by tuning the receiver #1 to λ_1. For th

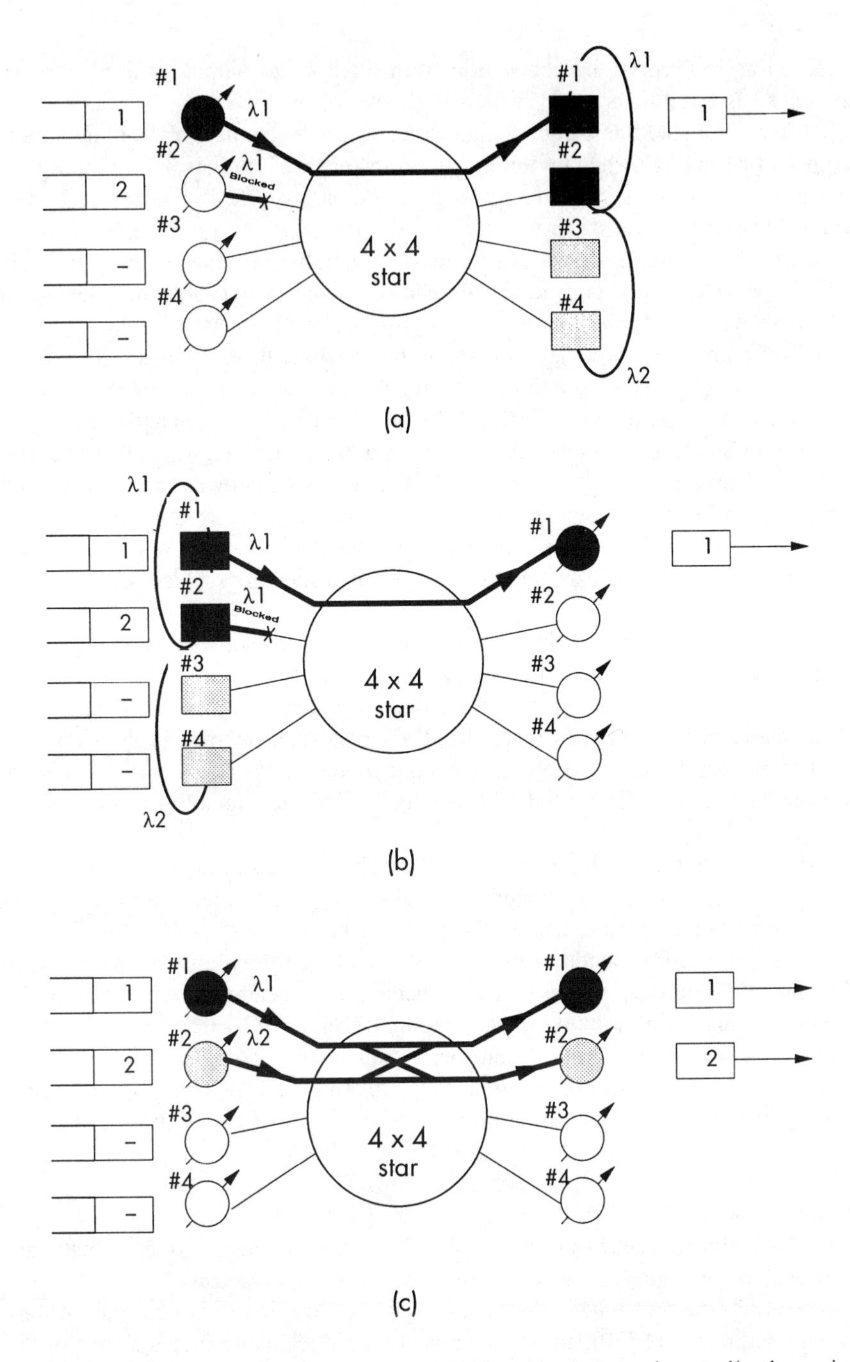

Figure 7.3 Illustration of the difference in network capability when the transmitters and/or the receivers are made tunable or are fixed-tuned.

same reason as in Case a, the connection from input #2 to output #2 is blocked since transmitter #2 is fixed-tuned to λ_1, which is already busy.

In Case c (tunable transmitters/tunable receivers), the connection from input port #1 to output port #1 is established by tuning both transmitter #1 and receiver #1 to λ_1. In the meantime, the connection from input port #2 to output port #2 is established by tuning transmitter #2 and receiver #2 to the free wavelength λ_2.

This simple example shows that the case in which both transmitters and receivers are tunable provides a higher potential utilization of the W available wavelengths than does the case where only the transmitters or only the receivers are tunable.

The difference in wavelength utilization between the three options has been investigated quantitatively in [Ramaswami 1990, Lu 1992]. The analysis assumes that each node has complete knowledge of the status (i.e., busy or idle) of all the wavelengths in the system. There is no buffering at the node such that upon a packet arrival, it is either transmitted or lost immediately depending on whether the connection is allowed or not, respectively. The propagation latency is omitted, and the analysis is therefore appropriate to either circuit-switched networks or centralized packet-switched networks (e.g., ATM optical switch) provided that the packet length is longer than the geographical coverage of the switch.

The number of nodes attached to the broadcast star network is equal to N and the number of available wavelengths is equal to W, with $W<N$. A packet arriving at node i is addressed to node j with a probability $1/N$ independent of i and j. The packet length is exponentially distributed with mean $1/\mu$ (in seconds/packet) and is the same for all nodes. The packet arrival at each node's input obeys a Poisson distribution with an average rate of λ packets per second. The average load of each input link is therefore equal to $\rho = \lambda/\mu$.

For a given ρ, the number of busy wavelengths w varies randomly according to the statistics of the input links. The dynamic behavior of w can be modeled by a *birth-death process* whose state transition diagram is shown in Figure 7.4.

Figure 7.4(a) corresponds to the case of wavelength tunability at the transmitters only or the receivers only (Cases a and b in the above example), while Figure 7.4(b) corresponds to tunability at both transmitters and receivers (Case c).

For all cases, a left-directed state transition results from the release of a busy wavelength after a successful connection. For a given state in which w wavelengths are busy, the left-directed state transition probability (per unit time) is simply equal to

$$\sigma_{w \rightarrow w-1} = w\mu \tag{7.1}$$

The right-directed state transition probability, which corresponds to the activation of an additional wavelength, depends on where the tunability is located.

When only the transmitters are tunable, such a transition can occur only if the two following conditions are fulfilled: (1) a connection request is generated at one of the (N - w) free transmitters, and (2) this connection is addressed to one of the free fixed-tuned

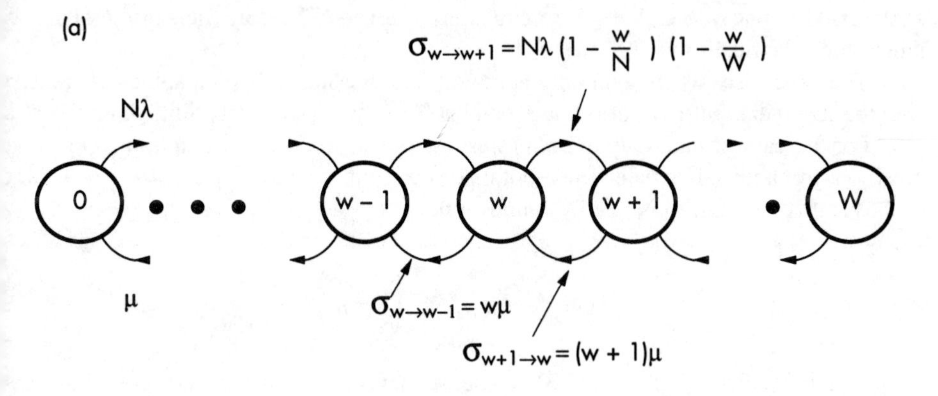

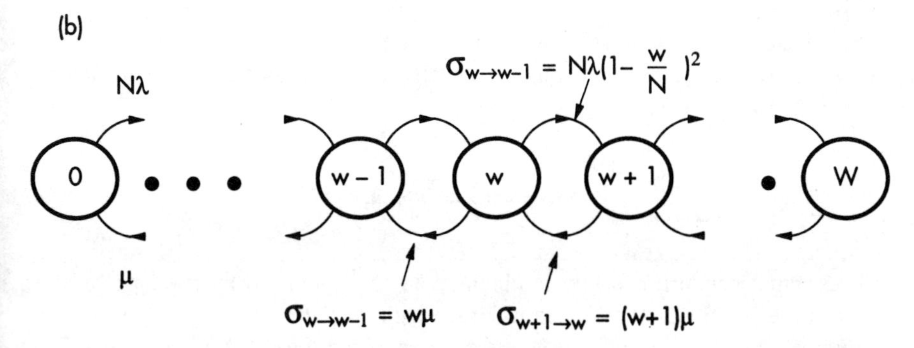

Figure 7.4 Transition-state diagram associated with the number of wavelengths in use: (a) tunable transmitters/fixed tuned receivers or fixed tuned transmitters/tunable receivers, (b) tunable transmitters/tunable receivers.

receivers. Since there are W available wavelengths, of which w are already busy, the probability that the second condition is fulfilled is $(1-w/W)$ (note that the destination probability of every input packet is uniformly distributed over all outputs). Therefore, a transition from state w to state $w + 1$ occurs with a probability

$$\sigma_{w \to w+1} = \lambda(N - w)\left(1 - \frac{w}{W}\right) \tag{7.2}$$

When only the receivers are tunable, a right-directed state transition can occur only if: (1) a connection request is generated at a transmitter whose fixed-tuned wavelength is not yet busy (the corresponding probability is equal to $(1-w/W)$), and (2) this request is addressed to one of the $(N - w)$ free tunable receivers. The state transition probability is therefore identical to (7.2).

Thus, the cases where tunability is provided at one side only (i.e., at the transmitters or at the receivers, but not both) have identical transition state probabilities.

For the case of tunability at both transmitters and receivers, the right-directed state transition probability is greater since both sides can tune. Following the same reasoning as above, this probability is readily found to be

$$\sigma_{w \to w+1} = \lambda(N - w)\left(1 - \frac{w}{N}\right) \tag{7.3}$$

Equations (7.1), (7.2), and (7.3) can subsequently be used to determine the steady-state probability p_w of being in state w (i.e., w busy wavelengths). This probability p_w can be obtained by applying classical techniques commonly used in the context of Markov chains [Kleinrock, 1975].

Since the average number of successful packets transmitted per unit time is also equal to the average number of busy wavelengths in the system, we can define the normalized network capacity S as

$$S = \langle w \rangle = \sum_{w=0}^{W} p_w w \tag{7.4}$$

The actual network capacity is obtained by multiplying S by the bit rate of each input link times the load ϱ (assumed to be the same for all nodes).

Figure 7.5 shows S as a function of the average load ϱ for $W = 25$, 50, and 125 when $N = 250$. It is seen that, except for very low loads ($\varrho \leq 0.1$), the case of both ends being tunable always provides a higher wavelength utilization, and hence a higher network capacity than the case where only transmitters or only receivers are tunable. The difference can be as high as 40% for the special case where $W = 50$ and $\varrho = 0.3$.

For small values of W (i.e., $W = 25$ and 50 in Figure 7.5), S saturates when ϱ increases. This saturation occurs because as the input link load ϱ increases, the average number of busy wavelengths rapidly approaches the small value of W.

For large values of W (i.e., $W = 125$), the average number of busy wavelengths is always less than W, even for ϱ approaching unity. It must be noticed that the maximum value of $\langle w \rangle$ depends on the traffic statistics assumed for the model.

The analysis has been extended to multiple transmitters and receivers per node [Lu 1992]. It has been shown that a small number of tunable transmitters and receivers at each node is enough to produce performance close to the upper bound achieved when $W = N$. This is because under uniform traffic assumption the probability that more than a few packets are going to the same destination at the same time is very small.

Indeed, let us suppose that all packets arriving at a node's input are independent and

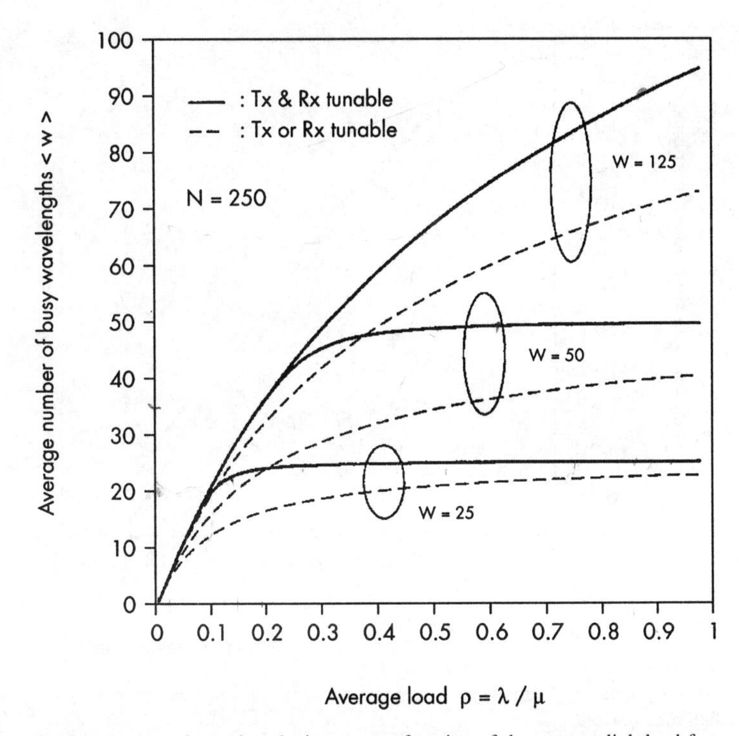

Figure 7.5 The mean number of wavelengths in use as a function of the average link load for various values of the maximum number of available wavelengths W. The advantage of transceivers having both transmitters and receivers tunable is clearly apparent.

equally likely to be destined to each of the N nodes in the network. With equal load ϱ for all the node's input streams, the probability p_k that k packets are simultaneously destined to the same node is binomially distributed such that

$$p_k = C_N^k \left(\frac{\varrho}{N}\right)^k \left(1 - \frac{\varrho}{N}\right)^{N-k} \qquad k = 0, 1, \ldots N \tag{7.5}$$

Figure 7.6 represents p_k as a function of k for several values of N and $\varrho = 0.9$. It is seen that, for example, the probability that 14 packets are simultaneously destined to the same output is less than 10^{-12} for arbitrarily large values of the network size N.

This property has been exploited in the *photonic knockout switch* discussed in Section 7.4.2.

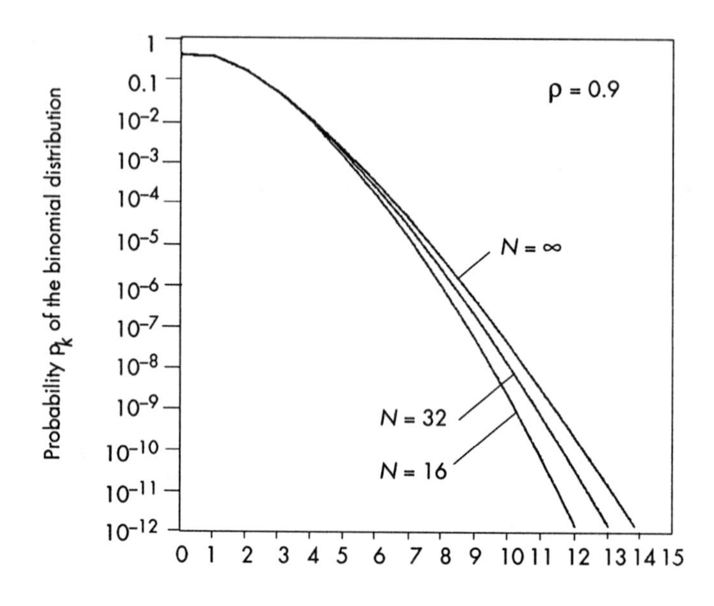

Figure 7.6 Probability of having k simultaneous packets destined to the same output for various values of N and a link load $\varrho = 0.9$.

7.1.2 Wavelength-Routing WDMA Networks

The second class of single-hop WDMA-networks are referred to as wavelength-routing networks. An example is illustrated in Figure 7.2. The network is composed of passive wavelength-selective (or routing) elements and a connection is uniquely determined by the wavelength of the transmitted signal and the node through which the signal is injected into the network.

For example, an $N \times N$ wavelength-routing network can be built with WDM elements interconnected with N^2 fibers, as shown in Figure 7.2(b) for $N = 3$. Each node is equipped with a tunable transmitter and a tunable receiver. By tuning the transmitter to a selected wavelength, the injected signal is passively routed to the addressed receiver, which is also tuned to the same wavelength for packet reception. It is seen that a complete N×N interconnection is possible with only N distinct wavelengths and that each receiver can be addressed by any one of the transmitters in a noninterfering way.

As already noticed in Section 6.6.3, the advantage of wavelength-routing networks over broadcast-and-select networks, which use wavelength-insensitive couplers or taps, is that they avoid splitting losses. However, their principal drawback when passive routing elements are used is that each node must be provided with both tunable transmitters and tunable receivers (or arrays of fixed-tuned elements) in order to achieve multiaccess.

Apart from this drawback, however, wavelength-routing networks make it possible to dynamically change the internal routing structure according to the traffic pattern within the network. This may prove attractive for unbalanced traffic between different nodes connected to the network. This dynamic change of the routing pattern can essentially be achieved in two distinct ways:

1. Using wavelength-selective space switches, which dynamically switch signals from one path to another by changing the WDM routing in the network;
2. Using wavelength converters, which transfer the signal from one wavelength to another wavelength.

In both cases, the interconnection pattern contains active elements that generally must be controlled electronically. The path between transmit and receive nodes may, however, remain optical depending upon the type of wavelength-selective space switches or wavelength converters chosen.

Figure 7.7 represents a wavelength-routing network using wavelength-selective space switches. Such switches, described in Section 7.2.4, can be viewed as three-port devices that have the remarkable ability to direct any wavelength at the input port to one of the two output ports. In other words, an arbitrary subset of the wavelengths $\lambda_1 \ldots \lambda_N$ on the input port of the device can be selected and directed to one of the two

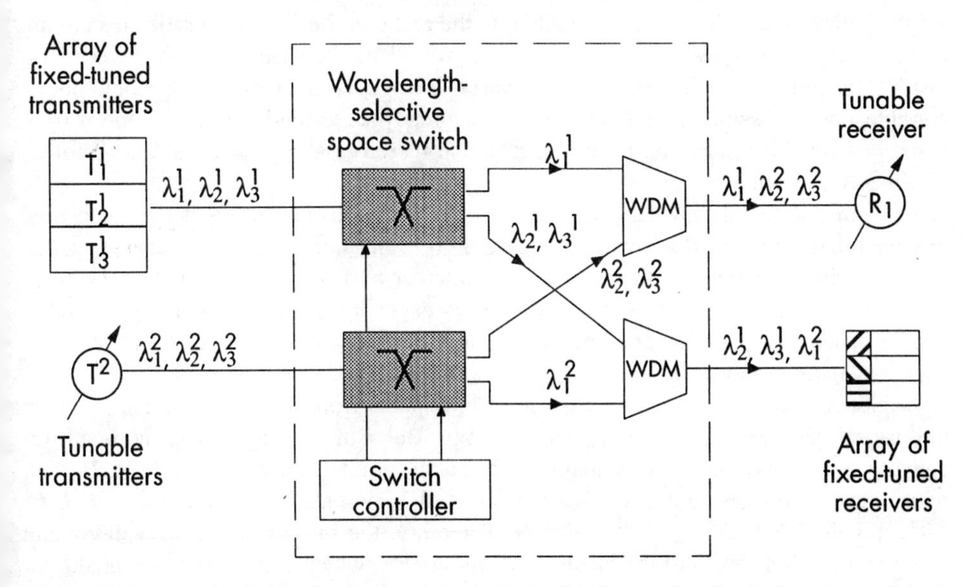

Figure 7.7 An example of a 2-node wavelength-routing network using wavelength selective space switches.

output ports. This selection is rearrangeable so that the path taken by any wavelength within the network can be changed whenever desired.

The second method to change the internal routing structure consists of using wavelength converters whose basic function is depicted in Figure 7.8(a). Applications of wavelength converters within wavelength-routing networks are similar to time-slot interchangers in a TDM digital electronic switch. An obvious way to perform wavelength conversion is shown in Figure 7.8(b). The input signal is first converted to electronic form by using a tunable optical receiver and the regenerated photocurrent is then used to modulate a tunable laser diode at the desired output wavelength. A nonblocking $N \times N$ optical switch has been proposed based upon this principle [Fujiwara, 1988]. Several other types of wavelength-converter devices based on multielectrode distributed Bragg reflector (DBR) laser structures, the *Y*-laser discussed in Section 2.3.5.5, or semiconductor optical amplifiers are being investigated in many research laboratories [Tada 1990, Ottolenghi 1993, Lach 1993, Stubkjaer 1993]. An example of a DBR-based structure is shown in Figure 7.8(c).

Recently, a new type of wavelength-routing network, the so-called *linear lightwave network* (LLN), has been proposed for circuit-switched traffic [Stern, 1990].

To illustrate the basic principle of LLNs, let us consider Figure 7.9 in which the nodes are interconnected with 2×2 wavelength-insensitive couplers whose coupling coefficients α_i are allowed to take any values between 0 and 1.

Each network node uses a distinct wavelength to establish the desired connection. For example, the connection from node 1 to node 1* is established on λ_1 via the path *A-B-C-F-G*. In the meantime, the connection from node 2 to node 2* can take place on λ_2 via the path *H-B-C*. This is made possible by the fact that the coupler coefficients are not restricted to taking only the values 0 or 1, as would be the case if switches were used instead. Indeed, with switches in place of variable couplers, the path *H-B-C* would not be possible since the switch at *B* would have been already set to send to *C* all the power from *A* ($\alpha_B = 1$ for the connection from 1 to *A**), and therefore also to send to *D* (but not *C*) all the power from *H*.

With additional wavelengths, other connections can take place at the same time provided that suitable values of α_i are chosen. The values of these coefficients α_i can be managed either centrally with a central controller or by using a distributed control protocol. In either case, each coupler coefficient is dependent upon the setting of all other couplers such that LLNs are appropriate for circuit-switched mode of operation but not for packet-switching networks.

Notice that to avoid reconvergence of multipaths from the same originating node, the distinct paths should follow a tree topology. Referring to Figure 7.9, multipath reconvergence exists at couplers *B* and *F* for the connection from node 1 to node 1*. Indeed, by setting the connection *A-B-C-F-G*, the signal at λ_1 will also take the routes *A-H-B-C-F-G*, *A-H-B-D-E-F-G*, and *A-B-D-E-F-G*. Therefore, the link from *F* to *G* will contain four time-delayed copies of the bit stream from node 1, which will degrade the quality of the connection through intersymbol interferences.

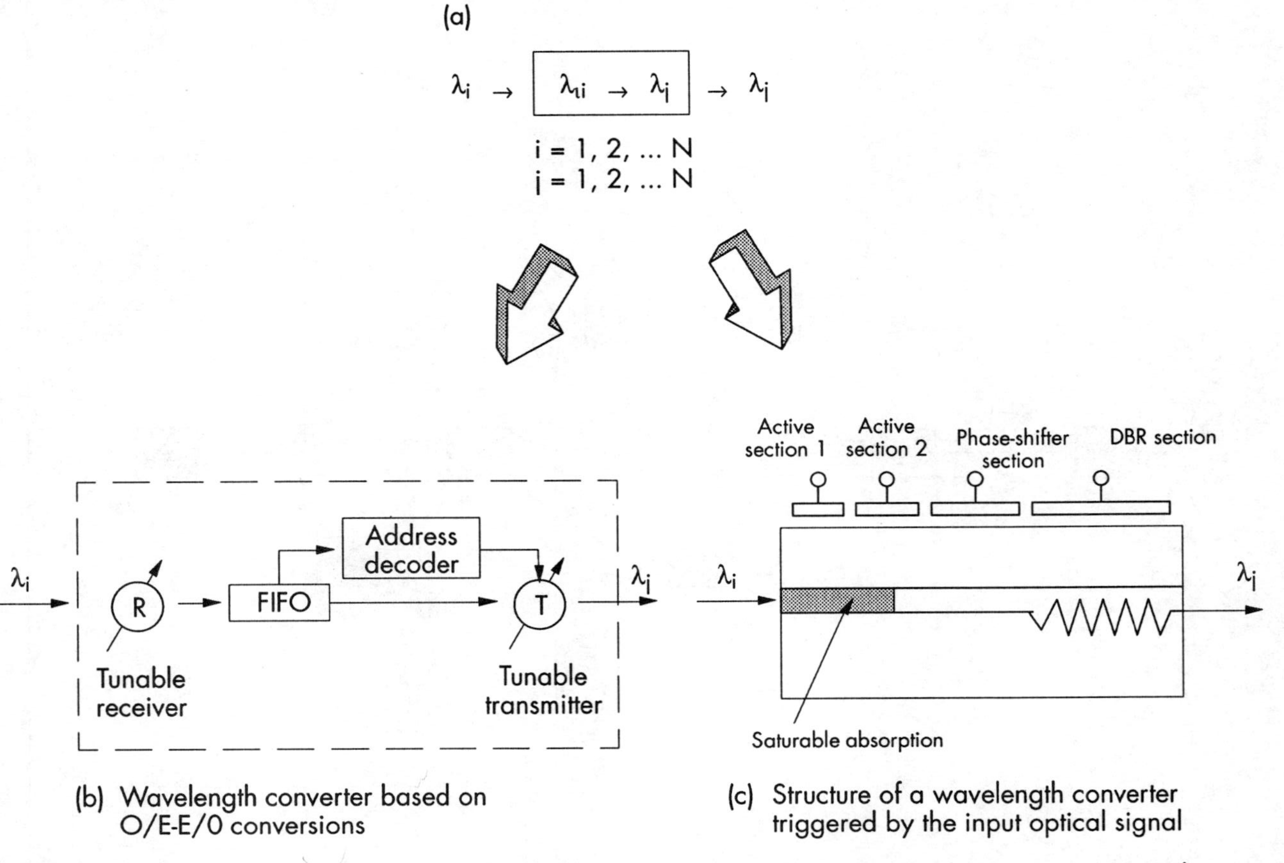

Figure 7.8 (a) Functional representation of a wavelength converter, (b) a λ converter using a tunable receiver and a tunable transmitter, (c) a λ converter using a multisection DBR-like laser structure.

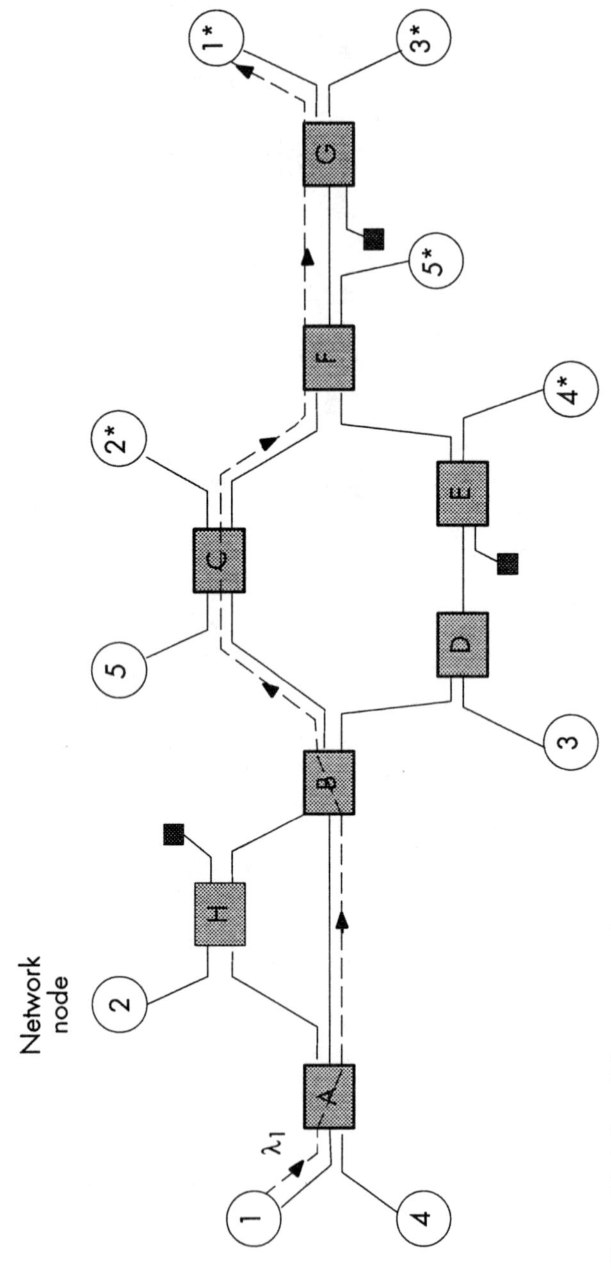

Figure 7.9 A linear lightwave network.

In [Bala, 1991], LLNs using wavelength-flat $\Delta \times \Delta$ couplers are studied. A solution is given for the coefficient settings that avoid multipath reconvergence.

Although the loss incurred by the signals within LLNs is higher than in wavelength-routing networks that use WDMs or switches, they offers the attractive feature of wavelength reuse. This is made possible if, after passing through several couplers, the signal at a specific wavelength is so attenuated that the same wavelength can be used at the same time for connections in some other parts of the network. This is illustrated in Figure 7.10, in which connections at λ_1 can take place simultaneously for communication between nodes 1 and 4* and between nodes 3 and 1*.

The possibility of wavelength reuse obtained by exploiting the attenuation of passive couplers has also been proposed independently in [Karol, 1991] to create large high-capacity LANs and MANs.

The quantitative study of LLNs is still incomplete and further work is required to better understand their performance and to determine the optimum tradeoff between the reduction in hardware complexity and the additional complexity of the networkwide control protocol.

7.2 TUNABLE OPTICAL FILTERS

A key component in WDMA networks is the tunable receiver, which can select the desired channel from a set of wavelength-multiplexed channels. Channel selection generally requires a tunable optical filter, whose basic function is represented in Figure 7.11. Many different channels at many different wavelengths appear at the input, but only one wavelength channel appears at the output.

The tunable optical filter can take different forms depending on the technology used. For all filters described in this section, optical interference effects are exploited to build very wavelength-selective devices. Some of these devices can also be useful within coherent optical receivers although receiver tunability in coherent detectors is usually obtained with a tunable local oscillator (i.e., a tunable laser) whose several design options have been discussed in Section 2.3.5.

From a system viewpoint, the critical issues involving tunable optical filters are:

- *The tuning range*: as shown in Figure 7.11, the filter-tuning range $\Delta\lambda$ is defined as the wavelength difference between the shortest and the longest wavelength that can be selected by the filter.
- *The maximum number of resolvable channels*: this number is the ratio of the tuning range to the minimum channel spacing required to guarantee a minimum crosstalk degradation. As we will see in Section 7.3.2, the channel spacing that guarantees < 0.5-dB crosstalk penalty varies, practically, between 3 to 10 times the channel bandwidth, and depends upon the modulation scheme and the optical-filter transfer function.
- *The tuning speed*: this is the speed with which a tunable optical filter can be reset from one wavelength to a new wavelength within the tuning range. For some

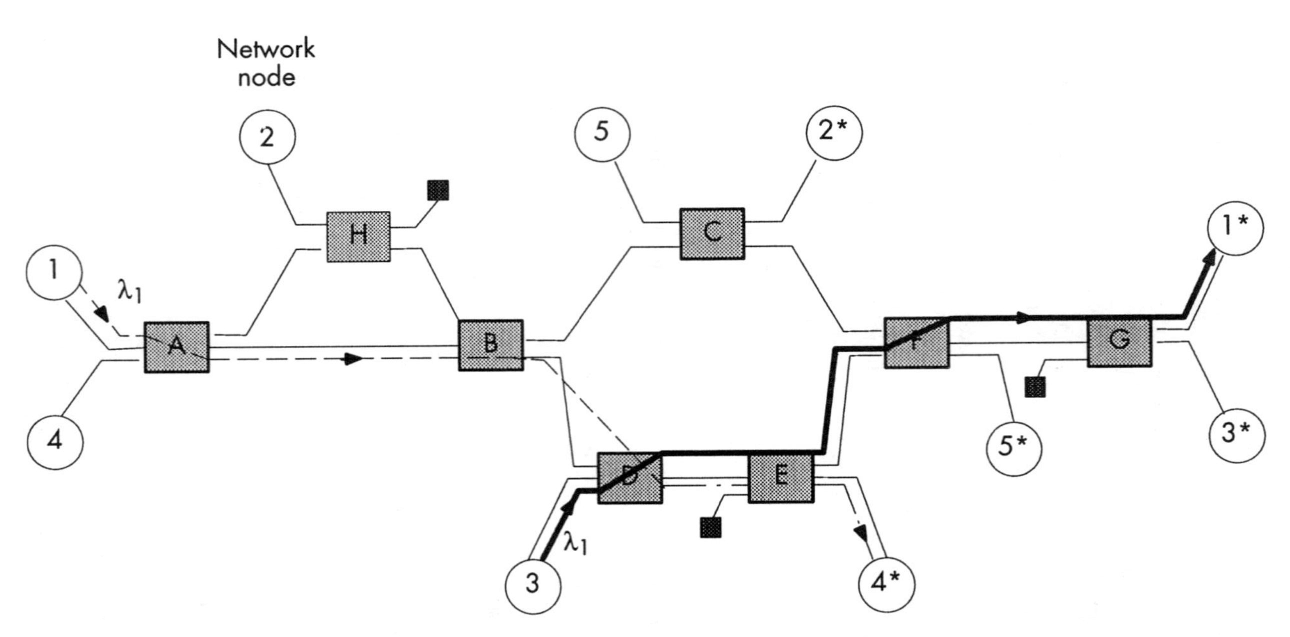

Figure 7.10 Wavelength reuse in linear lightwave networks. Connections node 1 → node 4*, and node 3 →. node 1* can take place simultaneously on λ_1.

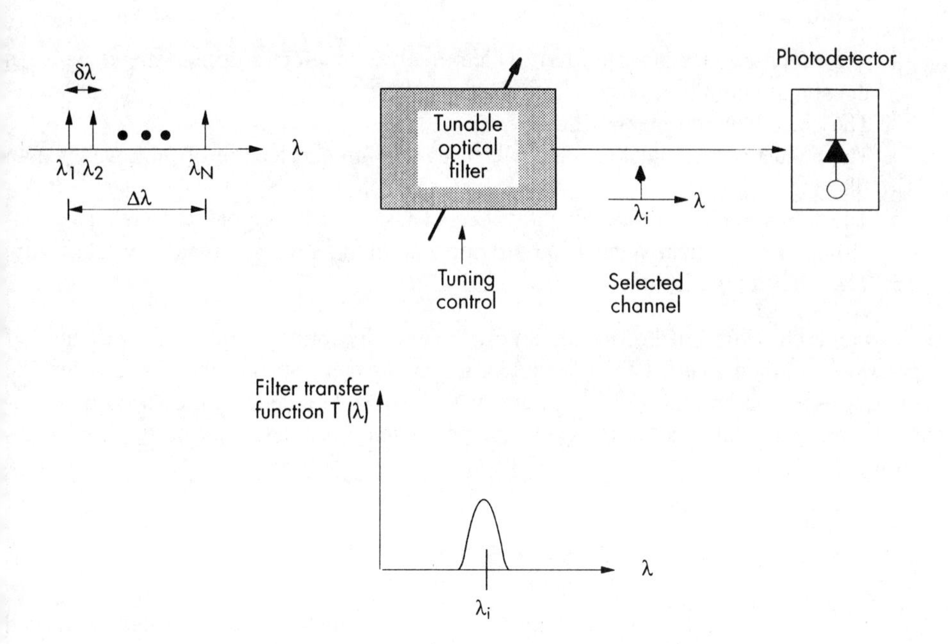

Figure 7.11 Block diagram of the basic function of a tunable optical filter.

circuit-switched applications, millisecond tuning time may be sufficient, while sub-microsecond tuning time is generally required for packet-switched applications (for example, at a 1-Gbps channel rate, the time to transmit a 1,000-bit packet is only 1 μs).

- *The attenuation*: the selected optical signal will in general incur some power attenuation because of insertion and internal losses of the filters. This attenuation must be minimized in order not to significantly reduce the already limited network power budget.
- *The polarization dependence*: the filter should preferably be polarization-independent (i.e., its transfer function is independent of all possible polarization states of the arriving light). In practice, this is almost always a necessary condition.
- *The stability*: thermal and mechanical factors should not cause the filter transfer function to drift more than a few percent of the channel bandwidth.
- *The size*: tunable optical filters and their associated control circuitry must be as compact as possible to meet user requirements.

The present section is intended to review the current state of practical tunable optical filters. Six basic designs will be considered separately:

1. The Fabry-Perot interferometer tunable filter, which has already been discussed in Section 5.1.1 in the context of optical amplifiers;

2. The Mach-Zender tunable filter, which involves feedforward interference between the signal and a delayed version of it;
3. The electro-optic tunable filter;
4. The acousto-optic tunable filter, which has the unique feature of multiwavelength filtering capability;
5. The semiconductor-based tunable filters. These devices are based on resonant amplification of optical signals and provide gain in addition to wavelength selectivity;
6. The fiber Brillouin tunable filter.

For each of these filter technologies, we will first briefly explain the principle of operation, and then quantify the relevant system parameters, such as the tuning range, the crosstalk isolation and the resulting number of resolvable channels, the speed of wavelength tuning, the attenuation (or gain), the polarization dependency, and the implementation complexity.

7.2.1 The Fabry-Perot Tunable Filter (FPF)

The basic physical mechanism of a Fabry-Perot interferometer filter was briefly described in Section 2.3.3 to introduce the longitudinal modes of a laser cavity. In Section 5.1.1 we discussed its main external characteristics in more detail, assuming that the internal medium provides gain (i.e., the device is a Fabry-Perot semiconductor amplifier). The FPF transfer function is given by (5.6) in which the gain is replaced by the loss A. This gives

$$T_{\mathrm{FPF}}(\nu) = \frac{A(1-R)^2}{(1-AR)^2 + 4AR\sin^2\left(\dfrac{2\pi(\nu-\nu_o)L}{c}\right)} \tag{7.6}$$

where A is the internal loss of the cavity.

This function is represented in Figure 7.12 for three values of the facet reflectivity R and $A = 1$ (i.e., lossless).

It is seen that $T_{\mathrm{FPF}}(\nu)$ is periodic with a period, called the free spectral range (FSR), given by

$$\mathrm{FSR} = \frac{c}{2nL} \tag{7.7}$$

The FSR is the frequency range that can be tuned freely without overlapping due to multiorder interference. If $\Delta\nu_{\mathrm{ch}}$ is the bandwidth of each of the N wavelength multiplexed channels, the FSR must be larger than the aggregate multichannel bandwidth; that is,

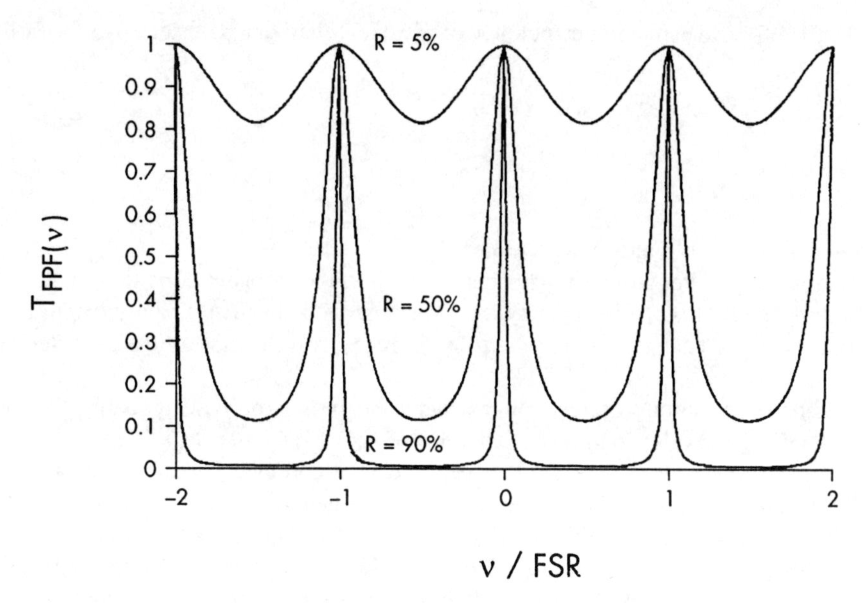

Figure 7.12 Power transfer function of a Fabry-Perot filter for various end-facet reflectivities R.

larger than $N\,\Delta\nu_{ch}$. The 3-dB bandwidth of the FPF (full width at half maximum, FWHM) is obtained from (7.6) and is given by

$$\Delta\nu_{FWHM} = \frac{c}{2nL}\frac{1-R}{\sqrt{R}} \tag{7.8}$$

The ratio of the free spectral range to $\Delta\nu_{FWHM}$ gives an indication of how many channels can be selected by the filter. This ratio is known as the finesse F of the Fabry-Perot filter and is given by

$$F = \frac{\text{FSR}}{\Delta\nu_{FWHM}} = \frac{\pi\sqrt{R}}{1-R} \tag{7.9}$$

Typically, F ranges from 20 to 100.

The maximum number of resolvable channels is set according to an allowed crosstalk penalty level. The crosstalk depends on the shape of the filter transfer function. For an FPF, the channel spacing can be as small as $3\,\Delta\nu_{FWHM}$ to keep the crosstalk below -10 dB [Mallison, 1987]. This crosstalk level corresponds to < 0.5 B power penalty [Hill,

1985]. Therefore, the maximum number of channels that can be selected by a FPF is restricted by

$$N_{\max} < \frac{F}{3} = \frac{\pi\sqrt{R}}{3(1 - R)} \tag{7.10}$$

With $F = 100$, (7.10) yields $N_{\max} < 33$.

To tune the filter from one channel to another, the resonant condition of the FPF must be changed. This can be done by fine-adjusting the Fabry-Perot cavity length L: the length needs to be changed by only half the operating wavelength to tune the filter over one entire FSR.

Among the various design options that have been proposed, probably the most successful design developed to date is the *tunable-fiber Fabry-Perot Filter* [Stone, 1987]. This filter is formed by a short gap between two fiber-end faces, which are coated with highly reflecting films to create the resonant cavity. The advantage of fiber-based FPF is that they can be integrated within the fiber system without incurring significant coupling losses. These filters are now commercially available with a finesse of over 150 and an insertion loss of 1 to 2 dB. The device is tuned by using a piezo-electric control in the configuration shown in Figure 7.13.

The tuning speed is fairly slow (tuning time ≥ 1 ms), and the filter is therefore only appropriate for circuit-switched applications.

To accommodate more channels, the finesse of the filter has to be improved. A useful approach to improve F is to use two FPF*s* in tandem with an isolator in between

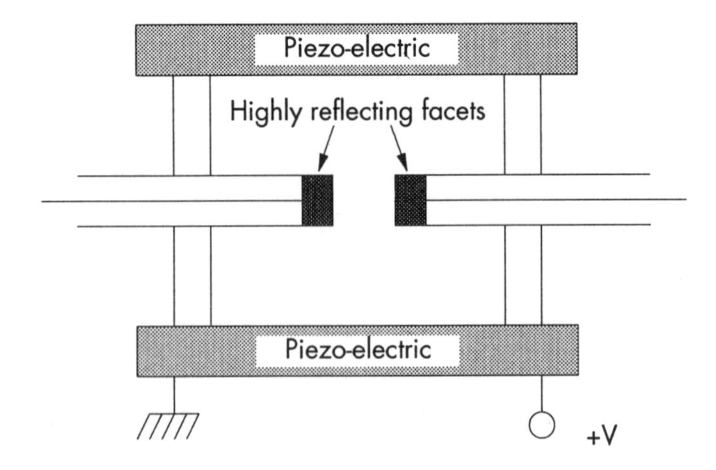

Figure 7.13 A tunable-fiber Fabry-Perot filter.

[Saleh, 1989]. This can bring the finesse to values in excess of $F \cong 1{,}000$ [Miller, 1992]. In addition, by sharpening the filter transfer function, the required channel separation is halved relative to the single-stage case, resulting in a doubling of the number of resolvable channels over the same FSR [Kaminow, 1988]. With $F = 1{,}000$, over 600 WDM channels can be supported with adequately low crosstalk. However, the loss is higher than the single-stage because of the need to isolate the two stages and the tuning is more difficult to control.

7.2.2 The Mach-Zender Tunable Filter (MZF)

The Mach-Zender interferometric filter is another potential tunable filter candidate for use in multiaccess networks. A single-stage Mach-Zender interferometer is represented in Figure 7.14.

The multichannel input signal is split into two equal parts by a 3-dB coupler. The two versions of the same signal traverse paths of slightly different lengths and merge together in another 3-dB coupler at the output.

The overall transfer function $T_{MZF}(\nu)$ of the single-stage Mach-Zender filter is obtained by multiplying the scattering matrix associated with the 3-dB couplers and that associated with the two different propagation paths. The scattering matrix of the 3-dB coupler was obtained in Section 6.1 (6.11) and is rewritten here as

$$[T_1] = \frac{1}{\sqrt{2}}\begin{bmatrix} 1 & -j \\ -j & 1 \end{bmatrix} \tag{7.11}$$

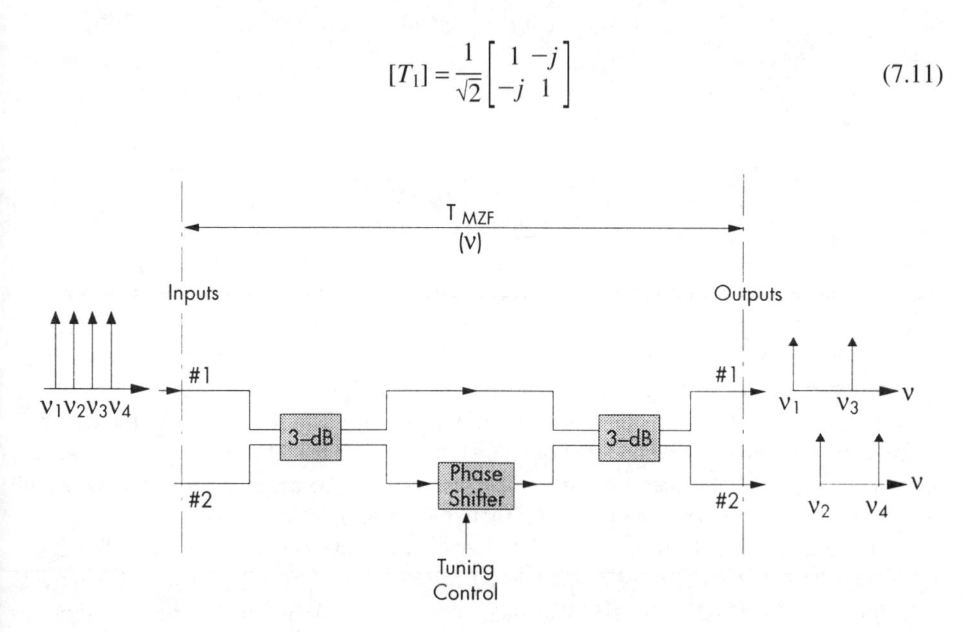

Figure 7.14 Structure of a Mach-Zender interferometric filter.

The scattering matrix associated with the two different paths, which differ by a delay τ, is simply given by

$$[T_2(\nu)] = \begin{bmatrix} 1 & 0 \\ 0 & \exp(-2\pi j\nu\tau) \end{bmatrix} \quad (7.12)$$

Therefore, $T_{MZF}(\nu)$ is given as

$$T_{MZF}(\nu) = \begin{bmatrix} T_{11}(\nu) & T_{12}(\nu) \\ T_{21}(\nu) & T_{22}(\nu) \end{bmatrix} = [T_1]\,[T_2(\nu)]\,[T_1]$$

$$= \frac{1}{\sqrt{2}}\begin{bmatrix} 1 & -j \\ -j & 1 \end{bmatrix}\begin{bmatrix} 1 & 0 \\ 0 & \exp(-2\pi j\nu\tau) \end{bmatrix}\frac{1}{\sqrt{2}}\begin{bmatrix} 1 & -j \\ -j & 1 \end{bmatrix} \quad (7.13)$$

$$= \frac{1}{2}\begin{bmatrix} 1 - \exp(-2\pi j\nu\tau) & -j[1 + \exp(-2\pi j\nu\tau)] \\ -j[1 + \exp(-2\pi j\nu\tau)] & -[1 - \exp(-2\pi j\nu\tau)] \end{bmatrix}$$

and the power transfer function is

$$\begin{bmatrix} |T_{11}(\nu)|^2 & |T_{12}(\nu)|^2 \\ |T_{21}(\nu)|^2 & |T_{22}(\nu)|^2 \end{bmatrix} = \begin{bmatrix} \cos^2(\pi\nu\tau) & \sin^2(\pi\nu\tau) \\ \sin^2(\pi\nu\tau) & \cos^2(\pi\nu\tau) \end{bmatrix} \quad (7.14)$$

Usually, the multichannel signal is injected at one of the two inputs of the MZF and (7.14) becomes

$$\begin{bmatrix} |T_{11}(\nu)|^2 \\ |T_{21}(\nu)|^2 \end{bmatrix} = \begin{bmatrix} \cos^2(\pi\nu\tau) \\ \sin^2(\pi\nu\tau) \end{bmatrix} \quad (7.15)$$

Equation (7.15) shows that the power transfer function of the MZF is periodic in frequency with a period of $1/\tau$. Such a filter is sometimes called a periodic filter. The filtering operation of the single-stage MZF with, for example, four input channels spaced in frequency by $\delta\nu = 1/2\tau$ can be understood as follows. Suppose that the resonant condition of the MZF is obtained at ν_1 such that $\cos^2(\pi\nu_1\tau) = 1$. Therefore, the channels at frequency ν_1 and ν_3 will appear at output #1, while channels at frequency ν_2 and ν_4 will appear at output #2. Further filtering of each output can be achieved by using a second MZF whose period is the double of the first stage; that is, $2/\tau$.

In general, the selection of one of $N = 2^M - 1$ channels equally spaced in frequency by $\delta\nu = 1/2\tau$ is achieved by a cascade of M MZFs whose periodicities are successively $2^m/\tau$ ($m = 1, \ldots, M$ refers to the MZF stage in the cascade). In other words, this requires that the delay introduced by the successive MZF stages is set according to

$$\tau_m = \frac{1}{2^m \delta\nu} \tag{7.16}$$

Figure 7.15 shows a four-stage MZF chain that can demultiplex 16 WDM channels. From (7.15) and (7.16), the overall power transfer function of the cascade of M MZFs is given by

$$T_{\mathrm{MZF}}^{M}(\nu) = \prod_{m=1}^{M} \cos^2(\pi\nu\tau_m) = \left[\frac{\sin(\pi\nu/\delta\nu)}{(N+1)\sin(\pi\nu/(N+1)\delta\nu)}\right]^2 \tag{7.17}$$

An eight-channel (M = 3) multiplexer/demultiplexer for an FDM system with 5-GHz spaced channels was demonstrated in 1987 by using silicon waveguide technology [Toba, 1987].

Tunability is achieved by thermal variation of the path-length difference in each MZF stage. A tunable 128-channel demultiplexer has been fabricated by using seven cascaded MZFs [Takato, 1990]. The crosstalk was -13 dB and the fiber-to-fiber attenuation was 6.7 dB.

The real advantage of MZF chains is that the filter can be realized using lithographic technology, leading to potentially low fabrication costs. Also, by designing a squared waveguide cross-section, these filters can be made polarization-insensitive. The main drawbacks, however, are the slow tuning speed due to thermal inertia (a few milliseconds) and the complexity of the multistage tuning control.

7.2.3 The Electro-Optic Tunable Filter (EOTF)

Another class of tunable optical filters makes use of wavelength-selective mode coupling mechanisms using electro-optic or acousto-optic effects. The present section discusses the electro-optic tunable filter (EOTF), while the next section will focus on the acousto-optic tunable filter (AOTF).

For both devices the filtering mechanism can be understood by referring to Figure 7.16.

The input multichannel with arbitrarily polarization state is split into two orthogonal polarization states by the input polarization splitter. In the TE/TM mode convertor, a periodic perturbation (i.e., an induced grating) is created through electro-optic or acousto-optic effect, depending on the particular material used. These perturbations change the polarization state of the TE and TM input modes. However, this change occurs only for specific wavelength channels for which the phase-matching condition (like the Bragg condition) is satisfied (channel at λ_3 in Figure 7.16). For those specific channels, a 90-deg rotation between the input and output planes of polarization can be achieved. The other wavelength channels that do not satisfy the phase-matching condition emerge from the

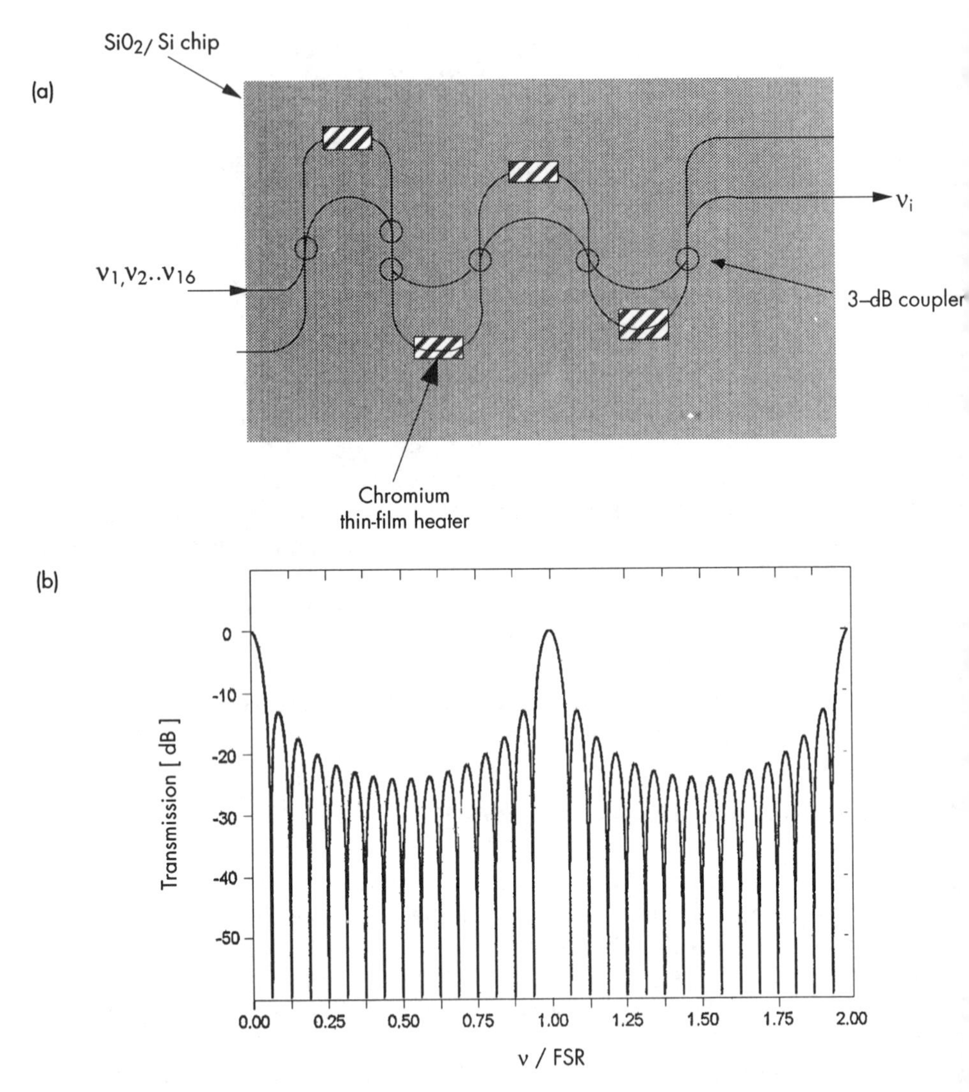

Figure 7.15 (a) A cascade of four Mach-Zender filters that select one of sixteen wavelength-multiplexed channels, and (b) the transmission characteristic as a function of the ratio ν/FSR for $M = 4$.

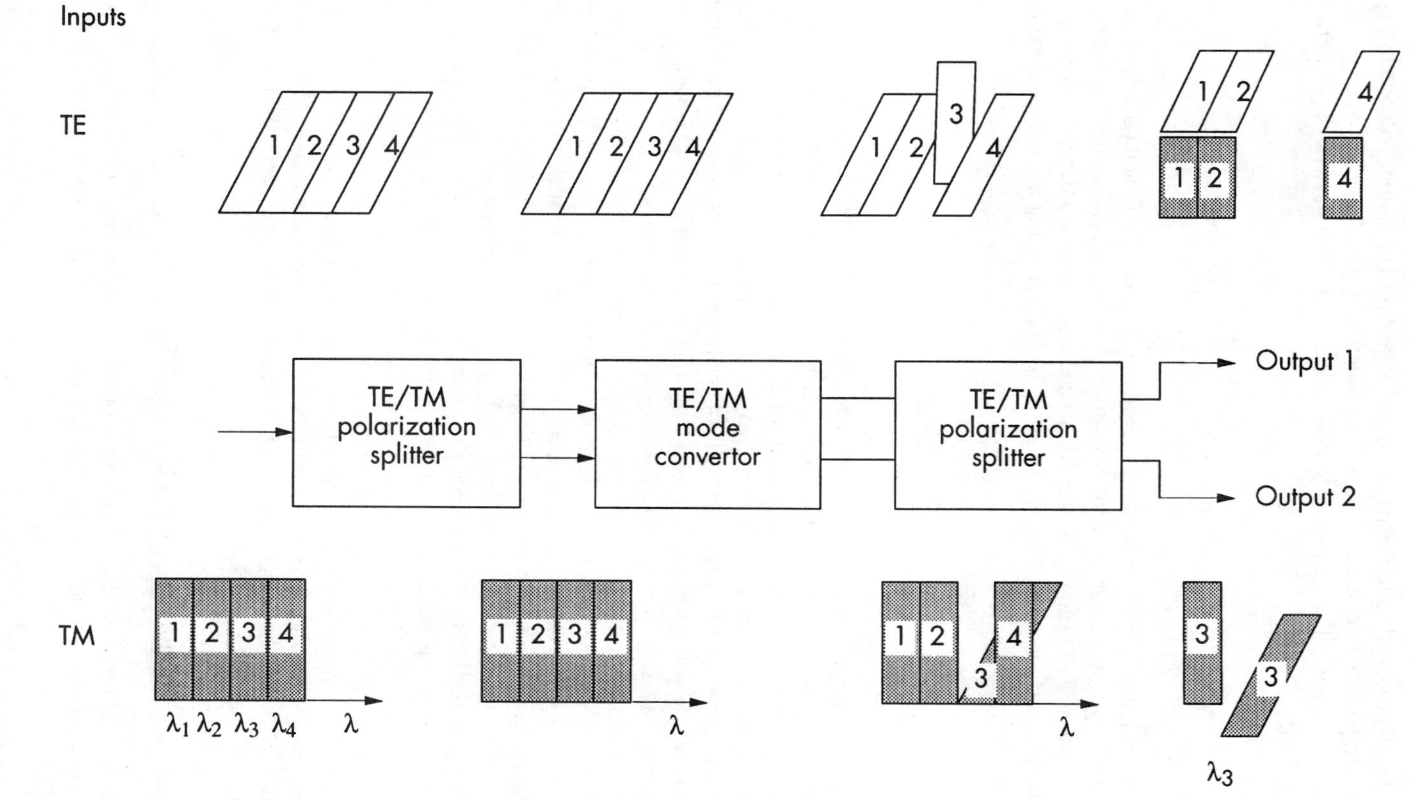

Figure 7.16 Schematic representation of the filtering mechanism used in both EOTF and AOTF. The figure illustrates the separation of the third channel at λ_3 out of four wavelength-multiplexed channels. [After Green, 1993.]

perturbation region with unchanged polarization states. The output-polarization splitter separates the polarization states, resulting in wavelength filtering.

It can be shown that the power transfer function of such devices is given, to a first order, by [Yariv, 1984]

$$|T(\nu)|^2 = \frac{\sin^2(\pi \cdot \Delta n \cdot L \cdot \nu / c)}{(\pi \cdot \Delta n \cdot L \cdot \nu / c)^2} \tag{7.18}$$

where L is the length of the perturbation region (i.e., the length of the induced grating) and Δn is the change of the refractive index induced by the electro-optic or acousto-optic effect. Equation (7.18) shows that the filter bandwidth $\Delta\nu_{FWHM}$ depends on the product $L \cdot \Delta n$. It can be shown that $\Delta\nu_{FWHM}$ is given by the following approximate expression [Harris, 1969]:

$$\frac{\nu}{\Delta\nu_{FWHM}} \cong \frac{L}{\Lambda} \tag{7.19}$$

where Λ is the perturbation period seen by the incoming signal.

Physically, (7.19) tells us that the larger the ratio L/Λ is, the narrower the filter bandwidth is. This is because a large value of L/Λ means a large number of corrugations within the grating length L, each acting as a scattering center. For the phase-matched wavelength, the scattered waves from all of these corrugations will interfere constructively, and the outgoing wave will have a large amplitude. For nonphase-matched waves, the scattered waves are out of phase and they cancel one another.

An EOTF fabricated with Ti-diffused waveguides on $LiNbO_3$ substrate is shown in Figure 7.17. Finger electrodes lithographed on the device surface are used to form the periodic grating of the refractive index through the electro-optic effect. Tuning is achieved by changing the voltage applied to the fingers (~ 100 V), which changes the grating period seen by the incoming signal (i.e., Δn).

Owing to the electro-optic effect, the tuning speed of EOTF is very fast (i.e., in the ns range). The tuning range, however, is limited to about 10 nm because of the small electro-optic effect (a 16-nm tuning range has been demonstrated in [Warzanskyj, 1989]).

A 3-dB bandwidth (FWHM) of 1 nm has been achieved. The number of resolvable wavelengths is therefore limited to about 10. A longer perturbation region would improve the filter bandwidth and thereby the number of resolvable channels within the 10-nm range, but this would be at the expense of increased attenuation, which is about 5-dB for the device shown in Figure 7.17.

7.2.4 The Acousto-Optic Tunable Filter (AOTF)

The basic principle of the EOTF can be applied to acousto-optic tunable filters (AOTF) which, as their name implies, use the acousto-optic effect instead of the electro-optic one.

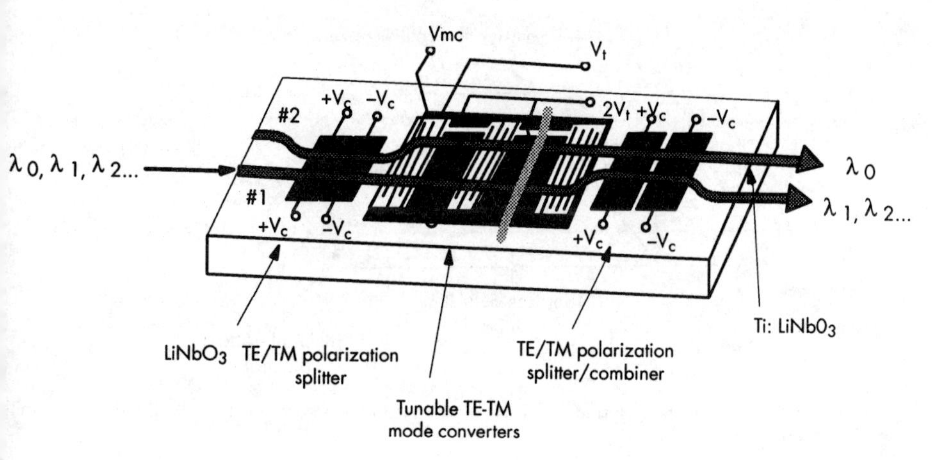

igure 7.17 Structure of an EOTF using Ti-diffused waveguides on $LiNbO_3$ substrate. [After Warzankyj, 1988.]

A schematic representation of an AOTF is shown in Figure 7.18. In spite of its milarity in appearance to the EOTF, the AOTF presents many different characteristics.

First, the diffraction grating is formed by surface acoustic waves (SAWs). Wavength tuning is accomplished by varying the SAW frequencies (usually between several ns to several hundreds of megahertz). The periodic perturbation induced by the acoustoptic effect can be viewed as a dynamic grating with a period equal to the acoustic avelength in the material of interest. As a result, the tuning range of the AOTF can be uch broader than that of the EOTF. In fact, the entire 1.3– 1.6-μm wavelength range can e covered [Cheung, 1989].

Notice first that since the light velocity is about 5 orders of magnitude higher than e sound velocity in the medium, the grating created acoustically remains in essentially e same position as light interacts with it (i.e., there is no Doppler shift of the diffracted ght).

Second, the tuning time of AOTF is limited by the time for the acoustic wave to fill the interaction length, usually in the μs range.

Third, the AOTF has the powerful and unique capability to select several wavength channels simultaneously. This is made possible thanks to the weak interaction tween different acoustic waves. Therefore, when multiple acoustic waves are present in e interaction length, multiple optical channels can be simultaneously and independently lected. Simultaneous selection of up to five wavelengths separated by 2.2 nm has been monstrated [Cheung, 1989]. The number of channels that can be selected simultaeously is limited by the maximum RF drive power that the transducer can tolerate ithout damage. With integrated-optic AOTF, a few milliwatts of drive power are reired per selected channel. It is expected that simultaneous selection of several tens of avelength channels can be achieved in the future.

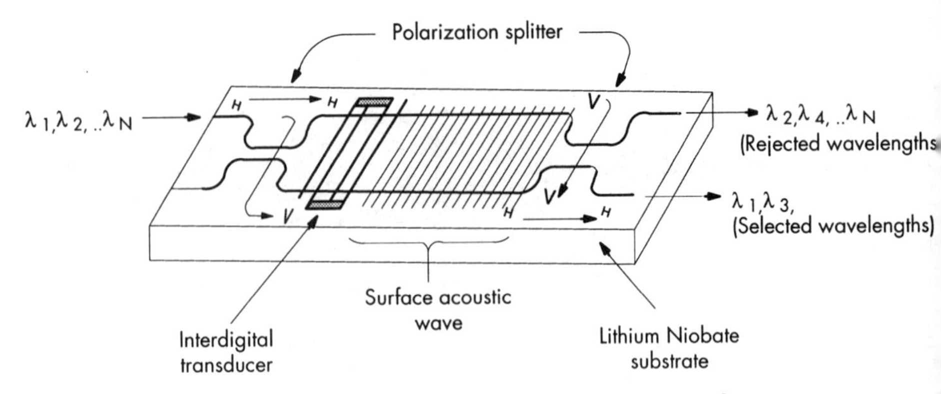

Figure 7.18 AOTF structure using surface acoustic waves over $LiNbO_3$ substrate. (© 1990 IEEE. [After Smit 1990.])

The 3-dB bandwidth of AOTF (FWHM) is similar to that of EOTF (i.e., ~ 1 nm) as is the insertion loss (i.e., ~ 5 dB) [Herrmann, 1992].

7.2.5 Semiconductor-Based Tunable Filters

Semiconductor laser and amplifier structures can also be exploited for wavelength filte ing. As discussed in Chapter 2, single-longitudinal mode (SLM) lasers have built-i gratings (e.g., DFB and DBR lasers) to provide for selection of a single optical frequency When biased below threshold, these devices can operate as resonant amplifiers th amplify only those input channels whose optical frequency coincides with those of th semiconductor amplifier structure.

Wavelength tuning is accomplished by varying the resonant frequency throug either current injection or temperature variation (temperature tuning leads to very slo response times, in the order of several milliseconds, and will not be further considere here).

With current injection, random tuning between any two frequencies within th laser/amplifier passband can be achieved in a few nanoseconds [Kobrinski, 1988].

Among the various structures that have been proposed and demonstrated, the mu tielectrode DFB laser amplifier with a phase-control section has attracted much attentio Such tunable filters are nothing but tunable multisection semiconductor lasers (discusse in Section 2.3.5) whose end-facet reflectivities have been made negligible by using a tireflection coatings.

Tuning is achieved by current injection in the phase-control section while simult neous gain is provided by injecting current into the active-gain section. The tuning rang of such filters is limited to about 5 nm at the most, while the channel bandwidth usuall ranges from 0.05 to 0.25 nm. The number of resolvable WDM channels is therefore restricted to a few tens.

The chief advantage of semiconductor-based tunable filters is that they can be monolithically integrated within the receiver, as they use the same semiconductor material.

Besides their limited tuning range, these filters are strongly polarization-dependent and would therefore require either the use of polarization control or polarization-diversity techniques such as those discussed in Chapter 4 in the context of coherent optical receivers.

It must be mentioned that a new wavelength channel selector based on the combination of a wavelength demultiplexer and a set of semiconductor optical amplifiers used as optical on–off gates has been proposed and demonstrated very recently [Chiaroni, 1994]. This device, which can be referred to as the *gated-wavelength demultiplexer,* is able to demultiplex up to 16 channels spaced by 0.6 nm with subnanosecond switching time and is therefore very promising for future fast dense-WDMA packet-switching applications.

7.2.6 The Fiber-Brillouin Tunable Filter

Nonlinear effects in optical fibers have been discussed in Chapter 3, and we will see in Section 7.3.2.2 that most of these nonlinear processes have detrimental effects in multiaccess optical-fiber networks.

The stimulated Brillouin scattering (SBS) in single-mode fibers can, however, be used as a selective amplification mechanism for wavelength-tunable filtering [Chraplyvy, 1986].

To use SBS amplification as a tunable optical filter, an unmodulated optical pump wave is injected from the receiver end of the optical fiber in the direction opposite to that of the incoming WDM channels. Selective amplification of one of the WDM channels is achieved by the SBS process, provided that the pump power ($>$ 10 mW) and the interaction length ($\sim$ several kilometers of fiber) are sufficient.

The selected channel frequency and the pump frequency must be separated by exactly the Brillouin shift ($\sim$ 11 GHz at 1.55 μm). Tunability is achieved by the use of a tunable pump laser. The filter bandwidth is limited by the SBS-gain bandwidth, which is about 100 MHz for an unmodulated narrow-linewidth pump. It can be increased up to several hundreds of megahertz by broadening the pump spectrum (e.g., by modulating the pump).

The switching time is determined by the time required for the pump to fill up the fiber. With a 10-km length of fiber, this provides a switching time of about 5 μs.

The potential of this scheme was demonstrated in an experiment in which one channel at 150 Mbps was selected from 128 WDM channels separated by 1.5 GHz [Tkach, 1989].

Table 7.1
Tunable Filter Characteristics

Technology	*Timing Range [nm]*	*3-dB Bandwidth [nm]*	*Number of Resolvable Channels*	*Loss [dB]*	*Tuning Speed*	*Applications*
FPF-Single Stage	50	0.5	10 s	2	ms	Circuit switching
FPF-Tandem	50	0.01	100 s	5	ms	Circuit switching
Mach-Zender	5–10	0.01	100 s	5	ms	Circuit switching
Electro-optic (EOTF)	10	1	10	≈5	ns	Packet switching
Acousto-optic (AOTF)	400	1	10 s	≈5	μs	Circuit & long packet switching
Active semiconductors	1–5	0.05	10 s	0 (gain possible)	ns	Packet & circuit switching
Fiber-Brillouin	10	< 0.01	100 s	0 (gain possible)	μs	Circuit & long packet switching

Table 7.1 summarizes the relevant system parameters of the wavelength-tunable filter technologies that have just been described.

It must be noticed that the design of tunable optical filters is still evolving and further development effort is under way not only to improve their performances but also to make them commercially available devices.

7.2.7 Multiwavelength Switched Receivers

To conclude this section on tunable receivers, it must be mentioned that another approach to wavelength-tunable receivers that does not require any optical-tunable filter has been proposed in the literature [Kirkby, 1990]. The basic idea, shown in an artistic view in Figure 7.19(a), consists of using a linear monolithic array of N photodetectors and a diffraction grating that separates the incoming WDM channels and directs them to distinct photodetectors. Channel selection (i.e., wavelength tuning of the receiver) is accomplished by using an electronic space-switching matrix, as shown in Figure 7.19(b).

Such a design can be referred to as the *multiwavelength switched receiver* (the same concept can be extended to *multiwavelength switched transmitters*). Note that unlike most tunable receivers using tunable optical filters, the multiwavelength switched receiver has the capability of simultaneous multiwavelength selection.

Proper design will enable all the required components to be incorporated into a module not much larger than today's single-channel hybrid receiver modules. Indeed, using planar technology a very compact grating-photodiode array has been fabricated recently [Cremer, 1992].

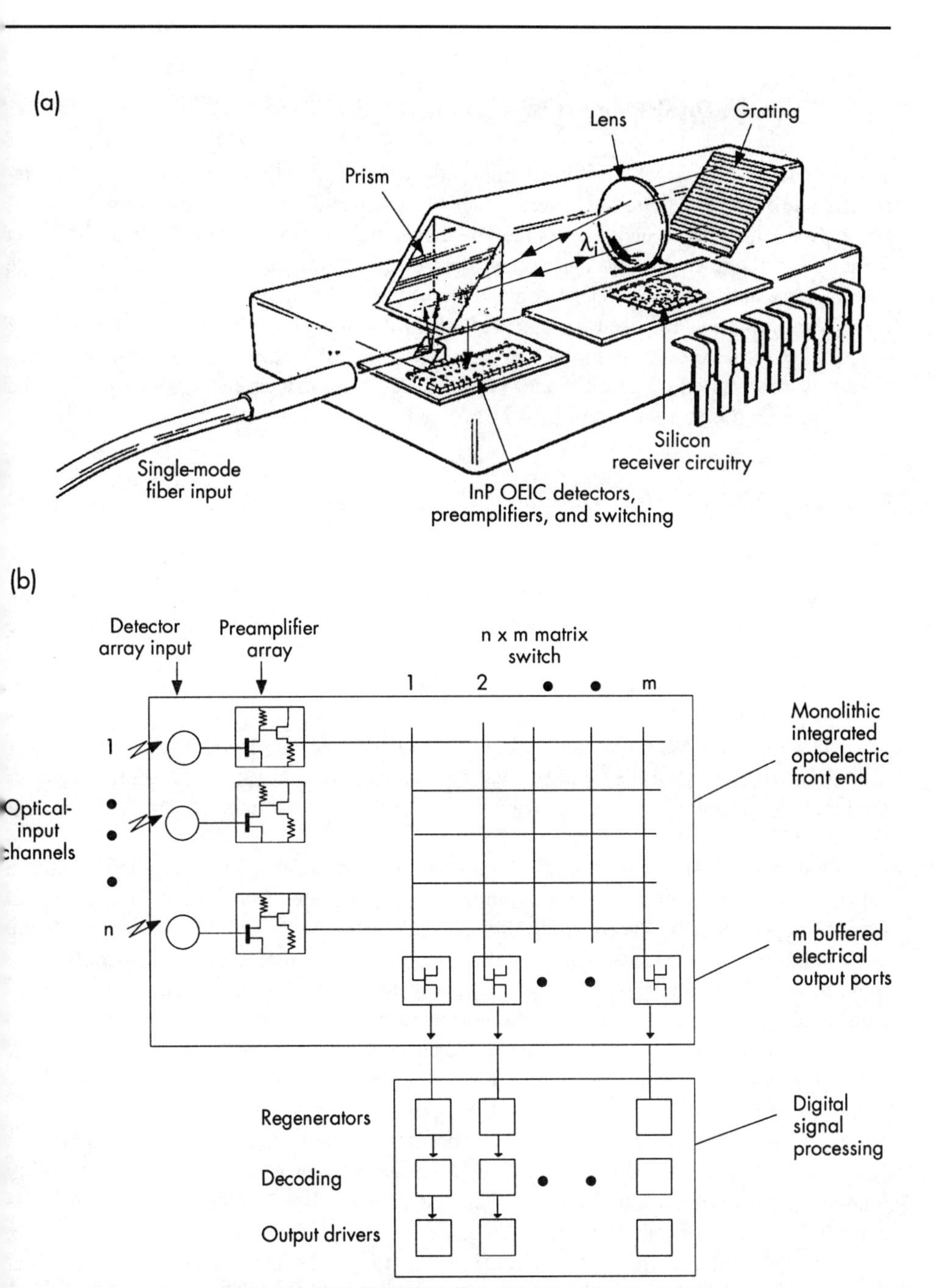

Figure 7.19 (a) Artistic representation of a multiwavelength switched receiver. (b) Electronic processing to select *m* signals out of *n* input signals. (© 1990 IEEE. [After Kirkby, 1990.])

7.3 NETWORK DESIGN AND PERFORMANCE ISSUES

Besides the technological constraints imposed by the limited range of tunability of transmitters and receivers, there are several other factors that affect the performance of single-hop WDMA networks. Such factors include the network protocol that coordinates the data flow within the network, the location of data buffers required to avoid packet loss, the level of crosstalk between WDM channels, and the wavelength stability.

The limitation imposed by the network protocol will be discussed in Section 7.4 along with some relevant examples of single-hop WDMA networks (the limitation imposed by the limited control-channel capacity has been discussed in Section 6.7). The other three factors are the subject of the following three sections.

7.3.1 Input Versus Output Queuing

In general, with arbitrary traffic within the network, the conflict of multiple (up to N) simultaneous packets destined for the same network node is unavoidable. The conflicting packets can either be denied or queued for later transmission. If they are denied, the information carried by them will be lost forever. Clearly, this is unacceptable unless the associated packet-loss probability is negligibly small. As a consequence, packet buffers are generally required to avoid this loss of data.

In single-hop WDMA packet networks, buffers can be placed at either the inputs (i.e., the transmitters) or the outputs (i.e., the receivers) of the system. The comparison of the network capacities obtained with input queuing and output queuing was quantified in [Karol, 1986].

Assuming first-in first-out (FIFO) buffers at each node, it has been shown that the capacity of an input-queued system (with $N \geq 20$) is about 58% of the capacity of an output-queued system. The reduction of capacity in input-queued systems is caused by what is known as the head-of-line (HOL) blocking phenomenon: of all the packets destined to the same output, only one has access to the network (i.e., selected for transmission) while the others are blocked in the buffers (assuming that the node's transmitters can send only one packet at a time). The buffered packets at the head of the FIFOs block network access to other packets in buffers, even if these packets are destined to idle nodes.

A possible way to reduce the HOL blocking is obtained by providing the ability to extract a packet that has access to the network among the first w ($w > 1$) queued packets [Hiuchyj, 1988]. It has been shown that this effectively increases the network capacity while also preserving all input-output packet sequences. The scheduling algorithm, however, is likely to be more complicated.

With output-queuing systems, the HOL blocking does not occur since all packets at the head of the FIFOs (i.e., at the transmitters) can be routed to their destination nodes, where they are buffered in the receiver FIFOs.

Notice that in addition to having high network capacity, it is also desirable to have minimal delay together with preservation of the FIFO packet sequence. Figure 7.20 shows the mean waiting time (normalized in number of packets all assumed to be of the same length) as a function of the input link load for input queuing and output-queuing systems. The advantage of output queuing is clearly seen from this figure.

7.3.2 Crosstalk

As already discussed in Section 7.2 concerning tunable optical filters, an important issue in the performance of WDMA networks is the level of crosstalk between the wavelength-multiplexed channels. Interchannel crosstalk may have several origins, which can be classified into two distinct categories:

- The so-called linear crosstalk, which is essentially induced by the nonideal characteristics of the channel-selection mechanism;
- The so-called nonlinear crosstalks, which are due either to the nonlinear effects in optical fibers or to gain-saturation effects in semiconductor optical amplifiers (when used of course). As discussed in Chapter 2, nonlinear effects in fibers are usually apparent only for signal propagation in long fiber lengths on the order of, say, 20 km

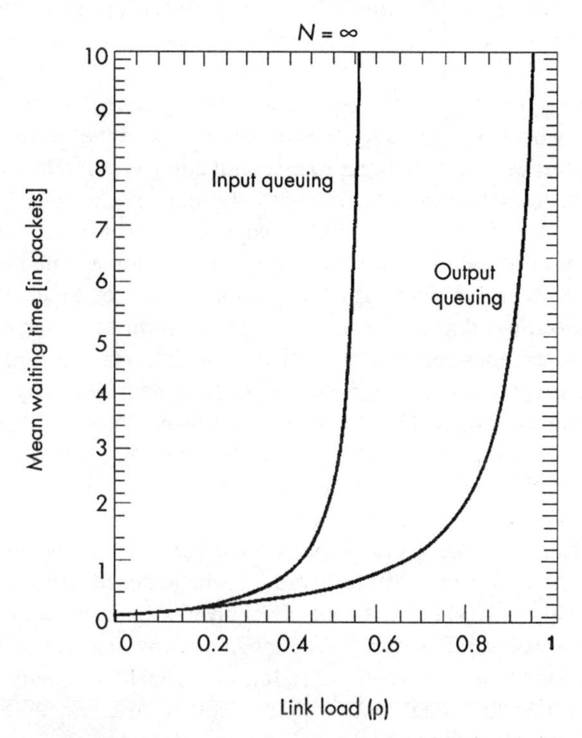

Figure 7.20 Mean waiting time in units of packets as a function of the link load for input and output-queuing networks. (© 1991 IEEE. [After Karol, 1991.])

or more. Therefore, unlike linear crosstalk, fiber-induced nonlinear crosstalks mus only be considered for WDMA networks having a large geographical coverage such as metropolitan area networks (MANs). Nonlinear crosstalks induced within semi conductor optical amplifiers have been discussed in Section 5.1.2.2.

The two following sections discuss the linear and fiber-induced nonlinear crosstalks separately.

7.3.2.1 Linear Crosstalk

Channel selection in WDMA-networks can be achieved using either a (tunable) optica filter with DD receivers or a (tunable) local-oscillator laser and an electrical bandpass filter (BPF) in coherent detection receivers.

Linear crosstalk depends essentially on the type of channel-selection device chosen as well as on the interchannel spacing. In fact, the interchannel spacing is determined by the characteristics of the channel-selection device and the allowed level of crosstalk.

The crosstalk characteristics of tunable optical filters commonly used with DD receivers have been discussed in Section 7.2.

In the case of coherent detection WDMA networks, channel selection is achieved by tuning the wavelength of the local-oscillator laser in the vicinity of the selected channe and passing the detected electrical signal through a fixed-tuned electrical BPF centered a the intermediate frequency (IF). The linear crosstalk is governed by the IF linewidth (i.e. $\Delta\nu = \Delta\nu_s + \Delta\nu_{LO}$ where $\Delta\nu_s$ and $\Delta\nu_{LO}$ are the linewidths of the transmitter laser and the local oscillator, respectively) and the transfer function of the electrical BPF. The IF linewidth-induced crosstalk can be made negligibly small by choosing $\Delta\nu/B \leq 0.1$ (B is the signal bit rate) and is therefore generally insignificant in properly designed coheren lightwave systems. The BPF-induced crosstalk depends on the signal-modulation forma chosen (i.e., ASK, PSK, or FSK). Notice that, as shown in Figure 7.21, the channe spacing in the electrical domain is generally not uniform even for equispaced optica channels. Usually, the nonselected channel that is closest to the local-oscillator frequency ν_{LO} provides the dominant source of crosstalk. If the center frequency of the nonselected channel falls inside the bandwidth of the BPF, its in-band power will appear as noise and will therefore increase the BER of the selected channel. A specified BER can be main tained by increasing the signal power. Such a power increase is referred to as the *linea crosstalk-induced power penalty.*

Figure 7.22 shows the power penalty as a function of the normalized channe spacing, $\Delta f/B$ (where Δf is the 3-dB bandwidth of the matched BPF), for ASK, PSK, and FSK modulation formats. The worst-case is obtained with ASK, while the lower powe penalty is realized when FSK is used (FSK modulation with a tone spacing equal to the signal bandwidth is assumed in Figure 7.22). It is also seen that a negligible power penalty below 0.5 dB is achieved in all three cases provided that $\Delta f \cong 5\,B$. With B = 10 Gbps, we have thus Δf = 50 GHz and the 150 nm (i.e., 20 THz) of optical bandwidth available around the 1.5-μm wavelength has therefore the potential to support 400 channels.

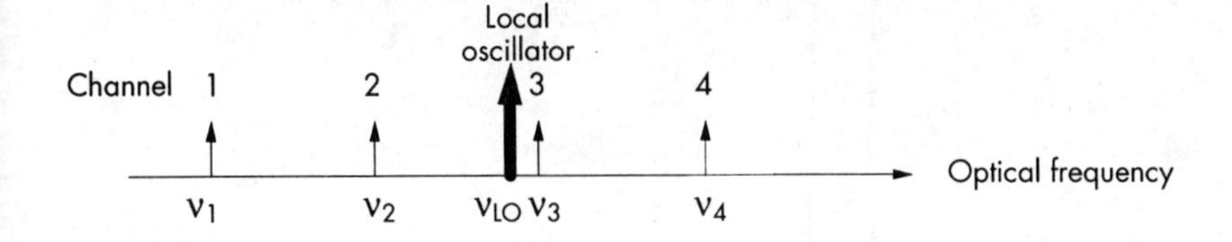

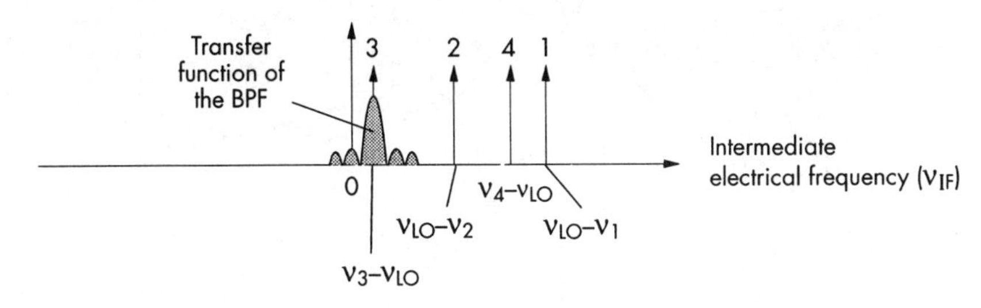

Figure 7.21 Channel selection in a coherent multiwavelength network.

7.3.2.2 Nonlinear Crosstalk

Most of the nonlinear effects in optical fibers discussed in Chapter 3 can produce interchannel crosstalk when multiple channels at different wavelengths propagate over long lengths of fiber. Each of these nonlinearities will affect WDMA networks in different ways. In all cases, however, nonlinear effects limit the power per channel in the common fiber where all the channels are present simultaneously. The channel multiplexing method will therefore have an important impact on the effects of nonlinear crosstalk. When an $N \times N$ star coupler is used for passive optical multiplexing, the power per channel injected into each output fiber (i.e., the "common fibers") is reduced by a factor N assuming zero excess loss for the star coupler. By contrast, with wavelength-selective multiplexers, the power of each channel injected into the common fiber is independent of the channel number. Consequently, such systems will be more susceptible to degradation by optical-fiber nonlinearities.

In the following paragraphs, the channel power limit imposed by each nonlinear effect must be understood as the power per channel present in the common fibers. This section describes the nature of the nonlinear crosstalk induced by each nonlinearity and discusses how the performance of WDMA networks is affected by them.

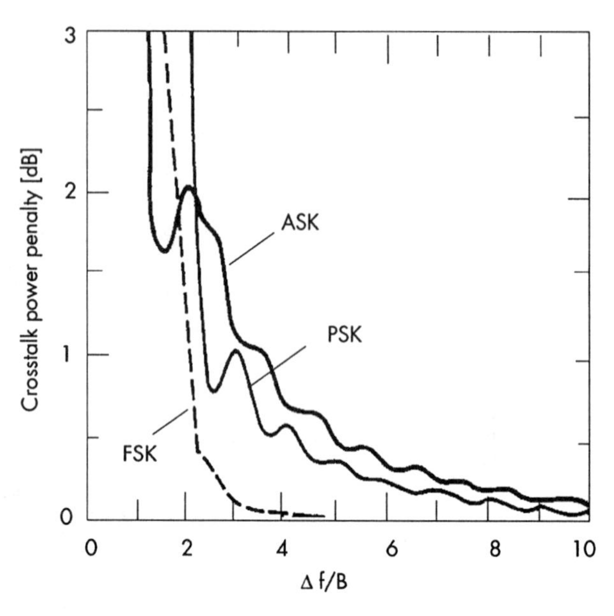

Figure 7.22 Crosstalk power penalty in coherent optical detectors for ASK, PSK and FSK modulation formats. (© 1988 IEEE. [After Kazovsky, 1988.])

Cross-Phase Modulation

As discussed in Section 3.5.1, the nonlinearity of the refractive index of an optical fiber is responsible for both self-phase modulation (SPM) and cross-phase modulation (XPM), which affect the phase of the propagating lightwaves. SPM affects the performance of each channel individually. In WDMA networks with many channels, XPM usually dominates by far over SPM. XPM converts optical-power fluctuations in a particular channel to phase fluctuations in the other channels and is therefore at the origin of a nonlinear crosstalk.

Since the XPM-induced crosstalk affects only the phase of the signals, it will degrade only systems that use phase-sensitive detection (i.e., coherent receivers) but not IM/DD lightwave systems. An exception to this rule is when an optical-phase-modulated signal is detected using a DD receiver. For example, an optical FSK signal can be converted to an intensity-modulated signal using an optical filter and then detected by a DD receiver [Kaminow, 1988].

The XPM-induced phase shift for the ith channel can be derived from (3.72) and is given by

$$\Delta\phi_i^{\mathrm{XPM}} = 2\gamma L_{\mathrm{eff}} \sum_{j \neq i}^{N} P_j \tag{7.20}$$

Two cases must be distinguished depending upon whether the signal is amplitude (ASK) or phase modulated (PSK, FSK).

Consider first the case of amplitude modulation (ASK) for which P_j takes one of the two values $P_j(0) = 0$ or $P_j(1) = 2\overline{P}$ ($\overline{P}$ is the mean optical power of the signal) to represent a logical data "0" or "1," respectively. From (7.20), the phase shift $\Delta\phi_i^{XPM}$ will vary from bit to bit depending on the bit pattern of the neighboring channels. In the worst case, the XPM-induced phase shift of channel i becomes

$$\Delta\phi_i^{XPM} = 4\gamma L_{eff}(N-1)\overline{P} \tag{7.21}$$

By imposing $\Delta\phi_i^{XPM} \leq 0.1$ rad to make the XPM effect negligible, (7.21) restricts the mean power in each channel to

$$\overline{P} \leq \frac{0.025}{\gamma L_{eff}(N-1)} \tag{7.22}$$

With $\gamma = 1 W^{-1} km^{-1}$ as a typical value and $L_{eff} = 22$ km ($L = 30$ km), this restriction implies P ≤ 0.1 mW even for N as small as $N = 20$.

The impact of XPM on coherent WDMA networks is less severe for phase-modulated signals since the power in each channel remains constant for all bits. The fundamental limitation for such systems stems from the intensity fluctuations associated with the transmitter lasers (i.e., the RIN). Assuming equal average power in each channel, the rms-induced phase fluctuations in a particular channel due to power fluctuations in the other channels is readily found from (7.20) and is given by

$$\sigma_\phi^{XPM} = 2\gamma L_{eff}\,\sigma_P\sqrt{N-1} \tag{7.23}$$

where σ_P is the rms power fluctuation of the transmitter lasers, which is typically less than $5.10^{-3}\,\overline{P}$ assuming an external PSK or FSK modulator. For $\overline{P} = 1$ mW, ($L_{eff} = 22$ km) and $\sigma_\phi^{XPM} \leq 0.1$ rad, (7.23) shows that the XPM effect on system performance is negligible even for very large number of channels.

Much larger values of σ_P are generated when the semiconductor laser transmitters are directly modulated. This is usually the case when the FSK modulation format is chosen, as we saw in Section 4.3.2.3. Values as large as $\sigma_P \cong 0.2\overline{P}$ are commonly observed in that case. It has been shown that σ_ϕ^{XPM} grows linearly with N rather than as $\sim\sqrt{N}$ [Chraplyvy, 1984]. In order to limit the XPM-induced power penalty to less than 1 dB, the power (in milliwatts per channel) must satisfy

$$\overline{P} \leq \frac{21}{N} \tag{7.24}$$

For $\overline{P} = 1$ mW, (7.24) provides $N \leq 20$.

The condition (7.24) is plotted in Figure 7.23 together with the power limitations imposed by four-photon mixing (FPM), stimulated Raman scattering (SRS) and stimulated Brillouin scattering (SBS), to be discussed in the next paragraphs.

FPM

The effect of FPM in an *N*-channel WDMA network is to generate new lightwaves at frequencies $\nu_{ijk} = \nu_i + \nu_j - \nu_k$, where *i, j,* and *k* can vary from 1 to *N*.

For equally spaced channels, these new frequencies generated at the expense of the useful channel power coincide with the channel frequencies and lead to crosstalk.

For the nonequispaced channels situation, the new FPM-generated frequencies can fall within the bandwidth of a modulated channel and appear as interference noise during the detection process.

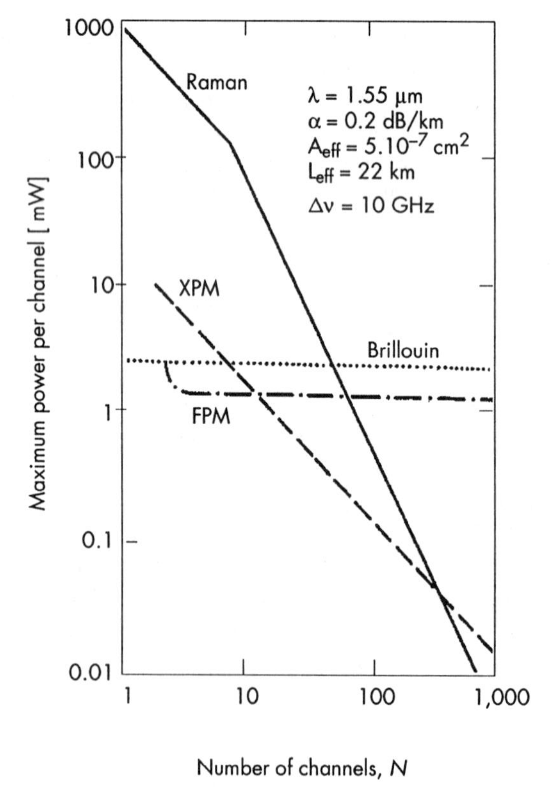

Figure 7.23 Maximum power limits per wavelength channel as a function of number of channels for the four main nonlinear effects in optical-fiber WDMA networks. [After Chraplyvy, 1991.]

Clearly, in both cases, FPM degrades the system performance by excess optical-power loss or crosstalk.

As seen from (3.74), the efficiency of FPM grows as P^3 if we assume equal power per channel. The power P in each channel must therefore be restricted to below a certain level to make the FPM-induced crosstalk negligible.

However, the efficiency of FPM depends also on the channel-frequency spacing and the fiber dispersion through the phase-matching condition. Because of fiber dispersion, the phase coherence among the interacting and generated lightwaves is destroyed, resulting in a reduction of power generated at the new frequencies. The larger the dispersion or the channel spacing, the lower the FPM efficiency will be. In general, FPM becomes important for systems whose operating wavelength is close to the zero-dispersion wavelength of the fiber.

It has been shown [Shibata 1987, Chraplyvy 1991] that the channel spacing must be larger than 20 GHz to make the FPM negligible in 1.5-μm WDMA networks that use standard single-mode fibers. With dispersion-shifted fibers, the required channel separation enlarges up to 50 GHz.

Figure 7.23 shows the maximum power per channel in a standard single-mode fiber for a channel spacing of 10 GHz. Clearly, FPM is the most limiting nonlinear phenomenon for small numbers of channels ($N \leq 10$). This limiting effect can be avoided in practice if the channel spacing is made larger than 100 GHz.

SRS

SRS is generally not a limiting nonlinear effect in single-channel lightwave systems since the required signal power to produce system degradation is above 500 mW.

In WDMA networks with many channels, SRS becomes much more important because it acts as an optical amplifier: the longer wavelength channels are amplified at the expense of the shorter wavelength channels as long as the wavelength difference falls within the bandwidth of the Raman gain. Since the Raman gain is about 200-nm broad (30 THz), the SRS effect will be noticeable in coarse or dense WDMA as well as FDMA networks. Channel amplification at the expense of power in other channels occurs only when there is power in all channels involved in the process. In other words, the amplification will be dependent upon the modulation format chosen (ASK, PSK, or FSK) since with ASK a bit ''0'' will correspond to the absence of light while for PSK or FSK there is always light irrespective of the content of the data pattern. In addition, for ASK the amplification will also be dependent upon the content and the synchronism of the bit pattern in each channel. The resulting signal-dependent amplification will produce power fluctuation in each channel and will therefore degrade system performance.

A simple model to account for the SRS-induced nonlinear crosstalk has been developed in [Chraplyvy, 1983]. This model considers the depletion of the shortest wavelength channel under the worst case in which all channels are synchronized and carry optical power associated with bit "1." Notice that this does not include the dependence of group velocity with wavelength (i.e., the fiber dispersion) since this would tend to desynchronize the bit pattern at the different wavelengths and, hence, reduce the impact of SRS on system performance.

The Raman gain profile has been assumed to increase linearly with the wavelength difference between the shortest wavelength channel and the $(N - 1)$ remaining channels up to a maximum of 120 nm. The result is that the SRS-induced crosstalk power penalty can be kept below 1 dB provided that the channel power is set according to [Chraplyvy, 1983]

$$P < \frac{500[\text{GHz} \cdot \text{W}]}{N(N-1)\,\Delta\nu} \tag{7.25}$$

The limit imposed by (7.25) is plotted in Figure 7.23 for a channel spacing of $\Delta\nu = 10$ GHz. It is seen that SRS becomes the major limiting factor for large numbers of channels ($N \geq 500$).

SBS

SBS has been described in Section 3.5.4. We have seen that its effect is similar to SRS; that is, short-wavelength channels transfer part of their energy to long-wavelength channels. Although the threshold of SBS (3.78) is much lower than the threshold of SRS, the SBS-induced nonlinear crosstalk can easily be avoided in a WDMA network by proper system design. The reasons are twofold:

1. SBS crosstalk does not occur when all channels propagate in the same direction along the fiber.
2. The SBS-gain bandwidth is extremely narrow (~ 10–100 MHz): hence, for SBS to be efficient, the channel spacing must match almost exactly the Brillouin shift, which is 11 GHz at 1.55-μm wavelength.

Condition (1) is automatically avoided for simplex communication links where separate fibers are used for each direction of propagations. Condition (2) is easily avoided by a proper choice of the channel wavelength allocation.

In addition, the efficiency of SBS decreases proportionately to $\Delta\nu_B/\Delta\nu_{\text{Laser}}$ where $\Delta\nu_B$ is the Brillouin-gain bandwidth (~ 10–100 MHz) and $\Delta\nu_{\text{Laser}}$ the transmitter laser linewidth. Therefore, the effect of SBS is negligibly small, unless transmitter lasers with extremely narrow linewidths use external modulators to impose the data on the optical carrier (note that even in this case one can always superimpose a slow modulation of the carrier frequency to increase the SBS threshold).

7.3.3 Wavelength Stability and Control

In WDMA networks, the wavelength separation between adjacent channels must be stable in order to avoid receiver-sensitivity degradation as a result of interchannel crosstalk [Toba,

1990]. In addition, even for stable wavelength separation, the components used for channels demultiplexing, such as optical filters, must be stable enough for the same reason.

Unfortunately, neither laser diodes nor optical filters have sufficient stability to be usable in dense WDMA or FDMA networks. Methods for device characteristic stabilization with regard to the wavelength are therefore required.

For laser diodes, the main sources of fluctuations of the emitted wavelength are related to variations of injected current (100 MHz/mA to 1 GHz/mA) and temperature ($\sim$ 10 GHz/°C). Many commercially available laser diodes include inside their package a thermoelectric cooler controlled by a thermistor and a feedback circuit to stabilize the temperature to within less than 0.1°C. This also makes it possible to keep the bias current within 0.1 mA. This simple method suffices to maintain the optical-carrier frequency to a few hundred megahertz around the center frequency.

Once means have been found to stabilize the wavelength of the emitted light, it is necessary to lock this wavelength to a reference wavelength. This reference wavelength can be that of a stable gas laser (e.g., HeNe laser) or, more generally, the wavelength associated to some atomic transitions. Simple and fairly inexpensive methods to lock the wavelength to atomic transitions using the optogalvanic effect have been demonstrated [Chung, 1990].

For optical filters, the main factor entering the picture is temperature. Temperature dependency of filter characteristics is a function of the material used to make the device. Particularly attractive materials with respect to temperature stability are silicon (for gratings and MZFs) and glass (FPFs) [Miller, 1990]. $LiNbO_3$ used, for example, for EOTFs and AOTFs has a much higher temperature coefficient of expansion and the same stabilization method as that used for laser diodes can be applied.

Besides the absolute wavelength-stabilization issue, relative wavelength stabilization must also be performed. In fact, the optical carriers may drift together as a group so long as they do not run into one another to create interference. Several methods for wavelength-separation locking have been demonstrated [Bachus 1985, Nosu 1987, Strebel 1987, Glance 1988, Shimosaka 1990]. The method described in [Shimosaka, 1990] has several attractive features. It converts frequency separation to time difference between adjacent pulses using a frequency-swept light source and a Fabry-Perot optical resonator. Details of the different techniques can be found in the references.

In general, these methods are simple to implement and have the potential of significant cost reduction.

7.4 EXAMPLES OF SINGLE-HOP WDMA NETWORKS

Many single-hop WDMA network designs with different potential applications have been proposed and demonstrated in research laboratories.

Recent reviews of advances in this field can be found in [IEEE 1990, IEEE 1993]. Application domains range from upgrading of existing optical-subscriber loop systems, dense WDMA passive optical networks, interoffice and metropolitan area networks, survivable networks, to circuit and, ultimately, high-speed packet-switching fabrics [Wagner 1989, Goodman 1989]. Furthermore, recent interests in WDMA networks have also extended their applicability to other new areas, such as optical interconnect in computer and neural networks. The purpose of this section is to provide the reader with an overview of some relevant WDMA networks that have been investigated so far. The section addresses the system issues, such as the mode of operation and the associated network protocol, the performance limitations, and the envisaged application domains.

The survey article [Mukherjee, 1992] and the references therein can be consulted for more detailed discussion of network protocols.

7.4.1 The Lambdanet

Bellcore's Lambdanet, shown in Figure 7.24, was the first single-hop WDMA network to be implemented at the prototype level [Kobrinski 1987, Goodman 1990].

It is an FT-FR^N broadcast-and-select star network. Each of the N = 18 nodes is equipped with a single-frequency DFB laser emitting at a unique wavelength in the 1,527- to 1,561-nm range with a spacing of ≈ 2 nm between adjacent wavelength channels. Each of the wavelengths is broadcast to every node receiver via an $N \times N$ star coupler constructed by cascading wavelength-flat 3-dB couplers as described in Section 6.2. At each node receiver, an 18-wavelength grating demultiplexer (the Stimax-type discussed in Section 6.3) separates the different wavelength channels, which are then independently converted to electrical form by a dedicated photodiode.

In order for every node to be able to communicate simultaneously to every other node, the data on each wavelength forms a time division multiplexed (TDM) frame. The TDM frame contains 18 slots (and a synchronization tag), each allocated to a predetermined destination node. Thus, the ith slot of the TDM frame transmitted on λ_j contains the bit stream from node j destined to node i.

Therefore, each node may receive and process, asynchronously and in parallel, transmission from all other nodes. Owing to this hybrid wavelength-time architecture, collisions are automatically avoided. In addition, the Lambdanet is an output-queued network, so its capacity is essentially determined by the number of wavelength channels and the bit rate of each channel within the network. A potential network capacity of 27 Gbps has been demonstrated with 18 wavelength channels, each operating at 1.5 Gbps.

Notice that the full connectivity provided by the hybrid wavelength-time Lambdanet architecture is achieved at the expense of high-speed opto-electronic conversion and electronic processing that are required to run at N times the bit rate of a single connection.

7.4.2 The Knockout Photonic Switch

The *knockout* principle for packet switching was first introduced in 1987 by Yeh and colleagues from AT&T Bell Laboratories [Yeh, 1987]. This principle was subsequently proposed to build a centralized photonic packet switch [Eng, 1988].

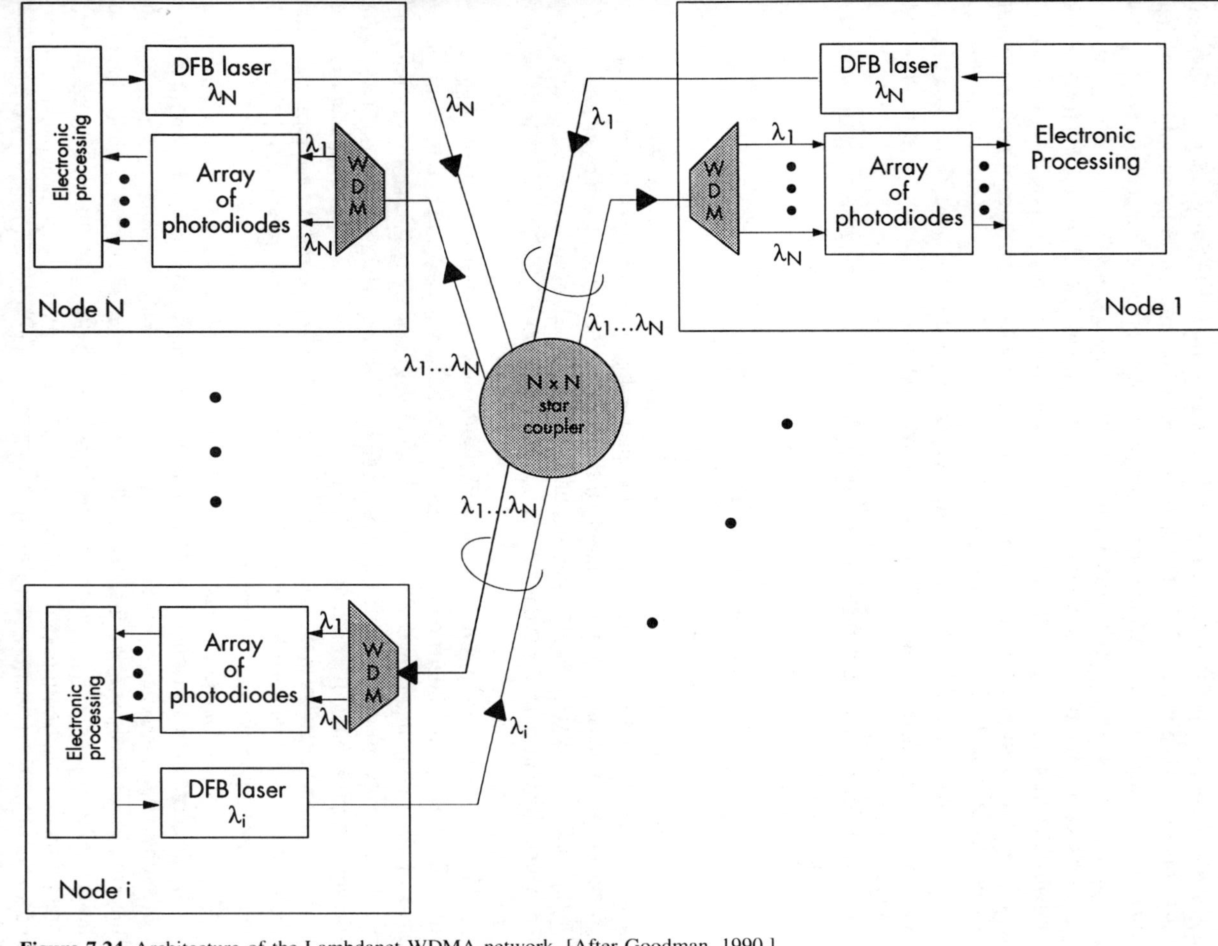

Figure 7.24 Architecture of the Lambdanet WDMA network. [After Goodman, 1990.]

The knockout switch is based on the observation, already mentioned in Section 7.1.1, that the probability p_k that k packets are simultaneously destined to the same node of an N node network rapidly decreases as k increases. This probability obeys a binomial distribution and is given by (7.5). This property can be exploited to build an output-queuing photonic packet switch in which the required number of optical receivers per node is minimized while the packet loss probability is maintained below a prescribed value.

The knockout photonic switch is illustrated in Figure 7.25. It is an FT-TR^L network in which each node receiver contains up to L tunable optical receivers with $L < N$. Thus, the maximum number of packets destined to the same node at the same time is restricted by construction to L.

If k packets (with $k > L$) are simultaneously destined to the same node, then only L packets out of k will be successfully received while $(k - L)$ packets will be dropped (i.e., lost).

It follows from (7.5) that the probability of a packet being dropped in the node receiver is given by

$$\text{Prob [packet loss]} = \frac{1}{\varrho} \sum_{k=L+1}^{N} (k - L) C_N^k \left(\frac{\varrho}{N}\right)^k \left(1 - \frac{\varrho}{N}\right)^{N-k} \tag{7.26}$$

Figure 7.26 shows a plot of the packet loss probability versus L for $\varrho = 0.9$ and for $N = 32$, 128, and infinity. It is seen that the packet-loss probability is not very sensitive to the switch dimension N, but is mainly influenced by the value of L. For example, if $L = 12$ is selected, a packet-loss probability of 10^{-10} can be achieved for any value of N. For each unit increment in L beyond 12, the packet-loss probability decreases by approximately an order of magnitude.

Notice that (7.26) has been obtained assuming uniform traffic. The performance of the knockout switch under nonuniform traffic has been analyzed in [Yoon, 1988].

Recognizing that packet loss is inevitable in any network (e.g., due to transmission errors, buffer overflow, etc.), the value of L can be chosen by design so that the associated packet-loss probability is substantially lower than that contributed by other factors.

The node receivers in the knockout switch must be informed as to which wavelengths their optical receivers must be tuned. This is achieved by means of the control channel C and a knockout contention controller.

The control channel C is issued by each node transmitter and is directed towards the knockout contention controller. The C channel may be composed of a 2-byte control packet where the first bit is an activity bit indicating the presence or absence of a valid data packet, the next 7 bits designate the address of the destination node, and the 7 following bits designate the source address (it remains one spare bit). In that case, the maximum switch dimension that can be supported is thus 128×128. The length of the control packet cannot be longer than that of a data packet since both must remain in synchronism (Figure 7.25(b)).

With ATM packets 53 bytes long and a data channel rate $D = 5$ Gbps, a data packet spans 85 ns, so that the 16-bit control packet must be transmitted and processed in the

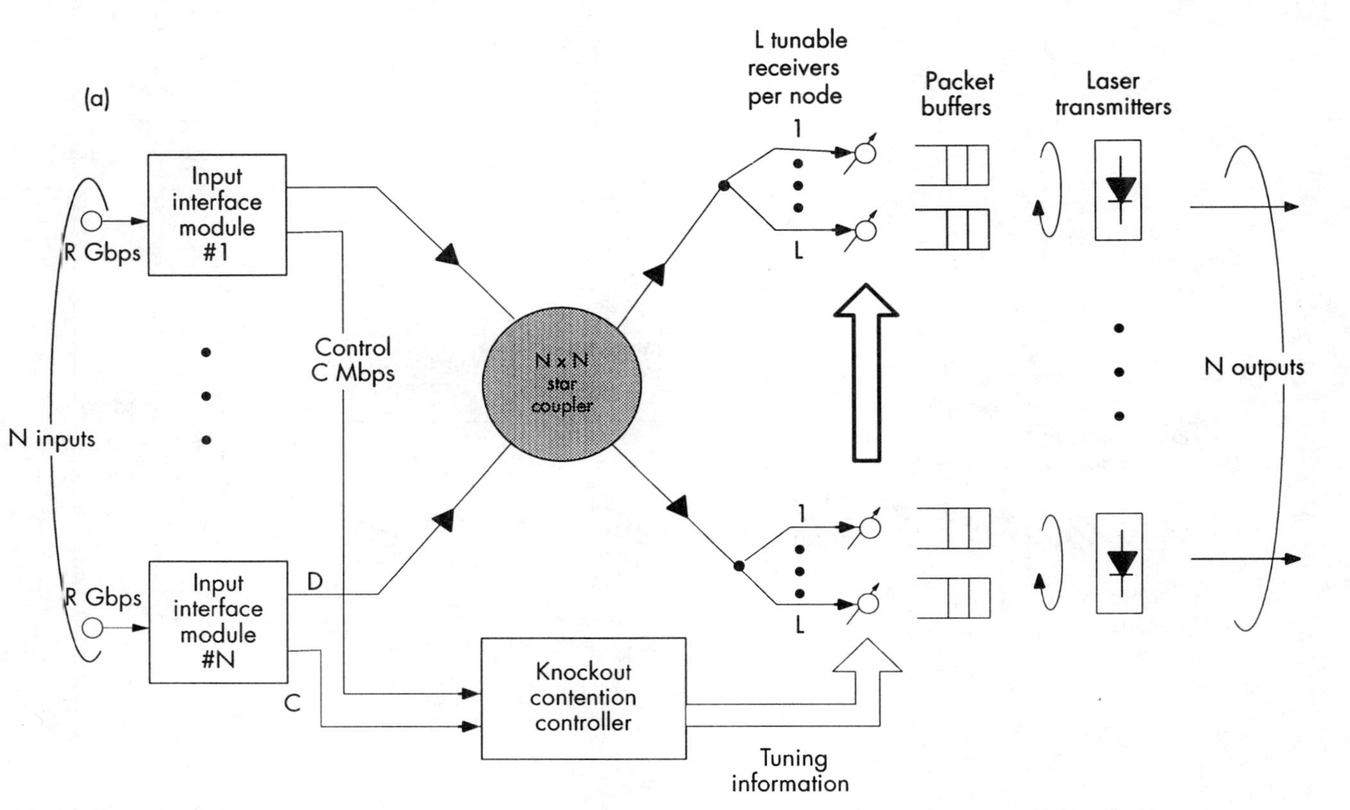

Figure 7.25 (a) The knockout photonic switch, and (b) functional block of the input-interface module included in each node. (© 1988 IEEE. [After Eng, 1988.])

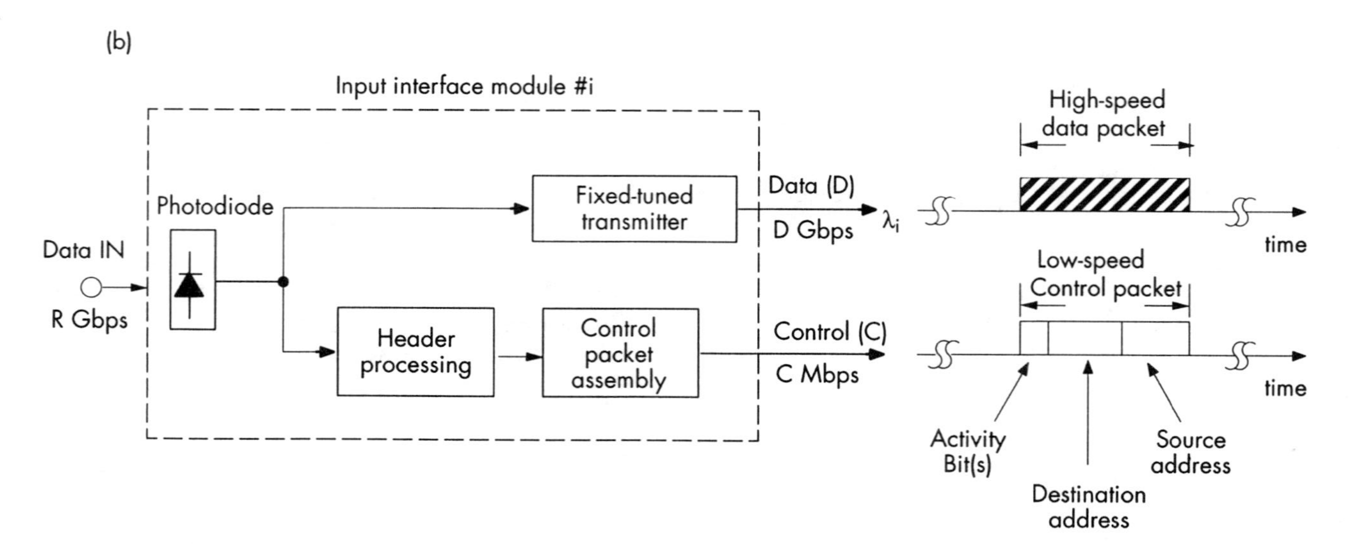

Figure 7.25 (continued)

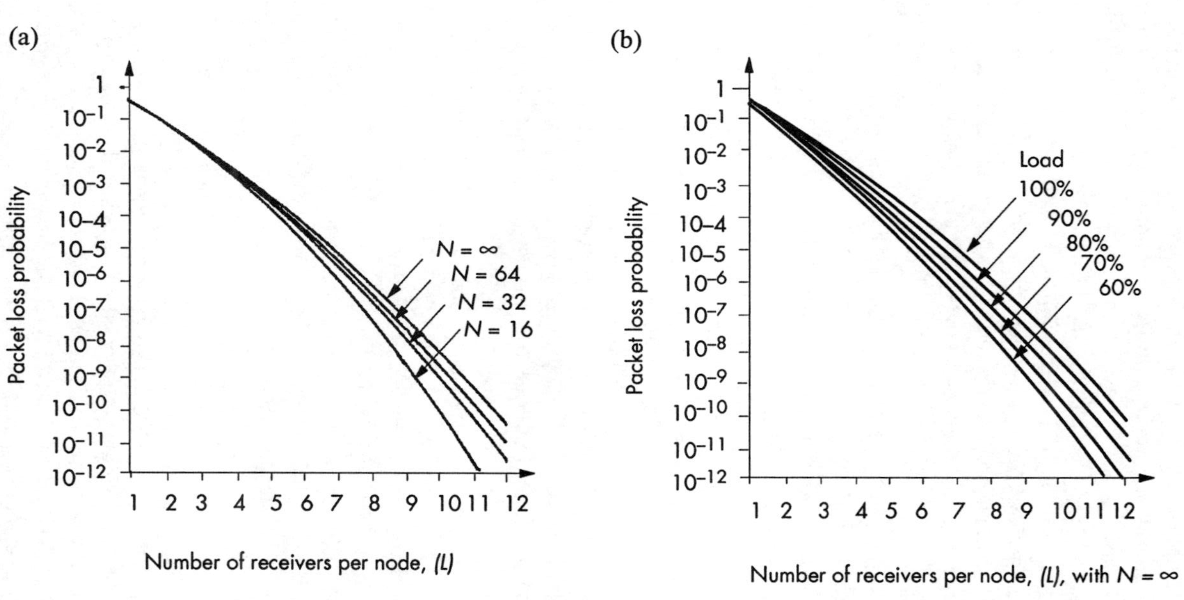

Figure 7.26 (a) Probability of packet loss versus L for 90% input link load. (b) Same as (a) but for various loads and $N = \infty$. [After de Prycker, 1991.]

knockout contention controller at a rate of about 200 Mbps. Such a rate can easily be processed using current VLSI circuits.

The switch contention is resolved within the knockout contention controller from which connection instructions (i.e., tuning information) are issued to each node receiver. Several distinct algorithms can be envisaged to decide which wavelengths the receivers must be tuned to. Packets that are successfully received can be called the "winners" while those that are dropped because of the lack of available tunable receivers can be called the "losers." Details of a specific algorithm are described in [Eng 1987] where contention resolution is achieved by organizing 2×2 contention switches, which issue one winner and one loser at each step (hence the name knockout given to this photonic switch).

7.4.3 The Passive Photonic Loop (PPL)

Figure 7.27 illustrates the basic elements of the passive photonic loop (PPL) as proposed by Bellcore in 1988 [Wagner, 1988]. The system is intended for subscriber loop applications in which substantial sharing of the fiber cable is achieved through the so-called double-star topology. The system provides each subscriber (i.e., each node), with a dedicated bidirectional link from the passive remote terminal (RT) while the feeder fiber, from the RT to the central office (CO), is shared by all subscribers. Each subscriber is assigned two unique wavelengths, one for reception (i.e., downstream) and one for transmission (i.e., upstream). The PPL relies solely on passive WDM devices at the CO and at the RT to perform multiplexing and routing functions.

Downstream and upstream transmissions are performed similarly in the way described below.

For downstream transmission, signals destined to different subscribers are modulated onto their assigned downstream wavelengths and multiplexed onto the common feeder. At the RT, the wavelength channels are passively demultiplexed onto the appropriate distribution fiber via a WDM device.

Similarly, for upstream transmission, different subscribers transmit onto their assigned upstream wavelengths, which are multiplexed at the RT, transmitted over the common feeder, and demultiplexed at the CO.

Note that since each subscriber transmits and receives only on unique wavelengths, the network intrinsically preserves security and privacy, which are particularly attractive features for subscriber loop applications.

The PPL architecture has been demonstrated experimentally using several types of light sources, ranging from the single-frequency DFB lasers to the broad-spectrum LEDs.

In one approach, wavelength-stabilized DFB lasers operating at both 600 Mbps and 1.2 Gbps were used. The WDM devices performed 20-channel multi/demultiplexing in the 1,550-nm band with wavelengths spaced at 2-nm intervals.

In a more cost-effective approach, LED light sources were used for upstream transmission. The use of LED in dense WDMA networks is made possible by the *spectrum slicing* concept illustrated in Figure 7.28 [Pendleton 1985, Blair 1993].

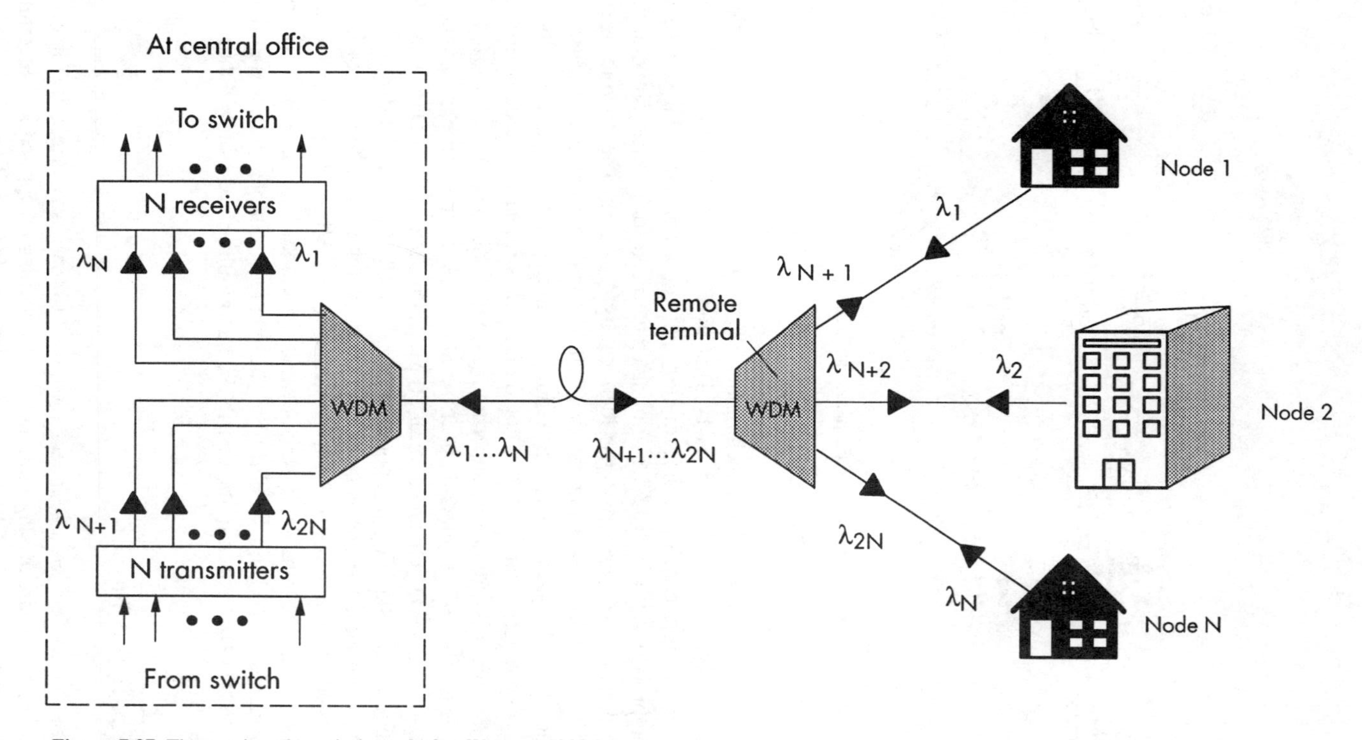

Figure 7.27 The passive photonic loop. [After Wagner, 1988.]

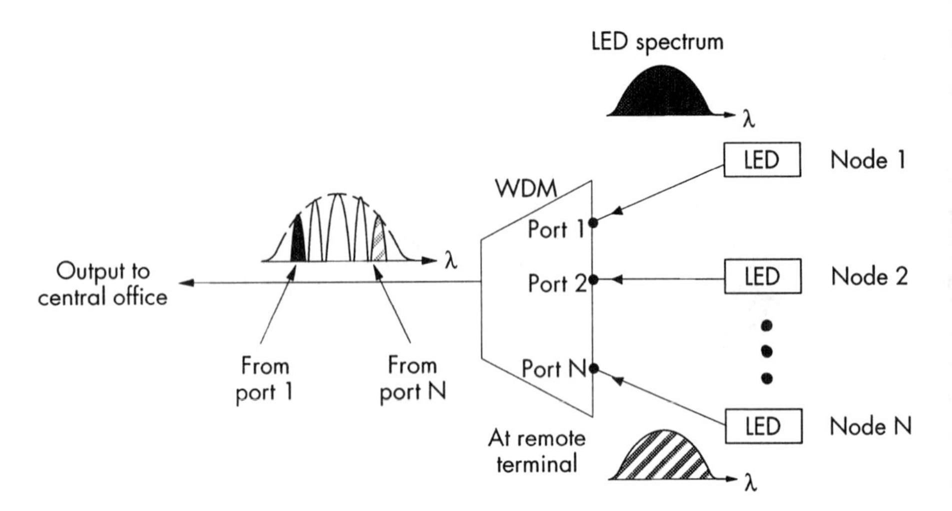

Figure 7.28 Principle of the spectrum-slicing technique using LED light sources.

When light from the LEDs reaches the WDM at the RT, the WDM selects and passes a narrow spectral band from each. This portion of the original LED spectrum nevertheless contains the original information and thus represents the subscriber data channel. The broad spectrum of LEDs eliminates the need for precise temperature stabilization and allows all subscribers to use identical light sources. The price to be paid is a reduction of the network power budget since only a small portion of each LED's power passes through the WDM at the RT. This reduction of the power budget limits the channel operating bit rate to less than 2 Mbps if standard LEDs with single-mode fibers are used.

To overcome this limitation, superluminescent diodes (SLDs) were used to demonstrate the feasibility of broadband spectrum slicing. Transmission experiments at 150 Mbps on each of 10 WDM channels, or at 50 Mbps on each of 16 WDM channels using spectrally sliced SLDs have been reported [Wagner 1990, Chapuran 1991].

A few years ago, the spectrum-slicing concept underwent field tests in London [Hunwicks, 1990]. The outputs of 8 LED*s* operating at 2 Mbps were spectrally sliced at the CO and transmitted to the end points located 3.5 km away. This experiment in a life environment demonstrated the feasibility of the relatively simple and inexpensive spectrum-slicing WDMA approach to the distribution and multiaccess of low data rate signals over modest distances.

7.4.4 The Fox/Hypass/Bhypass

The FOX, or fast optical cross-connect [Arthurs, 1988a], was originally proposed for applications in parallel-processing computers but may also be applied in telecommunication optical packet switching. The basic architecture of FOX is shown in Figure 7.29.

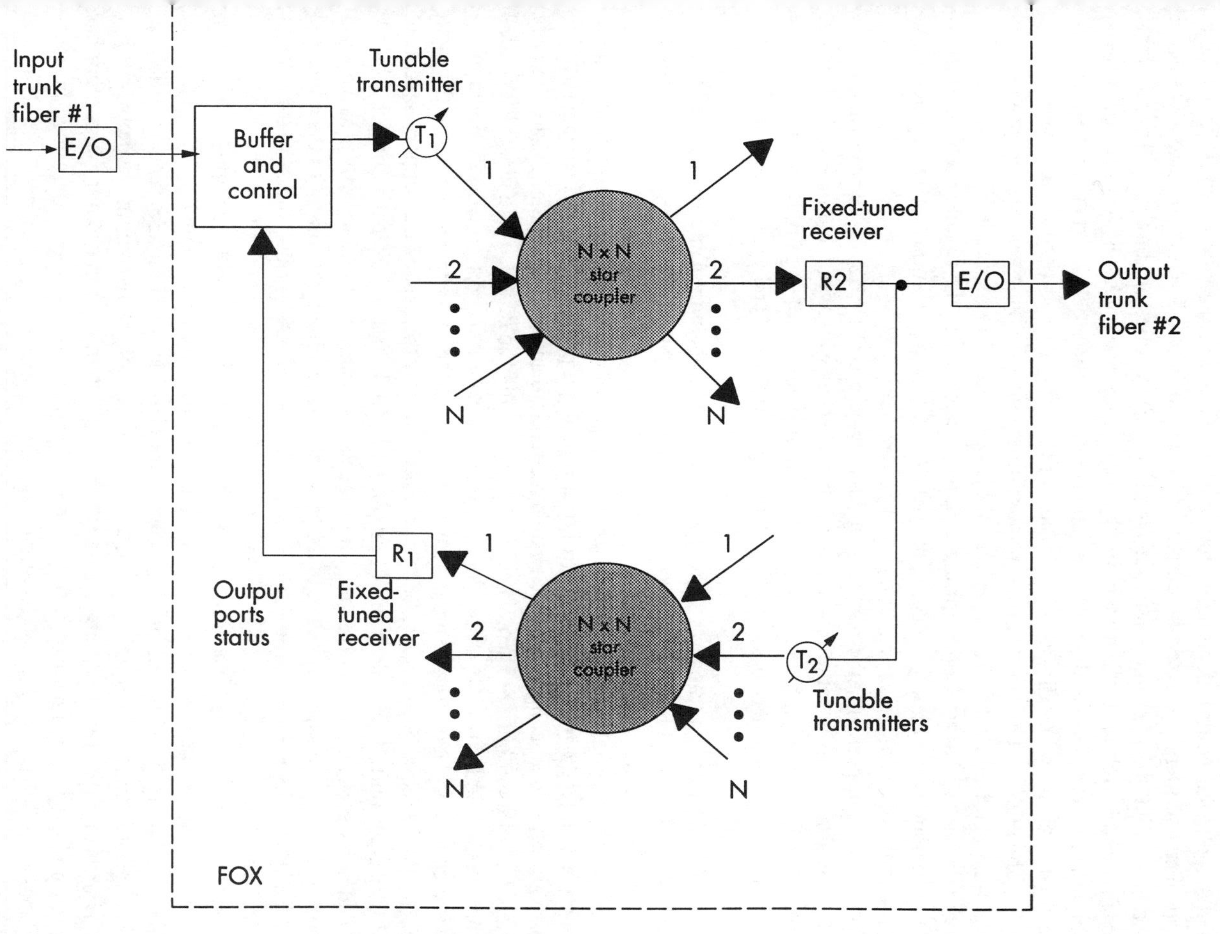

Figure 7.29 The FOX network architecture. [After Arthurs, 1988a.]

The system uses two $N \times N$ star interconnection networks: one for data packet transport and one for control information transport. Both interconnection networks use tunable optical transmitters and fixed-tuned receivers (i.e., it is a $(\text{TT-FR})^2$ broadcast-and-select network). Like the photonic knockout switch, the FOX relies on the fact that for uniform traffic there is only a small probability that more than one packet is destined to the same output port at the same time. If an output-port collision occurs, then the packets are retransmitted following a specific algorithm until they are successfully received.

The FOX system was the first proposal that required high-speed tunable laser diodes (i.e., that can be tuned within few tens of nanoseconds) in order to realize short packet-switching networks. Its limited throughput, due essentially to the required retransmissions when output port collisions occur, restricts FOX-like applications in which access to output ports is relatively low.

The high-performance packet-switching system (HYPASS) is an extension of the FOX system as it uses both high-speed tunable lasers as well as tunable receivers [Arthurs, 1988b]. The basic idea for HYPASS is shown in Figure 7.30. Again, there are two $N \times N$ star-interconnection networks, one for data packet transport and one for control information transport.

The data packet transport network uses tunable transmitters and fixed-tuned receivers (i.e., it is a TT-FR data network). Conversely, the control network uses fixed-tuned transmitters and tunable receivers (i.e., it is a FT-TR network).

The operation of HYPASS is based on an input-queued/output-controlled arbitration. In other words, the output ports control when the input ports are allowed to transmit. The packets arriving at an input port are first converted into an electrical signal and temporarily stored in an input buffer. The packet header containing the destination address is decoded and the tunable transmitter at the input port of the data-transport network is tuned to the wavelength corresponding to the output-port address. The stored packet waits in the input buffer until a request-to-send signal (i.e., a poll) issued by the desired output port is received at that input port. The polls from all output ports are broadcast over the control network to all input ports. To recognize the poll of the desired output port, the receiver at the input port is tuned to the same wavelength as that of the fixed-tuned transmitter of the addressed output port. Upon reception of a poll from the desired output port, the stored packet is transmitted through the data-transport network on the wavelength of the destination address. Simultaneously, packets from other input ports are being transmitted on different wavelengths to other output ports using the same data-transport network.

A variety of poll-generation protocols are possible for HYPASS. The specific protocol chosen may be strongly dependent on the type of traffic within the network. For (uncorrelated) packet switching, a control protocol based on a tree-polling algorithm has been proposed. This protocol is based on the dynamic-tree protocols described in [Capetanakis 1979, Hayes 1978] and assumes that the number of ports is a power of 2; that is, $N = 2^k$.

The basic tree-polling algorithm is based on collision detection to arbitrate demand-access requests. The first polling cycle is initiated by polling all input ports. If there is not

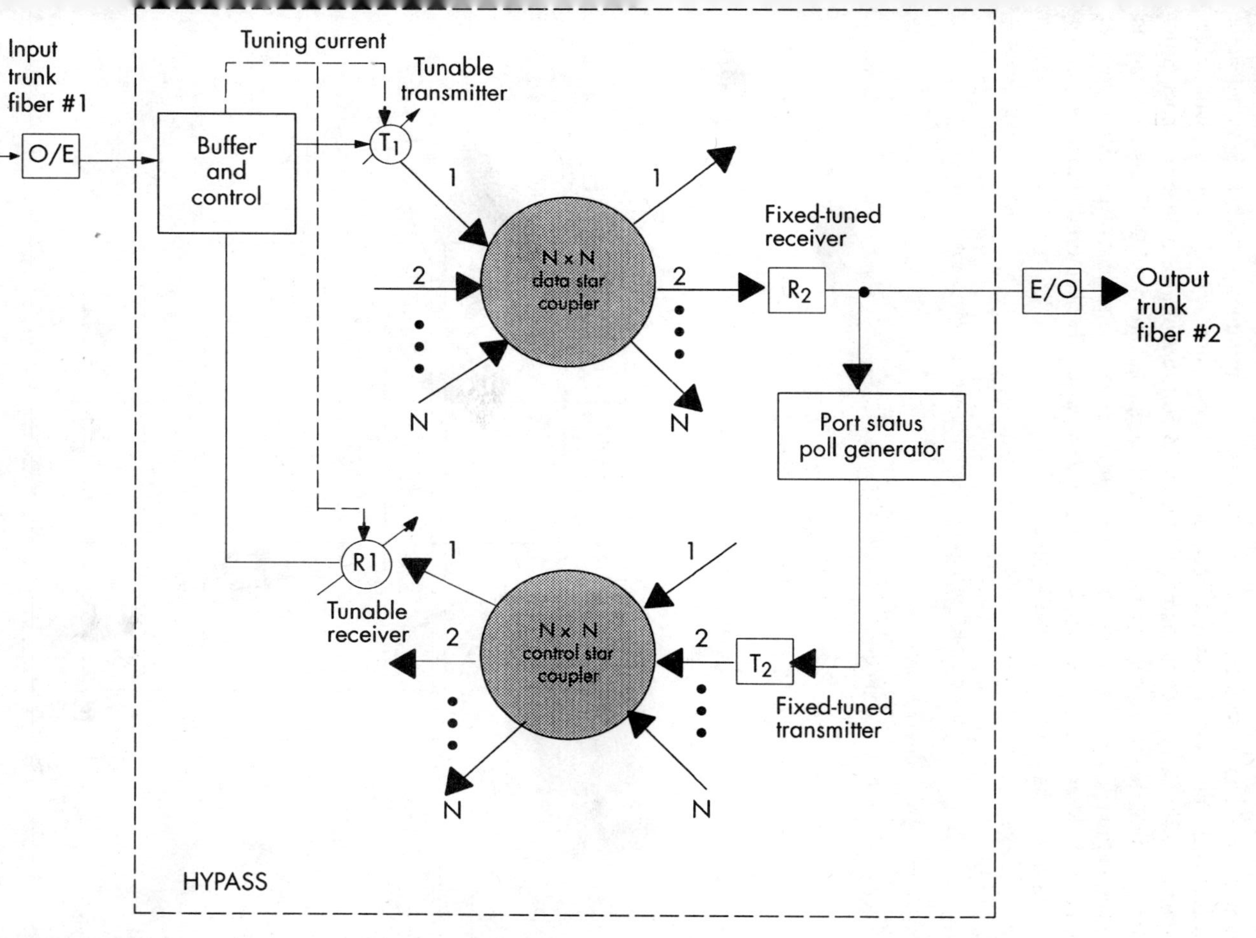

Figure 7.30 The Hypass WDMA network. (© 1988 IEEE. [After Arthurs, 1988b.])

more than one packet for the corresponding output port, then the cycle is completed in one round. If there is more than one packet destined to the output port, then the input ports receiving the poll will transmit their packets and a collision at the output port will occur. Upon collision detection, the output port issues a poll to a limited number of input ports. Group polling is repeated recursively until all collisions are resolved. This algorithm is illustrated in Figure 7.31 where there are four simultaneous requests (from input ports 1, 3, 7, and 8) to access a given output port of a 8×8 switch. The required number of polling cycles in this particular example is nine.

8 x 8 switch with four packets destined to a given output port

Input ports	Port 1	Port 2	Port 3	Port 4	Port 5	Port 6	Port 7	Port 8
	●	○	●	○	○	○	●	●
	Packet	Empty	Packet	Empty	Empty	Empty	Packet	Packet

	Input ports polled	Transmission from port
Cycle 1	1 to 8	Collision
2	1 to 4	Collision
3	1 to 2	1
4	3 to 4	3
5	5 to 8	Collision
6	5 to 6	–
7	7 to 8	Collision
8	7	7
9	8	8

Figure 7.31 Illustration of the tree-polling algorithm for the case in which there are four packets simultaneously destined to the same node.

Performance analysis of HYPASS with the dynamic tree-polling algorithm has been done in [Arthurs, 1988.b]. The model assumes uniform traffic so that any one of the output ports is addressed with equal probability ($1/N$). The buffers at the input ports are FIFOs and the throughput of the network will therefore suffer from HOL blocking. Figure 7.32 plots the calculated mean delay as a function of the input link load ϱ. It is seen that the practical maximum load is limited to about $\varrho = 0.25$. With 5-Gbps input links and 128×128 switch dimension, a peak throughput of about 150 Gbps can be achieved.

Although the HYPASS provides improved performance compared to the FOX, this is achieved at the expense of extra hardware complexity since tunability at both the transmitters and receivers is required. To overcome this drawback, a modification of the HYPASS, known as BHYPASS, has been proposed. In contrast to the HYPASS, the BHYPASS uses input control with a Batcher-Banyan [Acampora, 1994] electronic contention resolution circuit. The Batcher-Banyan electronic circuit processes only the packet headers and its operating speed is therefore much lower than if the whole packet must be switched through it, as is the case in actual electronic packet-switching fabrics.

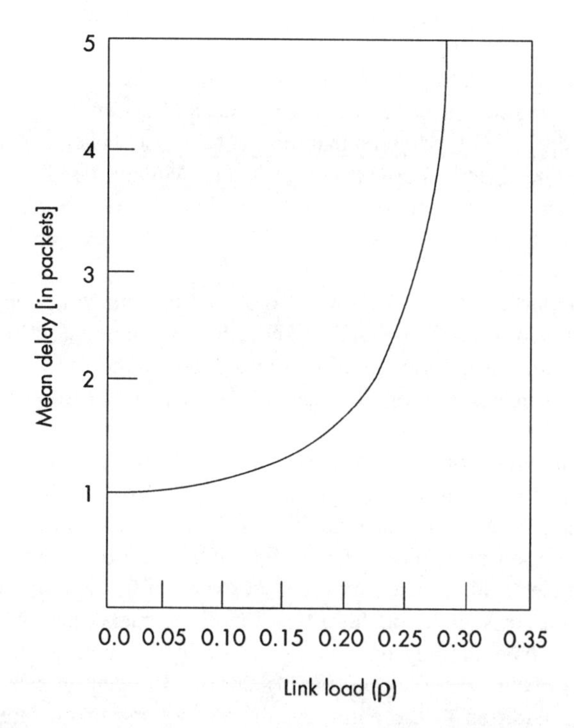

Figure 7.32 Mean delay versus ϱ for the Hypass WDMA network using the tree-polling algorithm for contention resolution. (© 1988 IEEE. [After Arthurs, 1988b.])

Although several experiments have demonstrated the feasibility of the photonic-switching approaches described above, none of the FOX, HYPASS, or BHYPASS approaches have been prototyped.

7.4.5 The Star-Track Network

With the exception of the Lambdanet, all previous designs have been intended for point-to-point traffic. This is essentially because they use fixed-tuned receivers within the data-transport network.

As already mentioned in Section 7.1.1, using tunable receivers within the data transport network one can take advantage of the inherent broadcast nature of the multi-wavelength optical star. This is the basic idea behind the star-track network shown in Figure 7.33 [Lee, 1990].

This network (used as a centralized-switching fabric) is formed by two distinct interconnection networks, an optical star for data transmission, and an electronic control ring (the "track") surrounding the star and sequentially connecting the input ports and output ports. The data-transport network uses fixed-tuned transmitters and tunable receivers (i.e., it is a FT-TR network).

The control ring uses a token consisting of a subtoken for each of the N output ports. An output-port reservation is done by writing the address of the input port into the corresponding subtoken field, provided that this subtoken field is empty. If it is not empty, the input port must wait until the next control cycle. Multicast reservation is done in a straightforward manner by writing one input-port address into several output-port subtoken fields. Prioritization schemes for some reserved traffic can also easily be implemented.

Each output port examines its own subtoken field and tunes its receiver to the appropriate input-port wavelength. Data-packet transmission and control cycles are pipelined so that the data-transmission cycle is executed during the succeeding control cycle. More details on the ring-reservation protocol and the switch performance can be found in [Lee, 1990].

A potential drawback of the star-track is that the ring-reservation scheme is sequential in nature so that the token must be processed at each port, whether there is a packet to be switched or not. Moreover, since the data-transmission and control cycles overlap in time, the control ring must complete its cycle within the data-transmission cycle. In other words, the limited electronic processing speed of the control ring may, in fact, determine the maximum number of switch ports as a function of the packet length and the input link bit rate. Such issues have already been discussed in Section 6.7.

Applications of the star-track are mainly confined to centralized packet-switching fabrics because, in addition to the above-mentioned factors, token-ring-based protocols are inefficient when large propagation delays (i.e., high latency) must be encompassed.

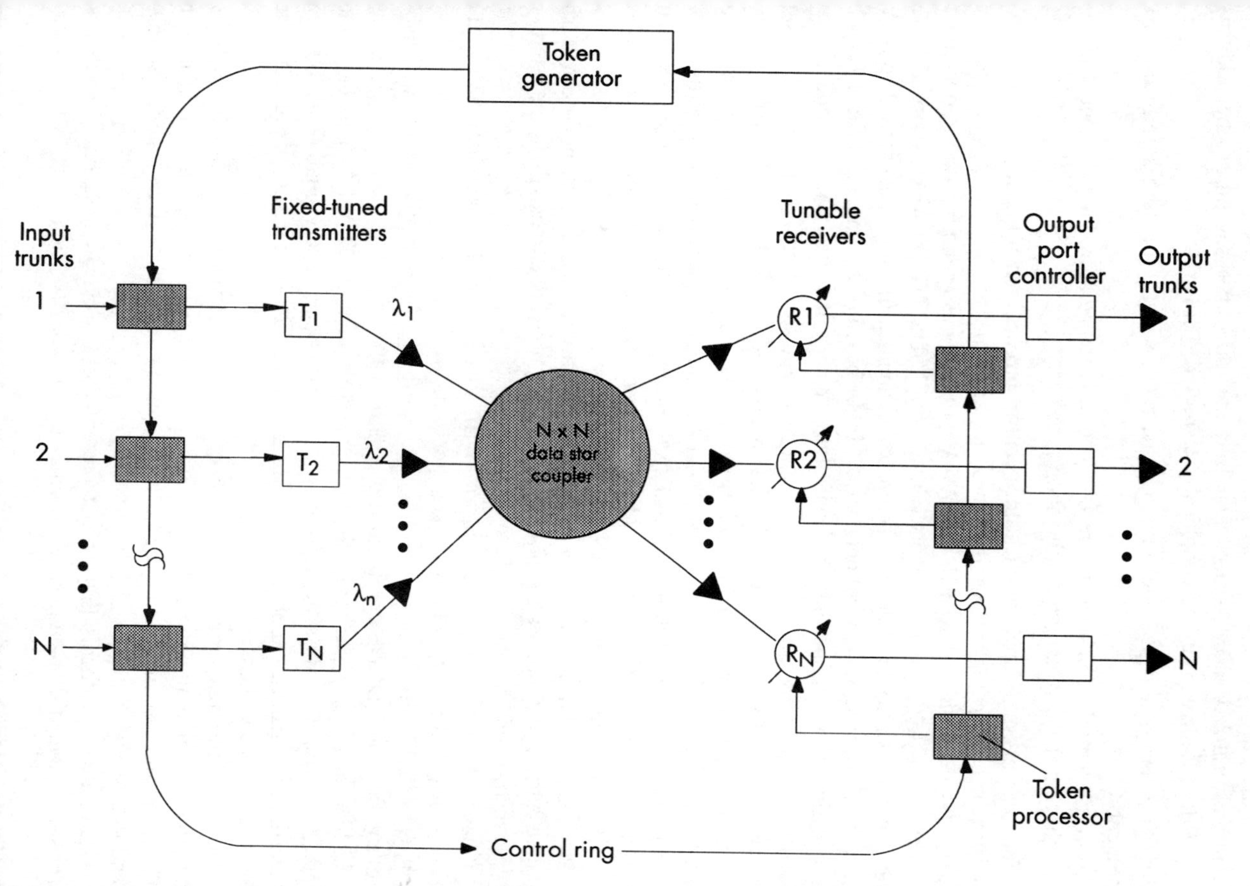

Figure 7.33 The star-track WDMA network for centralized photonic packet switching. [After Lee, 1990.]

Notice that centralized WDMA-based switching networks offer several practical advantages over decentralized (i.e., geographically dispersed) networks:

- Arrays of lasers and photodetectors can be used, which provide a significant cost reduction.
- All optoelectronic components are in a common temperature environment, thereby simplifying the wavelength stabilization problem.

For LAN or MAN applications, other control protocols than token-ring-based protocols must be used to make the FT-TR network efficient.

The dynamic time WDMA (DT-WDMA) protocol has been proposed to overcome the limitations of token-ring protocols [Chen, 1990]. The DT-WDMA uses a separate wavelength common to all nodes for the control channel, which is broadcast via an $N \times N$ star coupler. Access to the control channel is TDMA-based (see Chapter 9). Each TDMA frame contains N minislots, one for each node. A node that has a packet queued for transmission writes a transmission request into its dedicated minislot. The transmission request contains the source address, the destination address, and an additional field by which the source node can signal the priority of its queued packet. Upon reception of the TDMA frame, each node determines to which wavelength its receiver must be tuned. If more than one node want to transmit its packet to the same node at the same time, that addressed node checks the priority fields of the corresponding minislots of the TDMA frame and selects the one with the highest priority.

Provided that all nodes follow the same arbitration protocol, each node can learn about the status of data-packet transmission within the network and can therefore monitor its own transmission.

It has been shown that, under uniform traffic assumption, the maximum throughput per TDMA frame (or slot) is $0.6N$ packets per slot.

A DT-WDMA-like protocol has also been proposed for a multiwavelength distributed-queue dual bus (DQDB) protocol [Lu, 1993].

Unlike many other protocols, the principal advantage of DT-WDMA is that some packets (the "winners" of the arbitration protocol) experience a very small delay to reach their destination, namely about one propagation time across the network due to the "tell-and-go" mechanism.

However, the main drawbacks are twofold: First, they have a limited scalability because the maximum number of nodes N is fixed by the number of assigned minislots in the TDMA frame. Second, each node must be equipped with two (fixed-tuned) transmitters and two receivers.

Examples of protocols that do not present these disadvantages, but nevertheless present some other disadvantages, are discussed in the next sections.

7.4.6 The Rainbow Network

The Rainbow-1 is an FT-TR network that has been developed by IBM Research for MANs supporting circuit switching only [Janniello 1992, Dono 1990].

It uses the IM/DD technology with 1.5-μm DFB lasers and piezoelectric-tunable fiber Fabry-Perot filters [Miller, 1990]. The filters have a 3-dB bandwidth of ≈ 0.4 nm and a free spectral range of ≈ 40 nm, resulting in a finesse of 100. According to (7.10), the maximum number of nodes is approximately 33. The tuning time is about 10 ms, which restricts the Rainbow-1 applicability to circuit-switched networks and nonpacket-switched networks.

The protocol required to set up and disconnect full-duplex connections across the network uses a circular search scheme.

The essential operation of such a protocol can be understood as follows. Suppose a node A wishes to set up a bidirectional communication link with node B. Node A starts by transmitting connection requests repeatedly on its wavelength λ_A. Simultaneously, node A tunes its receiver to λ_B, which is the transmitter wavelength of node B.

The connection request on λ_A, which contains the source (node A) and destination (node B) addresses, is broadcast to all nodes by the $N \times N$ star coupler.

If destination node B does not already have a connection set up with some other nodes, it sweeps its optical filter over the entire range of transmitter wavelengths until a connection request intended for it is detected. Upon detection of such a request, the optical filter is locked to λ_A and a connection-accept message is sent repeatedly on λ_B. The full-duplex connection between nodes A and B is then established since both nodes have their optical filter tuned to the correct wavelength. A connection is released, for example, if no data is received for a specified period.

The Rainbow-1 system implemented using readily available components was demonstrated in the field in 1991. In this field trial, the system connected six nodes, four of them supporting 270 Mbps for digitized TV with the other two supporting 125-Mbps FDDI bitstreams.

7.4.7 The OASIS Photonic Switch

OASIS (optical ATM switching) is a single-hop WDMA-centralized ATM-cell switching fabric that has been prototyped in all its essential components by Alcatel Alsthom Recherches in 1993 [Chiaroni, 1993]. The switching fabric can be viewed as a TT-FR network, although the tunable transmitters are actually replaced by tunable wavelength converters. The system is depicted in Figure 7.34 and is composed of four different functional units:

- The ATM cell encoder, located at each of the N inlets, which performs wavelength conversion;
- The time-switching and cell-buffering unit based on optical on-off gates and fiber delay lines;
- The wavelength demultiplexer, located at each of the N outlets, based on fixed-tuned optical filters;
- The electronic control device, which controls the first two units.

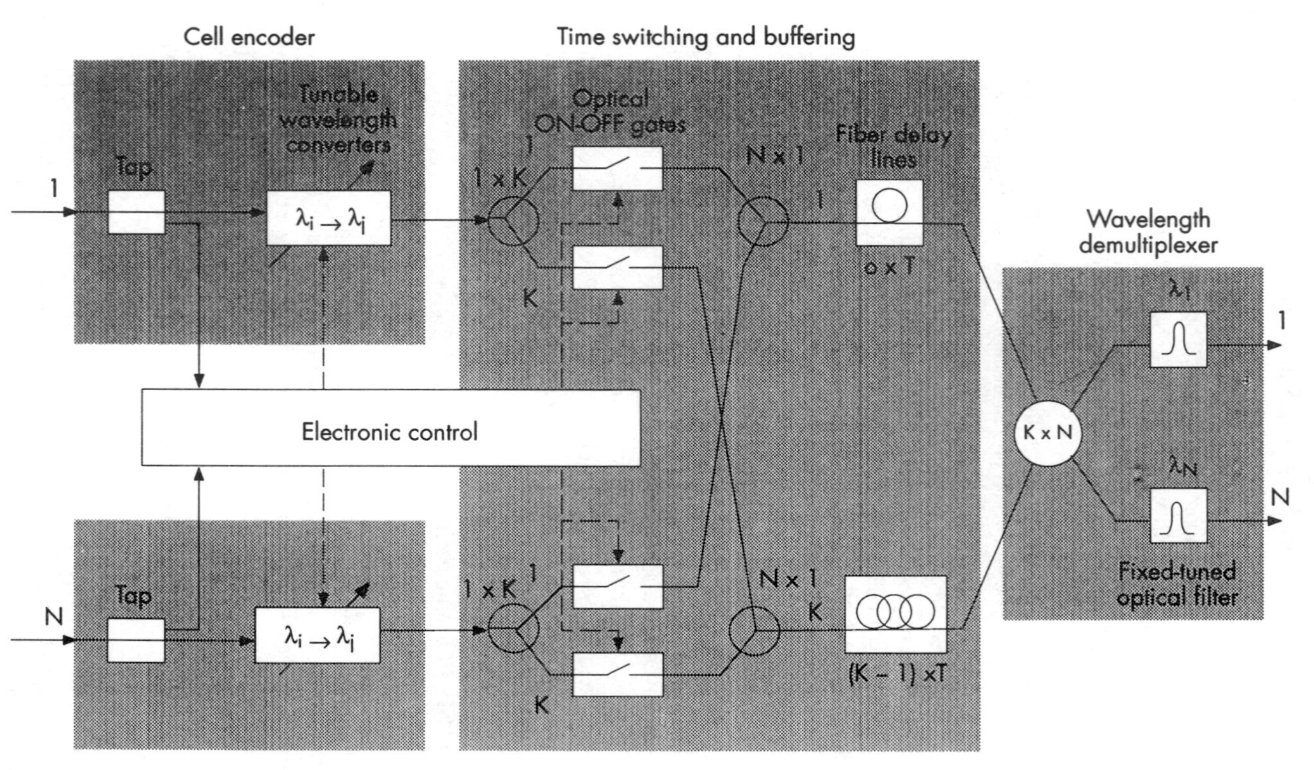

Figure 7.34 Structure of the OASIS photonic-switch preserving optical path continuity of input ATM cells. [After Jacob, 1993.]

The essential operation of the $N \times N$ photonic switch can be understood as follows. Suppose that all input streams are synchronized so that the ATM cells arrive in phase at each of the N inlets. A tap on each input link extracts an amount of optical power for conversion to electronic form while the remaining optical power is launched into a wavelength-converter device.

The electronic control device processes the cell headers and resolves the contention by issuing control parameters for the wavelength converters and the optical gates.

The output wavelength of each λ converter is determined by the destination output port (fixed-tuned receivers). Therefore, at the output of each cell-encoder unit, ATM cells have been wavelength-encoded.

Collisions between synchronous cells destined to the same output (i.e., cells having an identical wavelength) are avoided by means of the time switching and buffering unit. This unit comprises a set of K optical-fiber delay lines with increasing delays equal to $0 \times T$, $1 \times T, \ldots (K - 1) \times T$, where T is the duration of an ATM cell (i.e., 170.4 ns at 2.488 Gbps) [Gavinet-Morin, 1993].

Access to a delay line is commanded by the optical gates. The control device ensures that cells with the same wavelength (i.e., same output-port destination) do not enter the same delay line at the same time. In other words, cells destined to the same output are maintained in distinct delay lines as if they were queued in an output FIFO. The required number of delay lines depends on the traffic load at each inlet and the allowed cell-loss probability: for an input link load $\varrho = 0.4$ and a 10^{-12} cell-loss probability, $K = 15$ delay lines are required [Gabriagues, 1994].

The feasibility of a 4×4 OASIS matrix with input links operating at 2.488 Gbps was demonstrated in 1993 [Chiaroni, 1993]. Different types of wavelength convertors, such as those shown schematically in Figure 7.8, have been used [Durhuus, 1993]. The devices can switch between four wavelengths spaced by 0.6 nm around 1.548 μm in a maximum reconfiguration time of 9 ns. The optical gates are polarization-insensitive semiconductor optical amplifiers providing a 12-dB fiber-to-fiber gain with a response time of less than 200 ps.

Theoretical investigations have shown that the size of the switching matrix would be limited to $N = 16$ due to the accumulation of the amplified spontaneous emission within the optical gates and the corresponding degradation of the extinction ratio [Chiaroni, 1993]. Nevertheless, with improved device performance, further extension to larger values of N as well as cascading of such elementary matrices while preserving the optical-path continuity may be expected.

Besides the few examples we have described, many other single-hop WDMA networks have been proposed and demonstrated. To cite but a few alternatives, there are the UCOL (ultrawideband coherent optical LAN) [Fioretti, 1990], the PAC network (protection against collision) [Glance, 1992], the Symfonet [Westmore, 1991], Orion and Eros [Gautheron, 1992] and the multiwavelength survivable ring networks [Wagner 1992, Elrefaie 1993]. Details on these proposals can be found in the references.

7.5 MULTIHOP WDMA NETWORKS

Multihop lightwave networks were first proposed by Acampora in 1987 to obviate the need for both transmitter and receiver tunability in WDMA networks [Acampora, 1987]. Since then, several theoretical studies have been carried out that have resulted in a number of variants of the original proposal [Sivarajan 1991, Ayanogly 1989, Li 1992, Todd 1991]. A recent survey of multihop lightwave networks can be found in [Mukherjee, 1992b]. The present section focuses on the so called shufflenet as originally proposed in [Acampora, 1987]. We begin with the basic concept of multihop (shufflenet) WDMA networks. In Section 7.5.2, we evaluate their performance. Sections 7.5.3 and 7.5.4 give a description of the two multihop lightwave systems that have been implemented to date; namely, the teranet and the starnet [Gidron 1991, Kazovsky 1992].

7.5.1 Basic Concept

The basic idea behind multihop lightwave networks can be highlighted by referring to Figure 7.35, which shows a star network with eight nodes, each of them having two

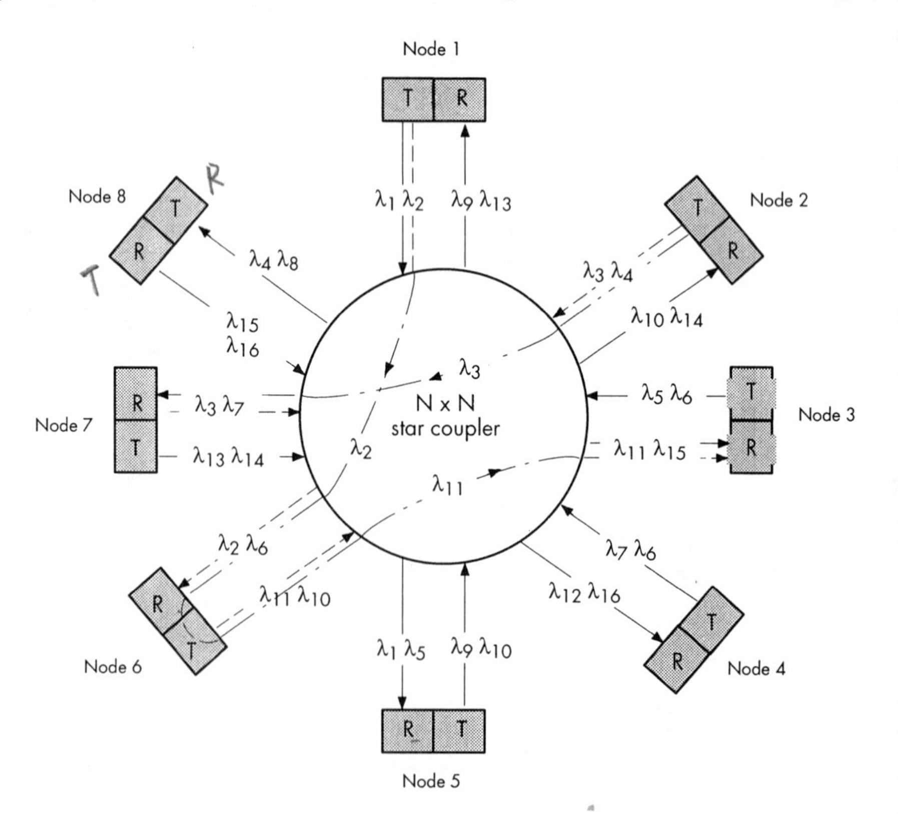

Figure 7.35 Example of an 8-node multihop WDMA star network.

fixed-tuned transmitters and two fixed-tuned receivers. We use the star topology for the purpose of illustration, although the physical topology can take a variety of forms (e.g., a bus or a tree).

The two transmit wavelengths of each node can only be received by two other nodes. In other words, although all channels at distinct wavelengths are broadcast, only fixed-assigned node-to-node connections can be supported within the network (i.e., the logical-path topology is point-to-point).

Nevertheless, since each node transmitter is connected to a different other node receiver, a connection from any given node to any other node can be achieved by allowing retransmission through one or more intermediate nodes.

To illustrate, consider Figure 7.35 and suppose node 2 has a packet destined to node 7. Node 2 can transmit on either λ_3 or λ_4. Since there is a direct connection (i.e., single-hop) from node 2 to node 7 on λ_3, node 2 will use λ_3 to transmit its packet in one hop to node 7. The decision as to which wavelength to transmit is based on a mapping of destination addresses to output ports, which can be implemented either as a hardware lookup table or as a logical circuitry executing a specific routing algorithm [Karol, 1991].

Suppose now that node 1 wishes to send a packet to node 3. Since there is no direct connection between node 1 and node 3, the packet must be relayed through one or more intermediate nodes. For instance, node 1 can transmit on wavelength λ_2 for reception at node 6, which forwards the packet on wavelength λ_{11} for eventual reception by the intended destination node 3. There is thus a two-hop path from node 1 to node 3. Note that if node 1 had used λ_1 instead of λ_2, the packet would have needed four hops to reach its final destination (i.e., node $1 \xrightarrow{\lambda 1}$ node $5 \xrightarrow{\lambda 10}$ node $2 \xrightarrow{\lambda 4}$ node $8 \xrightarrow{\lambda 15}$ node 3). Although larger numbers of hops lead to longer network delays and lower link efficiency (see Section 7.5.2), the multiplicity of potential routing paths can be exploited to route packets along less congested paths or when a network failure occurs (either node or link failures).

Another desirable feature of the multihop approach to lightwave networks is that, unlike the single-hop WDMA networks described previously, a control channel is no longer required. In fact, each network node serves as an active repeater and decides whether the received packet is intended for it or must be forwarded to another node.

7.5.2 Performance Analysis

Full connectivity between nodes in multihop lightwave networks can be achieved in different ways. The connectivity pattern can be represented by a directed graph in which arrows from each node transmitting on a particular wavelength are directed to all nodes receiving that wavelength. As an example, Figure 7.36 shows the directed graph associated with the multihop network of Figure 7.35.

For the purpose of generality, let us suppose that each network node has p fixed-tuned transmitters and p fixed-tuned receivers. Given p and some requirements on the number of hops from any given source node to any given destination node, the ideal

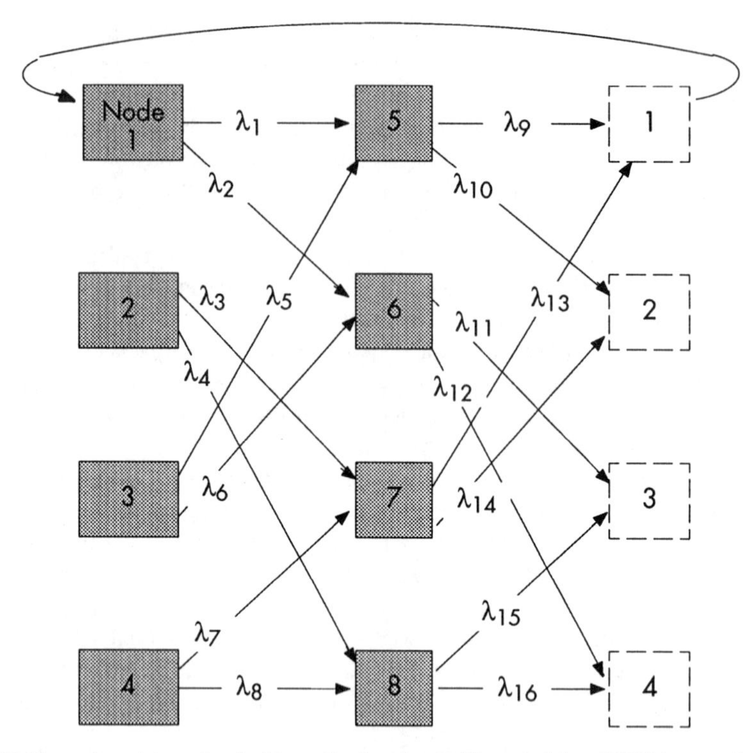

Figure 7.36 Directed graph associated with an $N = 8$ nodes shufflenet multihop WDMA network.

multihop directed-graph of degree p would maximize the number of nodes that can be connected within the network. In addition, it should be regular so that the network looks the same from any node. For such an ideal multihop lightwave network, each node would be one hop away from p nodes, which in turn would be one hop away from p other nodes, and so forth, until all nodes are reached without revisiting any node (i.e., the directed graph has the form of a p-ary spanning tree with each node appearing only once in the tree) [Acampora, 1989].

If h_{max} denotes the maximum number of hops from a source node to a destination node (h_{max} is sometimes called the diameter of the directed graph), then the following inequality is always verified

$$N \leq 1 + p + p^2 + \ldots + p^{h_{max}} = \frac{p^{h_{max}+1} - 1}{p - 1} \qquad (p \geq 2) \tag{7.27}$$

The equality in (7.27) corresponds to the *Moore bound* and represents the upper limit of the maximum number of network nodes, N_{max}, for a given diameter h_{max} (which is associated with the maximum delay a packet will experience). Unfortunately, the Moore bound cannot be realized, except for the extreme case $p = N—1$ [Bridges, 1980].

Besides this impossibility, two highly regular directed graphs have been suggested for multihop lightwave networks, namely the shufflenet graph [Hluchyj, 1988] and the deBruijn graph [Sivarajan, 1991]. For the sake of space, we will confine ourselves to a discussion of the shufflenet, while the deBruijn-graph is described in the cited reference.

The directed graph of a p shufflenet is constructed by arranging $N = kp^k$ ($k = 1, 2, \ldots$) nodes in k columns of p^k nodes each, with the last column wrapped around the first one in a cylindrical fashion. Figure 7.36 is an example of a $p = k = 2$ shufflenet directed graph, which can be viewed as the shuffling of two decks of cards (hence, the name given to these networks). This particular directed graph is also known as the perfect shuffle [Stone, 1971]. Another example is shown in Figure 7.37 and corresponds to an 18-node shufflenet with $p = 3$ and $k = 2$.

The figure shows also that the 8-node shufflenet directed graph (Figure 7.36) appears within the 18-node directed graphs. In fact, it can be shown that a (p, k) shufflenet is a subgraph of a $(p + 1, k)$ shufflenet. This property can be exploited for the modular growth of N in small increments [Acampora 1989, To 1993].

For the general (p, k) shufflenet directed graph, there are p arcs outgoing from and p arcs incoming to each node. In total, there are kp^{k+1} arcs and thus the same number of distinct wavelengths used within the network. From any node in any column (say the first column to fix the ideas), p nodes can be reached in one hop, p^2 additional nodes can be reached in two hops, and so on, until the p^k - 1 remaining nodes of the original columns are visited. From that point, any other nodes which were not visited within the first pass can be reached in a second pass (we assume a routing algorithm that minimizes the delay through the network; i.e., the number of hops [Karol, 1991]).

For the first hop in the second pass, there will be $(p^k - p)$ additional nodes that can be reached since p nodes of the second column were already visited during the first pass. Continuing on the same reasoning, the number of nodes that are h hops away from any given source node is indicated in Table 7.2.

It is seen that the maximum number of hops is limited to $h_{max} = 2k - 1$. The maximum number of nodes in a (p, k) shufflenet can therefore be expressed as

$$N_{max} = \frac{1}{2}(h_{max} + 1)\sqrt{p^{h_{max}+1}} \quad (7.28)$$

An important parameter in the performance evaluation of a (p, k) shufflenet is the mean delay experienced by a packet flowing across the network. This mean delay, $\overline{D}$, is related to the mean number of hops $\overline{h}$ by

$$\overline{D} = \overline{h} \cdot \frac{L}{v} \quad (7.29)$$

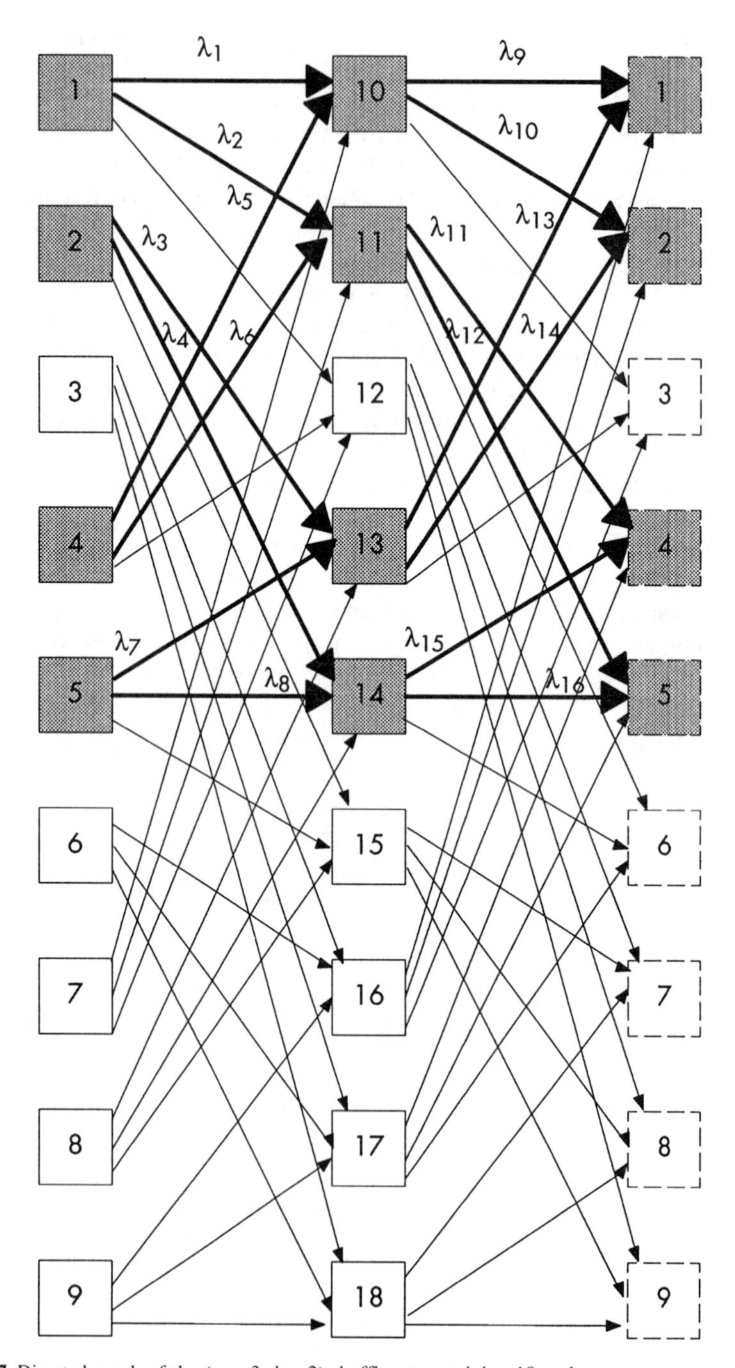

Figure 7.37 Directed graph of the ($p = 3$, $k = 2$) shufflenet containing 18 nodes.

Table 7.2
Number of Nodes That Are h Hops Away From Any Given Source Node for a (p, k) Shufflenet

Number of Hops h	*Number of Nodes h Hops From Source Node*
1	p
2	p^2
.	.
.	.
.	.
$k - 1$	p^{k-1}
k	$p^k - 1$
$k + 1$	$p^k - p$
$k + 2$	$p^k - p^2$
.	.
.	.
.	.
$2k - 1$	$p^k - p^{k-1}$

where L is the mean distance between nodes and v is the propagation speed inside the fiber $(= c/n)$. Notice that (7.29) does not include any additional delay that might arise from, for example, header processing or packet buffering in each node [Zhang, 1990].

From Table 7.2, the mean number of hops between two randomly selected nodes is given by

$$\overline{h} = \frac{1}{N - 1} \left\{ \sum_{j=1}^{K-1} jp^j + \sum_{j=0}^{k-1} (k + j)\,(p^k - p^j) \right\} \tag{7.30}$$

with $N = kp^k$.

The summations in (7.30) can be expressed in closed form with the result

$$\overline{h} = \frac{kp^k\,(3k - 1)\,(p - 1) - 2k\,(p^k - 1)}{2\,(p - 1)\,(kp^k - 1)} \tag{7.31}$$

Table 7.3 provides some representative parameters of (p, k) shufflenets and also indicates the Moore bound.

Due to multihopping, only a fraction of the link's capacity B is actually used to carry first-offered traffic (i.e., newly arriving traffic), while the remaining portion of the link's capacity is used for relayed traffic. In other words, each hop consumes a certain portion

Table 7.3
Representative Parameters of (p, k) Shufflenets

p	h_{max}	*Moore Bound* N_{max}	k	*Shufflenet* N	$\bar{h}$
2	3	15	2	8	2.0
3	3	40	2	18	2.2
2	5	63	3	24	3.3
3	5	364	3	81	3.6
4	5	1,365	3	192	3.7
6	5	9,131	3	684	3.8
8	5	37,449	3	1,536	3.9

of the total data rate of the network, which is equal to the product of the number of WDM channels (kp^{k+1}) times the bit rate per channel (B).

Assuming a uniform traffic pattern and a routing algorithm that balances the traffic load on the WDM channels, the network throughput is simply given by

$$S = \frac{kp^{k+1} \cdot B}{\bar{h}} \tag{7.32}$$

Equation (7.32) expresses that, even though there may be up to kp^{k+1} packets flowing across the network, on average only $1/\bar{h}$ of them constitute first-offered traffic.

Finally, the maximum throughput per node is obtained using (7.31) and (7.32) and is given by

$$\frac{S}{N} = \frac{2p\,(p-1)\,(kp^k - 1) \cdot B}{kp^p\,(p-1)\,(3k-1) - 2k\,(p^k - 1)} \tag{7.33}$$

Figure 7.38 shows the maximum achievable throughput per node for several values of p, obtained from (7.33). It is seen, for example, that with $N = 1{,}000$, $B = 1$ Gbps, and $p = 2$, an aggregate network throughput of about 200 Gbps with each node supporting 200 Mbps of first-offered traffic can be achieved.

The above results assume a uniform traffic pattern for which all the shufflenet links are uniformly loaded. In actual situations, however, the offered loading is expected to fluctuate and to be nonuniform. Depending on the routing algorithm used, it has been shown that the throughput per node under nonuniform traffic is reduced by a factor between 0.3 and 0.5 from that obtained for a uniform traffic pattern [Eisenberg 1988, Krishna 1990]. The use of tunable transceivers (i.e., slowly tunable lasers or tunable optical filters) allows the connectivity pattern to be changed in response to a changing traffic pattern (or network failures) [Labourdette, 1991].

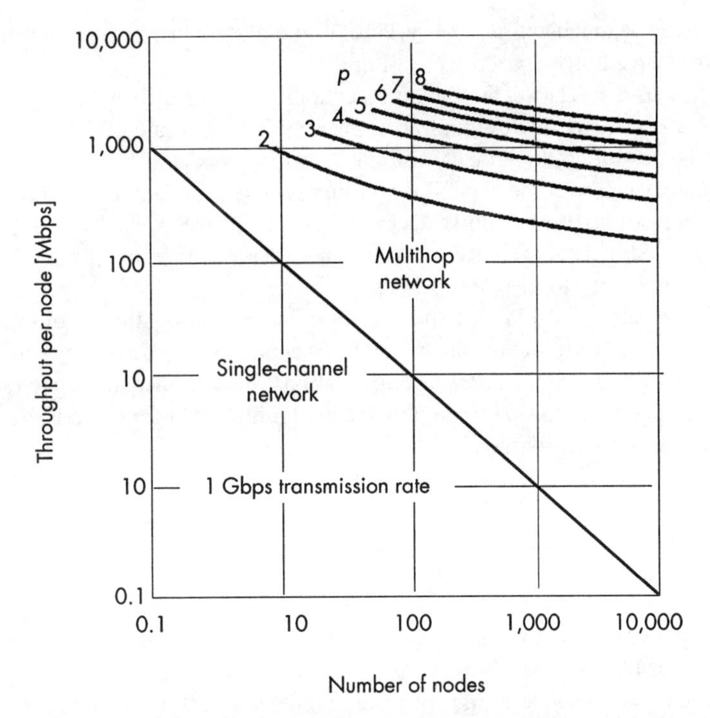

Figure 7.38 Throughput per node versus N for shufflenet WDMA networks with p transmitters/receivers at each node. (© 1989 IEEE. [After Acampora, 1989.])

Notice that the multihop approach is not restricted to the wavelength domain. For example, instead of distinct wavelengths, p separate transmit fibers and p separate receive fibers per node connected to a central patch-panel can be used. In this implementation, the patch panel, which replaces the star coupler, is used to make all the cross-connections according to the shufflenet directed graph.

Alternatively, subcarriers can be used in place of distinct wavelengths or separate fibers. An example of a multihop network that uses a subcarrier multiaccess technique will be described in Section 8.6.

7.5.3 The Teranet

Teranet is an experimental ($p = 2$, $k = 2$) shufflenet being developed at Columbia University to demonstrate 1-Gbps ATM cell switching or 1-Gbps circuit-switching with the multihop principle [Gidron, 1991]. Its connectivity graph is identical to that of Figure 7.36.

The optical transmitters are fixed-tuned DFB lasers while the optical receivers are tunable through a tunable fiber Fabry-Perot filter placed in front of each photodiode Wavelength channels are spaced by 1.5 nm.

The key to the ATM-cell switching capability of teranet is a 3×3 electronic ATM switching matrix whose input/output ports each operate at a rate of 1 Gbps. Figure 7.39 represents the functions implemented in each of the 8 nodes of teranet. The 3×3 ATM switching matrix provides the capability to route packets from any of the three input ports to one or more of its output ports according to the packet's final destination. Output buffers with a 186-packet capacity are shared between the three input ports to reduce the packet loss probability to very low values.

Besides electronic ATM switching used for packet routing, the use of tunable optical receivers allows reconfiguration of the network connectivity pattern in response to link congestion or failures. Since such reconfigurations are expected to occur rarely, transceiver tuning speed can be slow. The millisecond tuning time of tunable FPF*s* is therefore quite acceptable for such purposes.

7.5.4 The Starnet

Starnet is a coherent WDMA LAN being implemented at Stanford University [Poggiolini 1991, Hickey 1994]. The system supports both 3-Gbps circuit-switched traffic through single-hop WDMA and 100-Mbps packet-switched traffic through multihop WDMA.

Each node has a single transmitter and two receivers, of which one is tunable while the other is fixed-tuned. The system operates at a center wavelength of 1,319 nm over conventional single-mode fiber with a network diameter of 4 km and is designed to ensure a BER of 10^{-9} with a 10-dB power margin.

The initial experiment interconnected four nodes through a 4×4 star coupler. The 3-Gbps data stream is PSK modulated on the optical carrier while the 100-Mbps packet data (actually, this data is *4B/5B* encoded and uses modified FDDI hardware) is ASK modulated on the same carrier. The optical carrier spacing is 8 GHz ($\delta\lambda = 0.04$ nm), which corresponds to an extremely dense WDMA network (i.e., FDMA). Even with this very narrow channel spacing, less than a 2-dB crosstalk power penalty is foreseen.

Both receivers at each node use heterodyne coherent detection. The PSK receiver uses an ultralow linewidth thermally tunable Nd:YAG laser as the local oscillator (LO), while the ASK receiver uses the transmitter laser (also a Nd:YAG laser) as the LO.

The multihop WDMA approach is applied for the packet-switched traffic: the fixed-tuned receiver of node *i* is locked to the transmit wavelength of node (*i*—1). The receiver of the first node of the chain is locked to the transmit wavelength of the last node. In other words, a unidirectional store-and-forward packet ring network, similar to the FDDI, is formed. Notice that an alternate-routing capability, as provided with shufflenet, is not possible here.

An attractive feature of starnet resides in its novel approach to achieve relative optical-carrier stabilization of the whole network [Kazovsky, 1992]. This is accomplished

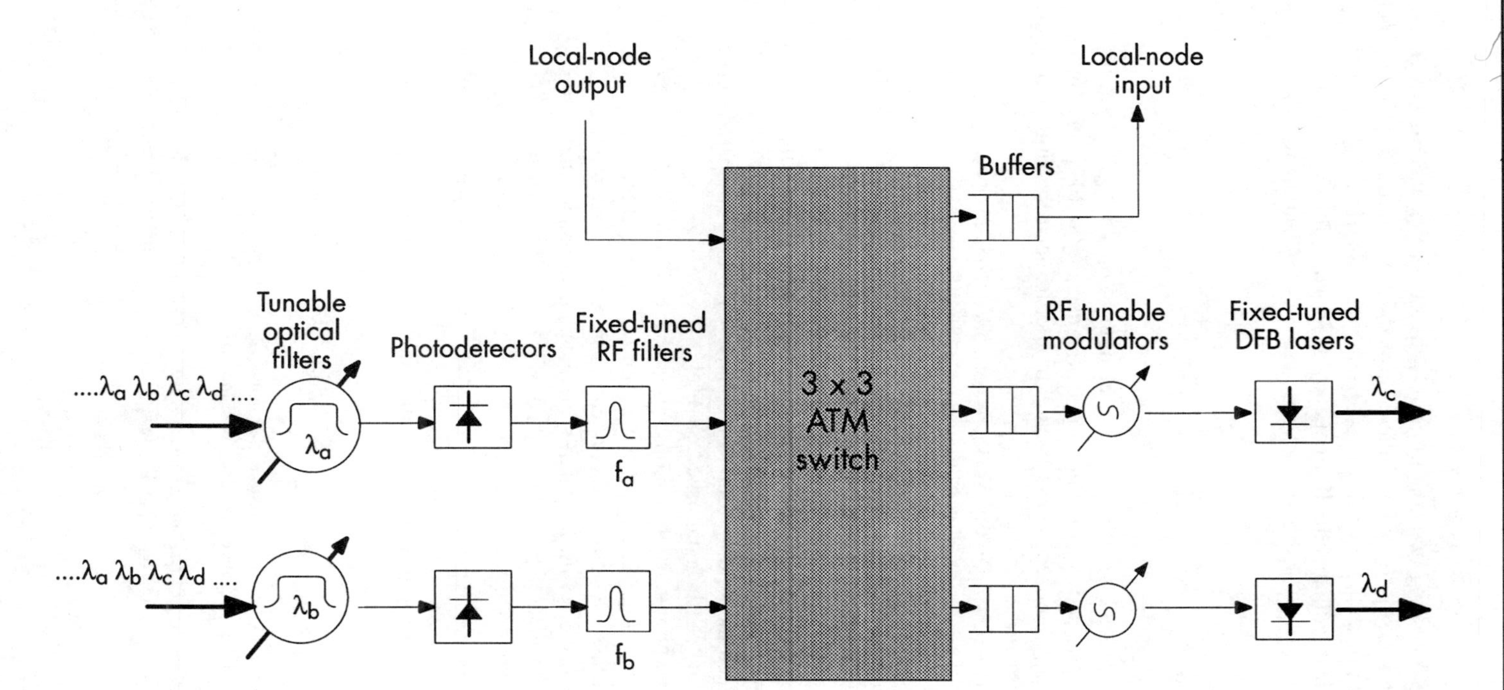

Figure 7.39 Functional block diagram of a teranet node.

by fulfilling the following two conditions: (1) the LO of the fixed-tuned receiver is also the node transmitter laser, and (2) the intermediate RF frequency of the heterodyne receiver is the same as the optical-carrier spacing (i.e., 8 GHz). Therefore, the frequency control circuit that maintains the IF frequency at 8 GHz will lock the LO on the previous node carrier, thereby establishing relative frequency stabilization among all nodes within the network.

7.6 SUMMARY

In this chapter, we have seen how the wavelength of optical signals can be exploited to construct fiber-optic networks having a potential capacity in the Tbps range.

Two distinct approaches have been explored; namely, the single-hop and the multihop WDMA networks.

The concept of single-hop WDMA networks (sometimes also referred to as all-optical networks) provides, in principle, very large network capacity by making efficient use of the wavelength domain as an additional degree of freedom to perform such functions as multiplexing, routing, and switching. Today, however, these systems suffer from technological constraints such as the limited tunability range and speed of optical transceivers. In addition, the performance of single-hop WDMA networks can be severely limited by the control protocol that manages the data flow within the network. These factors together restrict the applicability of the single-hop approach to circuit-switched networks, in which the holding time of a connection is long compared to the time during which connections can be established or reset.

Although exploratory lab demonstrations have shown the feasibility of fast packet switching using the single-hop approach, major technological breakthroughs must still be achieved before these systems find application in real-life environments.

Until transceivers with large tuning ranges and submicrosecond tuning speeds become routinely available, packet-switched WDMA networks have to be implemented through the multihop indirect approach.

It is noteworthy that, whether single-hop or multihop, WDMA networks that have been proposed so far use both optics and electronics for functions in which they perform best; that is, optics for transport and routing, and electronics for logics and memories.

REFERENCES

[Harris, 1969] Harris, S. E., et al. "Acousto-optic tunable filter," *J. Opt. Soc. America*, Vol. 59, 1969, p. 744.

[Stone, 1971] Stone, H. S. "Parallel processing with the perfect shuffle," *IEEE Trans. Computers*, Vol. C-20, No. 2, Feb. 1971, pp. 153–161.

[Kleinrock, 1975] Kleinrock, L. *Queuing Systems, Vol. I: Theory*, New York, NY: John Wiley & Sons, 1975.

[Hayes, 1978] Hayes, J. F. "An adaptive polling technique for local distribution," *IEEE Trans. Communication*, Vol. 26, 1978, pp. 1178–1186.

[Capetanakis, 1979] Capetanakis, J. I. ''Generalized TDMA : The multiaccess tree protocol,'' *IEEE Trans. Communication* , Vol. 27, 1979, pp. 1476–1484.

[Bridges, 1980] Bridges, W. G., et al. ''On the impossibility of directed Moore graphs,'' *J. Combinatorial Theory*, Series B, Vol. 29, 1980, pp. 339–341.

[Chraplyvy, 1983] Chraplyvy, A. R., et al. ''Performance degradation due to stimulated Raman scattering in wavelength division multiplexed optical fiber systems,'' *Electron. Letters*, Vol. 19, 1983, p. 641–642.

[Chraplyvy, 1984] Chraplyvy, A. R., et al. ''Carrier-included phase noise in angle modulated optical fiber systems,'' *IEEE J. Light. Technology* , Vol. LT-22, 1984, p. 6.

[Yariv, 1984] Yariv, Y., and P. Yeh. *Optical waves in crystals*, New York, NY: Wiley & Sons, 1984.

[Hill, 1985] Hill, A. M., et al. ''Linear crosstalk in wavelength division multiplexed optical fiber transmission systems,'' *IEEE J. Light Technology* , Vol. LT-3, 1985, pp. 643–651.

[Pendleton, 1985] Pendleton-Hughes, S., et al. ''Forty channel wavelength multiplexing for short-haul wideband communication networks,'' *Proc. European Conf. Opt. Communication, ECOC '85*, Venice, Italy, Oct. 1985, pp. 649–651.

[Bachus, 1985] Bachus, E. J., et al. ''Coherent optical fiber subscriber line,'' *Proc. European Conf. Opt. Communication, ECOC'85*, Vol. 3, 1985, pp. 61–64.

[Chraplyvy, 1986] Chraplyvy, A. R., et al. ''Narrowband tunable optical filter for channel selection in densely packed WDM systems,'' *Electron. Letters*, Vol. 22, 1986, pp. 1084–1085.

[Karol, 1986] Karol, M. J., et al. ''Input versus output queuing in a space-division packet switch,'' *Globecom '86, Conf. Rec.*, pp. 659–665. Also in *IEEE Trans. Communication*, Vol. COM-35, Dec. 1987, pp. 1347–1356.

[Mallison, 1987] Mallison, S. R. ''Wavelength selective filters for single mode fiber WDM systems using Fabry-Perot interferometers,'' *Applied Optics* , Vol. 26, 1987, pp. 430–436.

[Toba, 1987] Toba, H., et al. ''*5GHz*-spaced, eight-channel guided wave tunable multi/demultiplexer for optical *FDM* transmission systems,'' *Electron. Letters*, Vol. 23, 1987, pp. 788–789.

[Stone, 1987] Stone, J., et al. ''Pigtailed high-finesse tunable fiber Fabry-Perot interferometers with large, medium and small free spectral range,'' *Electron. Letters*, Vol. 23, 1987, pp. 781–783.

[Shibata, 1987] Shibata, N., et al. ''Phase-mismatch dependence of efficiency of wave generation through four-wave mixing in a single-mode optical fiber,'' *IEEE J. Quantum Electron.*, Vol. QE-23, 1987, p. 1205.

[Kobrinski, 1987] Kobrinski, H., et al. ''Demonstration of high-capacity in the *Lambdanet* architecture: a multi-wavelength optical network,'' *Electron. Letters*, Vol. 23, 1987, pp. 824–825.

[Yeh, 1987] Yeh, Y. S., et al. ''The *Knockout* switch: a single, modular architecture for high-performance packet switching,'' *IEEE J. Selected Areas of Communication*, Vol. JSAC-5, No. 8, Oct. 1987, pp. 1274–1283.

[Eng, 1987] Eng, K. Y., et al. ''A *Knockout* switch for variable-length packets,'' *IEEE J. Select Areas of Communication*, Vol. JSAC-5, Dec. 1987, pp. 1426–1435.

[Acampora, 1987] Acompora, A. S. ''A multichannel multihop local lightwave network,'' *Proc. Globecom '87*, 37. 5, Nov. 1987, pp. 1459–1467.

[Strebel, 1987] Strebel, B., et al. ''Multi-channel optical carrier frequency technique,'' Proc. *IEEE Globecom*, 1987, pp. 18. 3. 1–18. 3. 5.

[Nosu, 1987] Nosu, K., et al. ''Optical *FDM* transmission techniques,'' *IEEE J. Light. Technology*, Vol. LT-5, No. 9, 1987, pp. 1301–1308.

[Kobrinski, 1988] Kobrinski, H., et al. ''Wavelength selection with nanosecond switching times using DFB laser amplifiers,'' *Electron. Letters*, Vol. 24, 1988, pp. 969–970.

[Hluchyj, 1988] Hluchyj, M. G., et al. ''Queuing in high-performance packet switching,'' *IEEE J. Selected Area of Communication*, Vol. JSAC-6, Dec. 1988, pp. 1587–1597.

[Kaminow, 1988] Kaminow, I. P., et al. ''FDMA-*FSK* star network with a tunable optical filter demultiplexer,'' *IEEE J. Light. Technology*, Vol. LT-6, 1988, pp. 1406–1414.

[Warzanskyj, 1988] Warzanskyj, W., et al. ''Polarization-independent electro-optically tunable narrow-band wavelength filter,'' *Applied Phys. Letters*, Vol. 53, 1988, pp. 13–15.

[Fujiwara, 1988] Fujiwara, M., et al. "A coherent photonic wavelength division switching system for broadband networks," *Proc. European Conf. Opt. Communication, ECOC '88*, Brighton, England, Sept. 1988, pp. 139–141.

[Wagner, 1988] Wagner, S. S., et al. "A passive photonic loop architecture employing wavelength-division multiplexing," *Proc. Globecom*, Dec. 1988, pp. 48.1.1–48.1.5. See also "Experimental demonstration of a passive optical subscriber loop architecture," *Electron. Letters*, Vol. 3, 1988, pp. 325–326.

[Eng, 1988] Eng, K. Y. "A Photonic *Knockout* switch for high-speed packet networks," *IEEE J. Select Areas of Communication*, Vol. JSAC-6, No. 7, Aug. 1988, pp. 1107–1116.

[Yoon, 1988] Yoon, H., et al. "The *Knockout* switch under nonuniform traffic," *Proc. Globecom '88*, 49.5.1, 1988, pp. 1628–1634.

[Arthurs, 1988 a] Arthurs, E., et al. "Multi-wavelength crossconnect for parallel-processing computers," *Electron. Letters*, Vol. 24, No. 2, Jan. 1988, pp. 119–120.

[Eisenberg, 1988] Eisenberg, M., et al. "Performance of the multichannel multihop lightwave network under nonuniform traffic," *IEEE J. Selected Areas of Communication*, Vol. JSAC-6, Aug. 1988, pp. 1063–1078.

[Arthurs, 1988 b] Arthurs, E., et al. "*HYPASS*: An opto-electronic hybrid packet switching system," *IEEE J. Selected Areas of Communication*, Vol. JSAC-6, No. 9, Dec. 1988, pp. 1500–1510.

[Hluchyj, 1988] Hluchyj, M. G., et al. "Shufflenet: an application of generalized perfect shuffles to multihop lightwave networks" *Proc. IEEE Infocom*, 1988, pp. 4B4. See also IEEE J. Light. Technology, Vol. LT-9, No. 10, Oct. 1991, pp. 1386–1397.

[Kazovsky, 1988] Kazovsky, L. G., et al. *IEEE J. Light. Technology*, Vol. LT-6, 1988, p. 1353.

[Glance, 1988] Glance, B. S., et al. "Densely spaced *FDM* coherent star network with optical signals confined to equally spaced frequencies," *IEEE J. Light. Technology*, Vol. LT-6, No. 11, 1988, pp. 1770–1781.

[Wagner, 1989] Wagner, S. S., et al. "WDM applications in broadband telecommunication networks," *IEEE Communication Magazine*, March 1989, pp. 22–30.

[Goodman, 1989] Goodman, M. S. "Multiwavelength networks and new approaches to Packet Switching," *IEEE Communication Magazine*, Oct. 1989, pp. 27–35.

[Cheung, 1989] Cheung, K. W., et al. "Electronic wavelength tuning using acousto-optic tunable filter with broad continuous tuning range and narrow channel spacing," *IEEE Photon. Technology Letters*, Vol. 1, 1989, pp. 38–40. See also Cheung, K. W., et al. "Multi-channel operation of an integrated acousto-optic tunable filter," *Electron. Letters*, Vol. 25, 1989, pp. 375–376.

[Tkach, 1989] Tkach, R. W., et al. *IEEE Photon. Technology Letters*, Vol. 1, 1989, p. 111.

[Saleh, 1989] Saleh, A.A.M., et al. "Two stage Fabry-Perot filters as demultiplexers in optical *FDMA LANs*," *IEEE J. Light. Technology*, Vol. LT-7, 1989, pp. 323–330.

[Ayanoglu, 1989] Ayanoglu, E. "Signal flow graphs for path enumeration and deflection routing analysis in multihop networks," *Proc. Globecom '89*, Nov. 1989, pp. 1022–1029.

[Acampora, 1989] Acampora, A. S., et al. "An overview of lightwave packet networks," *IEEE Networks*, Jan. 1989, pp. 29–41.

[Brackett, 1990] Brackett, C. A. "Dense wavelength division multiplexing networks: principles and applications," *IEEE J. Selected Areas of Communication*, Vol. JSAC-8, No. 6, Aug. 1990, pp. 948–964.

[Ramaswami, 1990] Ramaswami, R., et al. "Tunability needed in multi-channel networks: transmitters, receivers, or both?," *IBM Research Report*, RC16237 (#72046), Oct. 1990.

[Takato, 1990] Takato, N., et al. "128-channel polarization insensitive frequency-selection-switch using high silica waveguides on *Si*," *IEEE Photon. Technology Letters*, Vol. 2, No. 6, 1990, pp. 441–443.

[Kirkby, 1990] Kirkby, P. A. "Multichannel wavelength-switched transmitters and receivers—New component concepts for broad-band networks and distributed switching systems," *IEEE J. Light. Technology*, Vol. LT-8, No. 2, Feb. 1990, pp. 202–211.

[Tada, 1990] Tada K. and Hinton H. S., *Photonic Switching II*, Part VI, Berlin, Germany: Springer Verlag, 1990, pp. 225–248.

[Stern, 1990] Stern, T. E. "Linear lightwave networks : How far can they go?," *Proc. Globecom*, 1990, pp. 1866–1872.

[IEEE, 1990] Special Issue on ''Dense Wavelength Division Multiplexing Techniques for High Capacity and Multiple Access Communication Systems,'' *IEEE J. Selected Areas of Communication*, Vol. JSAC-8, No. 6, Aug. 1990.

[Goodman, 1990] Goodman, M. S., et al. ''The *Lambdanet* multiwavelength network: architecture, applications and demonstrations,'' *IEEE J. Selected Areas of Communication*, Vol. JSAC-8, No. 6, Aug. 1990, pp. 995–1003.

[Kaminow, 1990] Kaminow, I. P. ''*FSK* with direct detection in optical multiaccess *FDM* networks,'' *IEEE J. Selected Areas of Communication*, Vol. JSAC-8, No. 6, Aug. 1990, pp. 1005–1014.

[Wagner, 1990] Wagner, S. S., et al. ''Broadband high-density WDM transmission using superluminescent diodes,'' *Electron. Letters*, Vol. 26, No. 11, May 1990, pp. 696–697.

[Hunwicks, 1990] Hunwicks, A. R., et al. ''A spectrally sliced, single-mode, optical transmission system installed in the *U. K.* local loop network,'' *Proc. Globecom*, 1990, pp. 1303–1307.

[Lee, 1990] Lee, T. T., et al. ''A broadband optical multicast switch,'' *XIII Intl. Switching Symposium, ISS'90*, Stockholm, Sweden, May 1990, pp. 7–13.

[Krishna, 1990] Krishna, A., et al. ''Performance of Shuffle-like switching networks with deflection,'' *Proc. IEEE Infocom '90*, June 1990, pp. 473–480.

[Smith, 1990] Smith, D. A., et al. ''Polarization independent electro-optically tunable filters for *WDM* networks,'' *IEEE J. Selected Areas of Communication*, Vol. JSAC-8, No. 6, 1990, pp. 1151–1159.

[Chen, 1990] Chen, M. S., et al. ''A media-access protocol for packet-switched wavelength division multi-access metropolitan area networks,'' *IEEE J. Selected Areas of Communication*, Vol. JSAC-8, Aug. 1990, pp. 1048–1057.

[Dono, 1990] Dono, N. R., et al. ''A wavelength division multiple access network for computer communication,'' *IEEE J. Selected Areas of Communication*, Vol. JSAC-8, No. 6, Aug. 1990, pp. 983–994.

[Miller, 1990] Miller, C. M., et al. ''Passively temperature-compensated fiber Fabry-Perot filter and its application in wavelength division multiple access computer networks,'' *Electron. Letters*, Vol. 26, No. 5, Dec. 1990, pp. 2122–2123.

[Zhang, 1990] Zhang, Z., et al. ''Analysis of multihop lightwave networks,'' *Proc. Globecom '90*, 903.7, Dec. 1990, pp. 1873–1879.

[Fioretti, 1990] Fioretti, A., et al. ''An evolutionary configuration for an optical coherent multichannel network,'' *Proc. Globecom '90*, Dec. 1990, pp. 779–783.

[Miller, 1990] Miller, C. M., et al. ''Passively temperature-compensated fiber Fabry-Perot filter and its application in wavelength division multiple access computer network,'' *Electron. Letters*, Vol. 26, No. 25, Dec. 1990, pp. 2122–2123.

[Chung, 1990] Chung, Y. C. ''Frequency locked 1. 3- and 1. 5- lasers for lightwave systems applications,'' *IEEE J. Light. Technology*, Vol. LT-8, No. 6, 1990, pp. 869–876.

[Shimosaka, 1990] Shimosaka, N., et al. ''Frequency separation locking and synchronisation for *FDM* optical sources using widely frequency tunable laser diodes,'' *IEEE J. Selected Areas of Communication*, Vol. JSAC-8, No. 6, Aug. 1990, pp. 1078–1086.

[Toba, 1990] Toba, H., et al. ''Factors affecting the design of optical *FDM* information distribution systems,'' *IEEE J. Selected Areas of Communication*, Vol. JSAC-8, No. 6, Aug. 1990, pp. 965–972.

[Chraplyvy, 1991] Chraplyvy, A. R. ''Nonlinear effects in Optical Fibers,'' in *Topics in Lightwave Transmission Systems*, Edited by Tingye Li, New York, NY: Academic Press, 1991, pp. 267–295.

[Bala, 1991] Bala, K., et al. ''A minimum interference routing algorithm for a linear lightwave network,'' *Proc. Globecom*, 1991, pp. 35–4. 1–4. 6.

[de Prycker, 1991] de Prycker, M. *Asynchronous Transfer Mode, solution for broadband ISDN*, Ellis Horwood, 1991.

[Karol, 1991] Karol, M. J. ''Exploiting the attenuation of fiber-optic passive taps to create large high-capacity *LAN's* and *MAN's*,'' *IEEE J. Light. Technology*, Vol. LT-9, No. 3, March 1991, pp. 400–408.

[Chapuran, 1991] Chapuran, T. E., et al. ''Broadband multichannel *WDM* transmission with superluminescent diodes and *LEDs*,'' *Proc. Globecom '91*, 1991, p. 18.4.1, pp. 612–618.

[Sivarajan, 1991] Sivarajan, K., et al. ''Multihop lightwave networks based on deBruijn graphs,'' IEEE Trans. Communication, Vol. 39, 1991. See also, Sivarajan, K., et al. ''Multihop lightwave networks based on deBruijn graphs,'' *Proc. IEEE Infocom '91*, April 1991, pp. 1001–1011.

[Todd, 1991] Todd, T. D., et al. ''Photonic multihop bus networks,'' *Proc. IEEE Infocom '91*, April 1991, pp. 981–990.

[Gidron, 1991] Gidron, R., et al. ''*Teranet*: a multihop multichannel *ATM* lightwave network,'' *Proc. Intl. Conf. Communication*, ICC '91, 20. 5, 1991, pp. 602–608.

[Labourdette, 1991] Labourdette, J-F. P., et al. ''Logically rearrangeable multihop lightwave networks,'' *IEEE Trans. Communication*, Vol. 39, No. 8, Aug. 1991, pp. 1223–1230.

[Karol, 1991] Karol, M. J., et al. ''A simple adaptive routing scheme for congestion control in shufflenet multihop lightwave networks,'' *IEEE J. Selected Areas of Communication*, Vol. JSAC-9, No. 7, Sept. 1991, pp. 1040–1051.

[Westmore, 1991] Westmore, R. J. ''*Symfonet*: interconnect technology for multinode computing,'' *Electron. Letters*, Vol. 27, No. 9, April 1991, pp. 697–698.

[Poggiolini, 1991] Poggiolini, P. T., et al. ''*Starnet*: an integrated services broadband optical network with physical star topology,'' *Proc. SPIE 1579*, 1991, pp. 14–29.

[Herrmann, 1992] Herrmann, H., et al. ''Integrated optical, *TE*- and *TM*-pass, acoustically tunable, double-stage wavelength filters in $LiNbO_3$,'' *Electron. Letters*, Vol. 28, No. 7, March 1992, pp. 642–644.

[Mukkerjee, 1992] Mukkerjee, B. ''WDM-based local lightwave networks—Part I: single-hop systems,'' *IEEE Network*, May 1992, pp. 12–27.

[Lu, 1992] Lu, J., et al. ''On the performance of wavelength division multiple access networks,'' *Proc. Intl. Conf. Communication, ICC'92*, 340.1.1, 1992, pp. 1151–1157.

[Miller, 1992] Miller, C. M., et al. ''Wavelength locked two stages fiber Fabry-Perot filter for dense wavelength division demultiplexing in an erbium-doped fiber amplifier spectrum,'' *Electron. Letters*, Vol. 28, No. 3, Jan. 1992, pp. 216–217.

[Cremer, 1992] Cremer, C., et al. ''Grating spectrograph integrated with photodiode array in *InGaAsP/InP*,'' *IEEE Photon. Technology Letters*, Vol. 4, No. 1, 1992, pp. 108–110.

[Janniello, 1992] Janniello, F. J., et al. ''A prototype circuit-switched multi-wavelength optical metropolitan-area Network,'' *Proc. Int. Conf. Communication, ICC '92*, 330.1, 1992, pp. 818–823.

[Mukherjee, 1992] Mukherjee, B. ''WDM-based local lightwave networks—Part II: Multihop systems,'' *IEEE Networks*, July 1992, pp. 20–32.

[Li, 1992] Li, B., et al. ''Virtual topologies for WDM star *LANs*—The regular structure approach,'' *Proc. IEEE Infocom '92*, May 1992, pp. 2134–2143.

[Kazovsky, 1992] Kazovsky, L. G., et al. ''WDM local area networks,'' *IEEE Light. Transmission Systems*, May 1992, pp. 8–15.

[Glance, 1992] Glance, B. S. ''Protection-Against-Collision optical packet network,'' *IEEE J. Light. Technology*, Vol. LT-10, No. 9, Sept. 1993, pp. 1323–1328. See also Karol, M. J., et al. ''Performance of the *PAC* optical packet network,'' *IEEE J. Light. Technology*, Vol. LT-11, No. 8, Aug. 1993, pp. 1394–1399.

[Wagner, 1992] Wagner, S. S., et al. ''Multiwavelength ring networks for switch consolidation and interconnection,'' *Proc. Intl. Conf. Communication, ICC'92*, 340.5, 1992, pp. 1173–1179.

[Lu, 1993] Lu, J. C., et al. ''A WDMA protocol for multichannel *DQDB* networks,'' *Proc. Globecom*, Houston, TX, Dec. 1993, pp. 149–153.

[To, 1993] To, P. P., et al. ''Multistar implementation of expandable Shufflenets,'' *Proc. Globecom '93*, Dec. 1993, pp. 468–473.

[Ottolenghi, 1993] Ottolenghi, P., et al. ''All-optical conversion with extinction ratio enhancement using a tunable *DBR* laser,'' *Proc. European Conf. Communication, ECOC'93*, TuC 5.5, Sept. 1993, pp. 141–144.

[Green, 1993] Green, P. *Fiber Optic Networks*, New York, NY: Prentice Hall, 1993.

[Lach, 1993] Lach, E., et al. ''*5Gb/s* wavelength conversion with simultaneous regeneration of extinction ratio using *Y*-lasers,'' *Proc. European Conf. Communication, ECOC'93*, TuC 5.5, Sept. 1993, pp. 137–140.

[Elrefaie, 1993] Elrefaie, A. F. ''Multiwavelength survivable ring network architectures,'' *Proc. Intl. Conf. Communication, ICC'93*, 1993, pp. 1245–1251.

[Chiaroni, 1993] Chiaroni, D., et al. ''Feasibility demonstration of a 2. 5*Gbit/s* 16x16 *ATM* Photonic Switching matrix,'' *Proc. Opt. Fiber Communication Conf., OFC'93*, WD2, Feb. 1993, pp. 93–94. See also, Chiaroni, D., et al. ''Rack-mounted 2.5*Gbit/s ATM* photonic switch demonstrator,'' *Proc. European Conf. Communication, ECOC'93*, ThP12.7, Sept. 1993, pp. 77–80.

[Gavinet-Morin, 1993] Gavinet-Morin, P., et al. ''Multiwavelength optical buffer based on fiber delay lines for gigabit packet switching,'' *Proc. Opt. Fiber Communication Conf., OFC'93*, WJ1, Feb. 1993, pp. 145–146.

[Durhuus, 1993] Durhuus, T., et al. ''Optical wavelength conversion and gating at 4*Gbit/s* using semiconductor optical amplifiers,'' *Proc. Opt. Fiber Communication Conf., OFC'93*, TuH6, Feb. 1993.

[Blair, 1993] Blair, L. T., et al. ''Impact of new optical technology on spectrally-sliced access and data networks,'' *British Telecom Technology J.*, Vol. 11, No. 2, April 1993, pp. 46–55.

[Jacob, 1993] Jacob, J. B., et al. ''Photonic broadband switching systems: count down to the new millenium,'' *Proc. 11th Conf. European Fiber Optic Communications and Networks*, The Hague, The Netherlands, June 1993.

[Stubkjaer, 1993] Stubkjaer, K., et al. ''Semiconductor optical amplifiers as linear amplifiers, gates and wavelength converters,'' *Proc. 19th European Conf. Optical Communication, ECOC'93*, Montreux (Switzerland), Sept. 1993, pp. 60–67.

[Acampora, 1994] Acampora, A. S. *An Introduction to BroadBand Networks*, Ch. 3, New York, NY: Plenum Press, 1994.

[Gabriagues, 1994] Gabriagues, J. M., et al. ''Design, modelling and implementation of the *ATMOS* project fiber delay line photonic switching matrix,'' *Optical and Quantum Electronics*, to appear, 1994.

[Chiaroni, 1994] Chiaroni, D., et al. ''Wavelength channel selector with subnanometric resolution and subnanosecond switching time,'' submitted to *20th European Conf. Optical Communication, ECOC'94*, Firenze (Italy), Sept. 1994.

[Hickey, 1994] Hickey, M., et al. ''The Starnet coherent WDM computer communication network: experimental transceiver employing a novel modulation format,'' *IEEE J. Light, Technology*, Vol. LT-12, No. 5, May 1994, pp. 876–884.

Chapter 8
Subcarrier Division Multiaccess Networks

Much of the technology for subcarrier optical-fiber communication systems is borrowed from microwave communication technology where multiple microwave carriers are employed for the transmission of a number of (electrically frequency multiplexed) channels over coaxial cables or free space. The name subcarrier comes from the fact that when the modulated channels are transmitted optically, the microwave carriers (10^7 to 10^{10} Hz) act as subcarriers for the optical carrier (10^{14} Hz).

There has been a growing interest recently in subcarrier multiplexing (SCM) for the distribution of analog television signals over passive optical networks [Olshansky 1989, Way 1989, IEEE 1990, Hill 1990, Gabla 1992].

The SCM technique is illustrated in Figure 8.1. The information from each channel modulates a dedicated subcarrier and all modulated subcarriers are summed using an RF/microwave power combiner. The composite signal is then used to intensity-modulate an optical carrier. The optical signal is transmitted over a point-to-point or point-to-multipoint fiber-optic transport system (possibly containing optical amplifiers) and received by a high-bandwidth photodetector, which reconverts the optical signal into electrical form. The desired channel at the receiver is selected with an RF/microwave bandpass filter or an electrical heterodyne receiver in a similar way as in radio tuners. The independence of each subcarrier means that both analog and digital signals can be simultaneously transmitted.

The basic concept behind SCM can be exploited as a means for multiple access by partitioning the subcarriers among a number of transmit/receive nodes [Darcie, 1986]. This results in the so-called subcarrier multiple access (SCMA) technique, which is the subject of the present chapter.

SCMA can be classified into two distinct categories: *single-channel* SCMA, in which a unique subcarrier channel is assigned to each optical transmitter, and *multichannel* SCMA, in which multiple subcarrier channels are combined together in the RF/microwave domain before driving an optical transmitter.

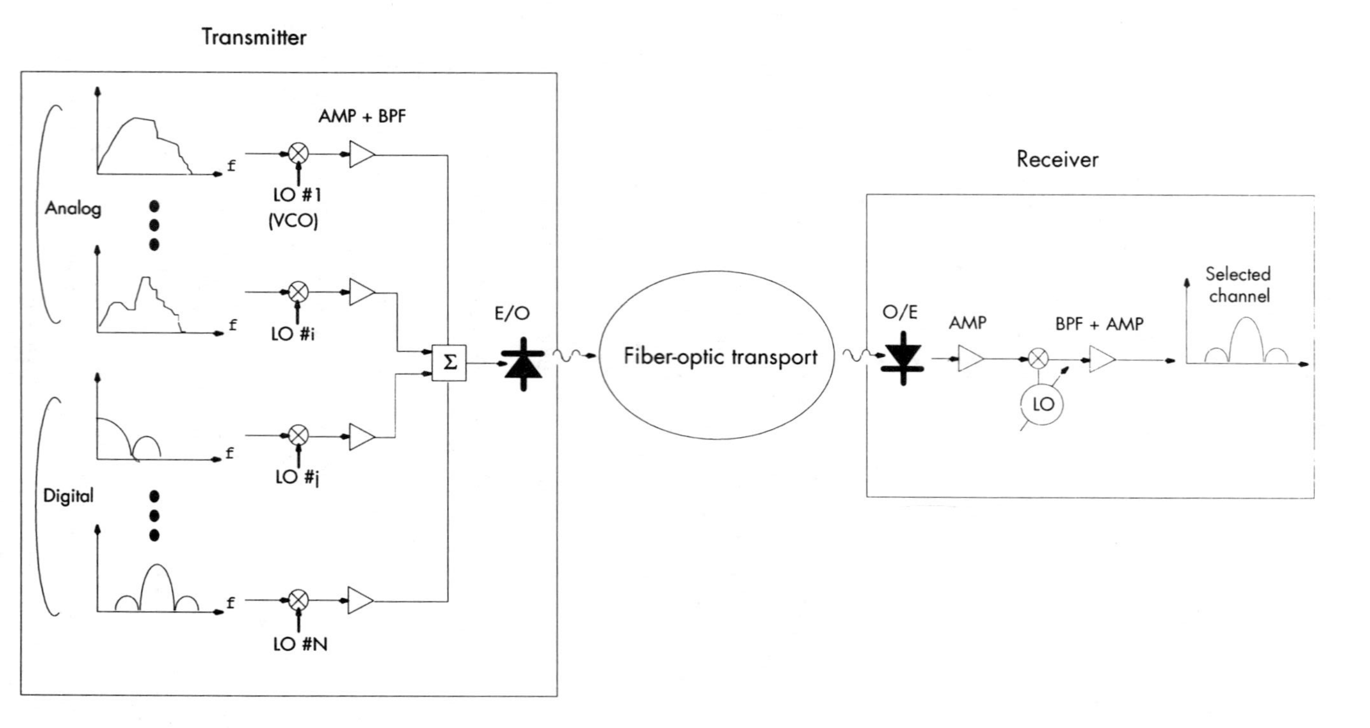

Figure 8.1 Basic SCM system configuration.

In Section 8.1 we introduce the basic concept of SCMA with focus on single-channel SCMA. In Section 8.2, the network performances achievable using single-channel SCMA technique will be analyzed. The physical mechanisms that can lead to the limitation of network capacity include shot noise, receiver electronic noise, laser-intensity fluctuations, and optical-beat interference effect; each will be discussed in separate Sections 8.2.1 to 8.2.4. Experimental single-channel SCMA systems are described in Section 8.3. Section 8.4 is devoted to multichannel SCMA, which is also called *hybrid* SCMA/WDMA. Issues related to system performance will be discussed in Section 8.5; they include intermodulation product distortions and clipping and will be analyzed separately in Sections 8.5.1 and 8.5.2. Section 8.6 describes two multichannel SCMA systems that have been prototyped in research laboratories. Finally, Section 8.7 summarizes the main results obtained in this chapter.

8.1 BASIC CONCEPT

Consider Figure 8.2 showing an $M \times M$ physical-star network that uses SCMA as a means for multiple access (other physical-network topologies may be envisaged as well). The data from each node is used to modulate a unique RF/microwave subcarrier frequency, which is then used to modulate the intensity of a semiconductor light source (a laser or a light-emitting diode, LED) directly by adding it to the bias current. The optical channels from all nodes are combined and broadcast to each node by the $M \times M$ star coupler. At the receiver, channel selection is accomplished with an RF/microwave bandpass filter or an electrical heterodyne receiver.

The transmitter or receiver RF/microwave local oscillators (or both) can be made tunable, in which case a dynamic multiple access capability is provided.

SCMA with single-channel per subcarrier and per light source (which we shall call single-channel SCMA, as already mentioned in the introduction) is thus very similar to WDMA, except that with single-channel SCMA a connection is set up in the microwave subcarrier frequency domain, while with WDMA it is performed in the wavelength (optical-frequency) domain. Although SCMA and WDMA are very similar in their principles, they differ in many respects, essentially with regard to the required technology.

First, SCMA uses RF/microwave technology that has already reached maturity and low cost levels.

Second, the stability of RF/microwave sources and filters allows for tighter channel spacing and efficient utilization of electrical bandwidth.

Third, RF/microwave filters or heterodyne receivers can be tuned much faster than today's commercially available tunable optical filters.

Finally, signals on RF/microwave carriers can easily be processed using electronic components. This can be exploited to implement highly efficient modulation formats (i.e., having more than one bit per hertz efficiency, such as quadrature amplitude modulation,

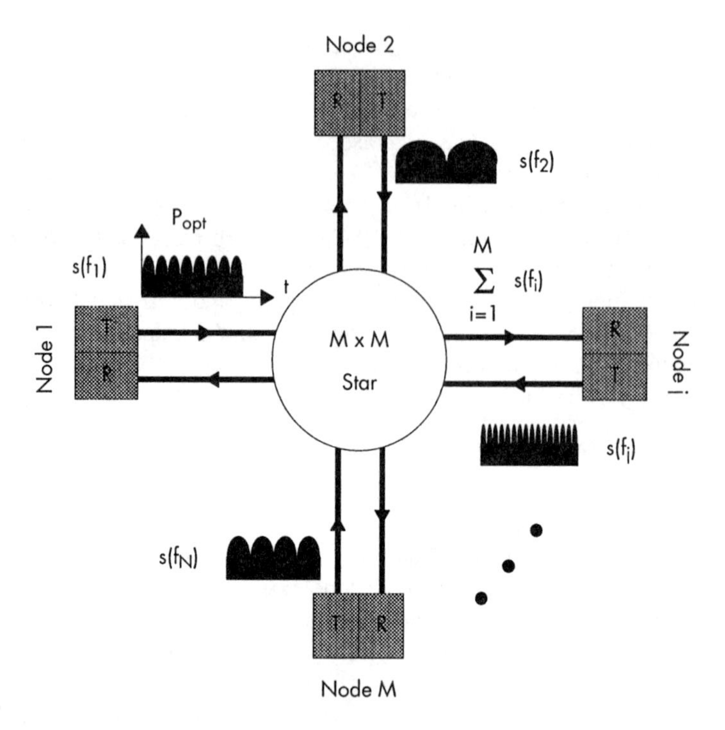

Figure 8.2 SCMA over a physical-star network topology. $s(f_i)$ represents the modulated optical signal using f_i as the subcarrier frequency.

QAM, or quadrature phase shift keying, QPSK), which are much more difficult to achieve at optical frequencies [Runge 1989, Otsuka 1991].

Besides these advantages, the subcarrier frequencies in single-channel SCMA networks must be distinct for each active connection on the network to avoid collisions in the electrical domain. This means that, in contrast to WDMA, SCMA requires both wideband optical transmitters and receivers that cover the full subcarrier frequency range used within the network. Since the modulation bandwidths of semiconductor light sources are limited to about 15–25 GHz, there will exist a maximum usable subcarrier frequency and hence a limitation on the number of channels (or nodes). This fact may turn the balance back in favor of WDMA if maximum network capacity is the objective.

Nevertheless, although the photodetectors detect all subcarrier channels over the total system bandwidth, only the desired channel(s) has to be amplified and demodulated, so that the receiver sensitivity is determined by the bandwidth of this channel only.

In order to quantify the network performance that can be obtained with single-channel SCMA, we must analyze the different impairments that arise with this multiple-access technique. This is a subject to which we will look in the next section.

8.2 SINGLE-CHANNEL SCMA: NETWORK PERFORMANCE

SCMA network performances can be analyzed in terms of the signal-to-noise ratio (SNR), defined as the ratio of root-mean square (rms) subcarrier power to rms noise power at the output of the photodetector. In dealing with RF/microwave subcarriers, the carrier-to-noise ratio (CNR) is often used in place of SNR. SNR and CNR may differ for analog-modulation formats, but are the same for digital signals [Way, 1989]. Since we consider only digital signals, we shall use the term SNR to keep consistency with other chapters.

Let us assume, for the moment, light sources with linear *L-I* curve (Figure 8.3). Then the optical output power from node i can be written as

$$P_i(t) = P_{\text{bi}}\{1 + m_i a_i \cos(2\pi f_i t + \phi_i)\} \tag{8.1}$$

where P_{bi} is the optical-output power at the bias current, m_i is the optical-modulation index defined as $(P_i^{\max} - P_{\text{bi}}) / P_{\text{bi}}$, and a_i (0 or 1), f_i, ϕ_i are the (normalized) amplitude, the frequency, and the phase of the subcarrier associated with node i, respectively.

The values a_i, f_i, ϕ_i or a combination of them is modulated to impose the data, depending on whether amplitude modulation (ASK), frequency modulation (FSK), phase modulation (PSK), or other well-known modulation formats are used [Sklar 1988, Gitlin 1992].

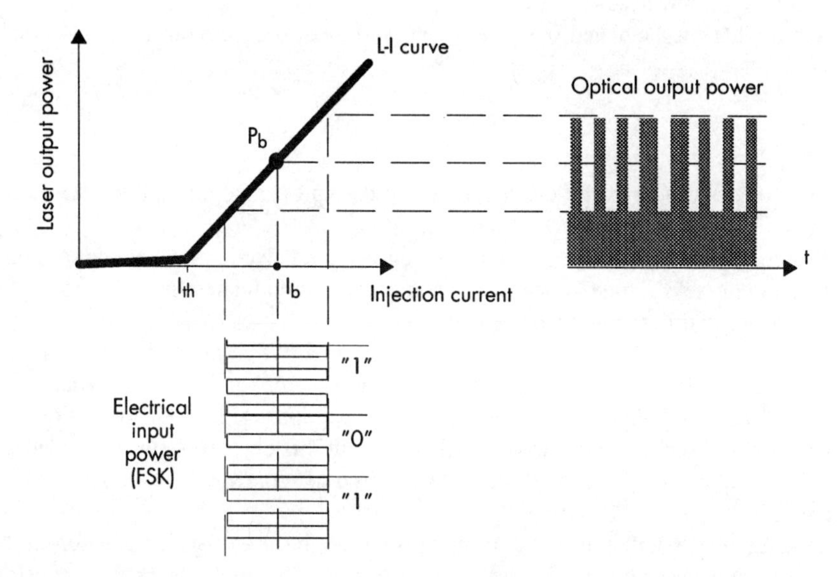

Figure 8.3 Subcarrier modulation for semiconductor laser. The laser is biased at the bias point (I_b, P_b). The modulated subcarrier (in the figure, FSK modulation format is illustrated) is added to the bias current I_b and owing to the linear nature of the *L-I* curve an optical replica of the electrical signal is produced.

Most single-channel SCMA experiments to date have used either FSK or QPSK [Darcie 1989, Mestdagh 1991, Bates 1991]. The frequency deviation of voltage-controlled oscillators (VCOs) that are commercially available today is only about 100 MHz, and FSK modulation using VCO is therefore limited to data rates ≤ 100 Mbps. At higher data rates, QPSK can be used but the corresponding receiver is likely to be more complex than the FSK receivers. For very high channel bandwidths, the simple ASK modulation format can be chosen, but this requires a 3-dB increase in SNR compared with FSK to achieve the same BER [Proakis, 1989].

In practice, the choice of modulation format will result from compromises between many factors, such as the allowable RF/microwave circuit complexity and its related cost and reliability, the required modulation efficiency (spectral occupancy), as well as network-related parameters (e.g., bit rate per node, network size, etc.).

It is noteworthy that nodes connected to the network can use different modulation formats and data rates since the subcarriers are all distinct and independent of each other. This may, however, make the receiver very complex if it has to be data-rate and data-format adjustable.

Although SCMA can be realized through a variety of physical-network topologies, we shall concentrate on the star topology as shown in Figure 8.2. The advantages of star networks have been highlighted in Chapter 6. Alternative physical-network topologies can be treated with the same considerations presented here.

Assuming a linear communication channel, the power reaching each node receiver, P_{rec}, is related to the launched power P_i (8.1) (assumed to be the same for all nodes) by

$$P_{\text{rec}} = L \cdot \sum_{i=1}^{M} P_i \tag{8.2}$$

where L is the loss incurred by the signals when they propagate through the star network (also assumed to be the same for all nodes).

The receiver at each node converts the aggregate received optical power P_{rec} into an electrical signal, selects one subcarrier channel, and demodulates the selected signal to recover the original data intended for it.

At the input of the demodulator, the SNR must be high enough to achieve a low BER. The required SNR depends on the modulation format and the demodulation technique [Sklar 1988, Proakis 1989, Gitlin 1992]. For example, with FSK modulation format and synchronous demodulation, which will be used in this chapter to illustrate theoretical results, SNR values of 15.6 and 18.0 dB are required to achieve a BER of 10^{-9} and 10^{-15}, respectively.

Several effects will contribute to the noise at the input of the demodulator and will degrade the SNR accordingly. Essentially four of these impairments must be considered for single-channel SCMA: (1) *Shot noise,* which determines the ultimate limit of SCMA systems; (2) *Receiver electronic noise (thermal noise)*; (3) *Laser-intensity noise,* expressed by the RIN parameter defined in Chapter 2; and (4) *optical beat interference* (OBI) noise, which results from the beating between the optical channels simultaneously received by the photodetector.

Additionally, other sources of noise such as harmonic distortion, subcarrier intermodulation, and clipping might contribute to the degradation of the SNR. These noise phenomena arise because of a departure from linearity of the light sources. Harmonic distortion in single-channel SCMA can be avoided by restricting the subcarrier frequency range to within one octave of bandwidth (from f_{min} to $f_{max} \leq 2f_{min}$). Intermodulation distortion arises only when more than one subcarrier is transmitted by the same light source and will be considered in Section 8.4 when we will deal with multichannel SCMA. Clipping results from the threshold characteristic of semiconductor lasers and from the fact that optical power cannot be negative. As will be explained in Section 8.5.2, clipping leads to a limitation of the modulation index per subcarrier, resulting in a reduction of the optical power budget. For single-channel SCMA, this effect is negligible provided that the modulation index $m \leq 1$.

In the following sections, we shall consider the four noise sources for the single-channel SCMA separately, assuming a PIN photodetector, and conclude by computing the overall SNR. Extension of the analysis presented here to the case of APD photodetectors can easily be done using the results of Chapter 4.

8.2.1 Shot Noise

Shot noise has been analyzed at length in Chapter 4, and (4.23) is directly applicable here provided that the mean photocurrent I_0 is replaced by

$$I_0^{\text{Tot}} = \Re \cdot L \cdot \sum_{i=1}^{M} P_{\text{bi}} \tag{8.3}$$

where $\Re$ is the responsivity of the photodiode, L is the network loss and P_{bi} is the optical output power at the bias current. Indeed, the time average of the sum in (8.2) is equal to $\sum_{i=1}^{M} LP_{\text{bi}}$ resulting in a total average photocurrent given by (8.3).

From (8.1), the rms signal current associated with channel i is given by

$$i_{\text{rms}} = \sqrt{<i_i^2>} = \Re \cdot L \cdot \frac{m_i P_{\text{bi}}}{\sqrt{2}} \tag{8.4}$$

Using (8.3), (8.4), and (4.23), one can express the ratio of subcarrier-signal power to shot-noise power, $\text{SNR}_{\text{shot,}}$ for channel i as

$$\mathrm{SNR_{shot}} = \frac{\Re \cdot (m_i P_{\mathrm{bi}})^2}{4e \cdot (\sum_{i=1}^{M} P_{\mathrm{bi}}) \cdot B_i} \cdot L \tag{8.5}$$

where B_i is the RF/microwave receiver filter bandwidth associated with channel i.

Equation (8.5) expresses the ultimate shot-noise limit SNR [Darcie, 1987]. To see clearly how this limit determines the maximum network capacity, we shall simplify (8.5) by assuming identical bias optical power, modulation index, and channel bandwidth for each node on the network; that is, $P_{\mathrm{bi}} = P_b$, $m_i = m$ and $B_i = B$.

Rearranging (8.5), we therefore obtain

$$M \cdot B = \frac{\Re m^2 P_b}{4e \cdot \mathrm{SNR_{shot}}} \cdot L \tag{8.6}$$

Equation (8.6) gives the shot-noise-limited network capacity (or the total usable bandwidth) of single-channel SCMA. Using (6.13) expressing L for the star network, we finally have

$$M \cdot B = \frac{\Re m^2 P_b}{4e \cdot \mathrm{SNR_{shot}}} \cdot L_k \cdot \frac{\beta^{\log_2 M}}{M} \tag{8.7}$$

where L_k includes transmission losses such as input and output coupling losses, intrinsic fiber losses, and splice and connector losses.

Table 8.1 lists the shot noise capacity limit calculated using (8.7) with $\mathrm{SNR_{shot}} = 16$ dB, $P_b = 1$ mW, $\Re = 0.8$ A/W, $m = 0.7$, $L_k = 5$ dB, and $\beta = 0.2$ dB. Given that the bandwidth per channel is approximately twice the bit rate for FSK modulation format, this type of network could support, for example, over 200 nodes at data rates of 40 Mbps per node, providing a network capacity of 8 Gbps in the shot-noise limit.

Although the allowed bit rates or the number of nodes can be increased by increasing the laser power or the modulation index, the resulting network capacity will always be far lower than those that could be obtained with the WDMA access scheme. Thus, where maximum network capacity is the objective, single-channel SCMA is certainly not the best multiple access scheme.

However, as we saw in Chapter 7, network capacities potentially achievable with WDMA cannot at present be reached because of the lack of suitable optical components. By contrast, SCMA can be implemented with today's commercially available RF/microwave components, but the achievable network capacity is rather limited.

The ultimate shot-noise-capacity limit is expressed by (8.6), but other noise sources that will be discussed in the following sections will further limit the network capacity that can be obtained in practice.

Table 8.1
Single-Channel SCMA Star Network; Shot-Noise Limit

Network size M	*Network Capacity* $M \cdot B$ *[GHz]*	*Channel Bandwidth* B *[MHz]*
200	17.1	85.5
500	6.4	12.8
1,000	3.1	3.1
2,000	1.5	0.75

It turns out that by combining SCMA with WDMA, the advantages of both multiple-access schemes can be exploited while avoiding both the inherent limitations of SCMA and the current technological bottleneck of WDMA. The resulting hybrid SCMA/WDMA multiple-access scheme has been proposed and demonstrated [Shankaranarayanan 1991, Ramaswami 1993, Choy 1993], and will be discussed in Sections 8.4 and 8.5.

8.2.2 Electronic Noise

Under most operating conditions met in practice, the dominant noise source is likely to be electronic noise (i.e., thermal noise) developed at the receiver front-end. As we saw in Chapter 4, the electronic-noise current density increases rapidly as the bandwidth increases. For PIN-FET receivers, the mean power density of this noise is given by (4.34) which is recalled here for convenience (the term in K_1 can be neglected here if we assume a large total bandwidth):

$$<i_{\text{el}}^2> = \frac{4F_n kTB_{\text{Tot}}}{R_L} \cong K_2 B_{\text{Tot}}^2 + K_3 B_{\text{Tot}}^3 \tag{8.8}$$

where T is the absolute temperature, k is the Boltzman's constant, F_n is the amplifier noise figure, R_L is the load resistance, K_2 and K_3 are constant factors that depend on the receiver circuit design and technology, and B_{Tot} is the total receiver bandwidth. The quadratic term in (8.8) is mainly determined by the transimpedance resistance, while the cubic term mainly shows the influence of the FET channel-conductance noise [Hein, 1990].

Assuming the same bandwidth, B, for each of the M channels, we have $B_{\text{Tot}} = MB$ (note that we do not restrict the total bandwidth to within one octave), and (8.4) and (8.8) provide

$$\text{SNR}_{\text{el}} = \frac{(\Re LmP_b)^2}{2[K_2 + K_3 MB] \cdot M^2 B^2} \tag{8.9}$$

It is clear from (8.9) that with an increasing number of channels, the maximum channel bandwidth decreases rapidly. For large product MB, the second term in the bracket of the denominator of (8.8) dominates over the first term and together with (8.7) and (6.13), the SNR_{el} becomes

$$SNR_{el} = \frac{(\Re m P_b)^2}{2K_3} \cdot L_k^2 \cdot \frac{\beta^{2 \log_2 M}}{M^5 B^3} \tag{8.10}$$

where L_k is the link loss including all losses except splitting loss.

PIN-HEMT receivers with an equivalent input noise of 8 $pA/\sqrt{\mathrm{Hz}}$ (= $\sqrt{<i_{el}^2>}$) over the 0 to 14-GHz frequency range have been reported [Ohkawa, 1988]. Using this value in (8.8) with B = 80 MHz and M = 175 ($M{\cdot}B$ = 14 GHz), we obtain $SNR_{el} \cong 96$ dB.

Thus, for this particular example the shot noise dominates over the electronic noise by several orders of magnitude.

Note that using commercially available receivers, the shot-noise limit will, in general, never be reached. In Section 8.2.5, we shall calculate the overall SNR and a realistic example will show that electronic noise becomes the dominant impairment for large network capacities.

8.2.3 Laser-Intensity Noise

As we saw in Chapter 2, light emitted by any semiconductor laser or LED exhibits power fluctuations, referred to as laser-intensity noise and captured by the RIN parameter [Cunningham 1989, Way 1989, Blauvelt 1992].

Assuming that the RIN is uniform within the receiver bandwidth, $B_{Tot} = MB$, the rms noise power at the receiver induced by the laser-intensity fluctuations is given by

$$\sigma_{RIN}^2 = RIN \cdot (\Re L P_b)^2 \cdot MB \tag{8.11}$$

so that, with (8.4), the SNR due to laser-intensity fluctuations only is given by

$$SNR_{RIN} = \frac{m^2}{2 \cdot RIN \cdot M \cdot B} \tag{8.12}$$

RIN typically runs between -130 dB/Hz to -160 dB/Hz, depending on the type of laser and how well it is isolated from reflections. To illustrate (8.12), consider the following representative values: B = 80 MHz, M = 200, m = 0.7, and RIN = -130 dB/Hz. Then, (8.12) gives $SNR_{RIN} \cong 21.8$ dB. Comparing (8.12) with (8.6), it is seen that SNR_{RIN} is inversely proportional to M while SNR_{Shot} is inversely proportional to M^2. Keeping the same parameters as those used in the above example, the condition $SNR_{RIN} = SNR_{shot}$ is obtained for $M = 60$, above which shot noise dominates over the RIN.

8.2.4 Optical Beat Interference (OBI) Noise

The presence of M simultaneous optical carriers at the square-law photodetector results in a photocurrent containing beat notes (cross-mixing terms) at the difference frequencies corresponding to each pair of optical fields. If some beat-note frequencies overlap an active subcarrier channel, the SNR of that channel is degraded. The SNR degradation will depend on the type of light source used within the network (incoherent LEDs, Fabry-Perot or narrow linewidth laser diodes (e.g., DFB), as well as on the signal modulation bandwidth).

Optical-mixing effects within photodetectors have already been analyzed in Chapter 5 when we considered the combination of optical amplifiers and photodiodes. The light sources were assumed monochromatic and beat noise appeared in the photocurrent because of the mixing of the optical signal with the spontaneous emission generated by the optical amplifier. Here, the analysis must take into account the linewidth of the light sources because it is precisely the beating of all optical-frequency components of the light sources that will determine the system performance degradation.

The OBI effect in SCMA systems has been analyzed in [Desem 1988, 1990, Shankaranarayanan 1991 a, b]. The treatment outlined here follows that of [Desem, 1990].

Consider, for the moment, the simple case where two optical signals $S_1(t)$ and $S_2(t)$ with the same polarization states are detected by a single photodetector. The combined electric field $E(t)$ incident on the photodetector can be written as

$$E(t) = \sqrt{P_1(t)} \cdot E_1(t) + \sqrt{P_2(t)} \cdot E_2(t) \tag{8.13}$$

where the $P_i(t)$ is given by (8.1) and the $E_i(t)$ represents the normalized electric fields associated with the two optical carriers.

The photocurrent generated by the photodetector is given by (4.57) and following (8.13) it can be expressed as

$$i(t) = \Re\{P_1(t) + P_2(t) + 2\sqrt{P_1(t)P_2(t)} \cdot \cos(2\pi\delta\nu t + \varphi_2(t) - \varphi_1(t))\} \tag{8.14}$$

where $\delta\nu$ is the difference of the center optical frequency of the two light sources, and $\varphi_i(t)$ are the phases of the optical fields.

The first two terms in (8.14) are the modulated subcarrier signals and are assumed to occupy different frequency bands. The third term in (8.14) results from the mixing process in the photodetector and represents interference noise if it lies within the bandwidth B of the desired signal. In order to determine the amount by which this interference term degrades the SNR, it is necessary to determine the frequency spectrum of the interference term in (8.14).

It can be shown that this spectrum is essentially determined by the convolution of the frequency spectra of the two light sources, provided that B and the subcarrier frequencies are much smaller than the source linewidths [Nazarathy, 1989]. This interference spectrum is centered at $\delta\nu$; that is, the center frequency difference of the two light sources.

Let us first consider single-longitudinal-mode semiconductor lasers (SLM lasers) with Lorentzian line shape given by [Yariv, 1989]

$$g(\nu) = \frac{\Delta\nu_{\mathrm{FWHM}}}{2\pi\left[(\nu - \nu_0)^2 + \left(\frac{\Delta\nu_{\mathrm{FWHM}}}{2}\right)^2\right]} \tag{8.15}$$

where ν_o is the center optical frequency and $\Delta\nu_{\mathrm{FWHM}}$ is the full width at half-maximum (FWHM) of the power spectrum. It can easily be verified that $g(\nu)$ is normalized such that $\int_{-\infty}^{+\infty} g(\nu)d\nu = 1$.

Following [Nazarathy, 1989], the normalized power spectrum of the interference term in (8.14) is given by

$$g_{\mathrm{int}}(f) = \frac{\Delta\nu_{\mathrm{FWHM}}}{\pi[(f - \delta\nu)^2 + \Delta\nu_{\mathrm{FWHM}}^2]} \tag{8.16}$$

which also has a Lorentzian line shape but with FWHM = $2\Delta\nu_{\mathrm{FWHM}}$.

Assuming $f_i \ll \Delta\nu_{\mathrm{FWHM}}$ (this is a reasonable assumption provided that the linewidth of the SLM laser diode is significantly broadened by the modulating signal), the amount of interference power spectrum that falls within the signal bandwidth can be approximated by the dc component of (8.16) irrespective of $\delta\nu$, as shown in Figure 8.4(b, c).

The signal power is obtained from (8.4). Using (8.14) and (8.15) with $f = 0$ (i.e., the dc component of the interference spectrum), it can easily be shown that the signal-to-OBI-noise ratio, SNR_{OBI}, is given as

$$\mathrm{SNR_{OBI}}\,(M = 2, \delta\nu) = \frac{\pi}{8} m^2 \frac{\Delta\nu_{\mathrm{FWHM}}}{B} \cdot \left[1 + \left(\frac{\delta\nu}{\Delta\nu_{\mathrm{FWHM}}}\right)^2\right] \tag{8.17}$$

Figure 8.5 plots the $\mathrm{SNR_{OBI}}$ as a function of $\delta\lambda = -\lambda^2\delta\nu/c$ for various linewidths $\Delta\nu_{FWHM}$ and for $B = 5$ MHz. Laser sources with Gaussian line shape are also shown for comparison.

For a small wavelength difference between the two sources, $g_{\mathrm{int}}(f)$ is essentially centered at *dc* and, therefore, a large amount of $g_{\mathrm{int}}(f)$ falls within the bandwidth of the desired channel. This gives rise to the low values of $\mathrm{SNR_{OBI}}$.

As $\delta\nu$ increases, the $\mathrm{SNR_{OBI}}$ improves because the center of $g_{\mathrm{int}}(f)$ moves away from dc, and the portion of the tail of $g_{\mathrm{int}}(f)$ that falls within B around f_i reduces.

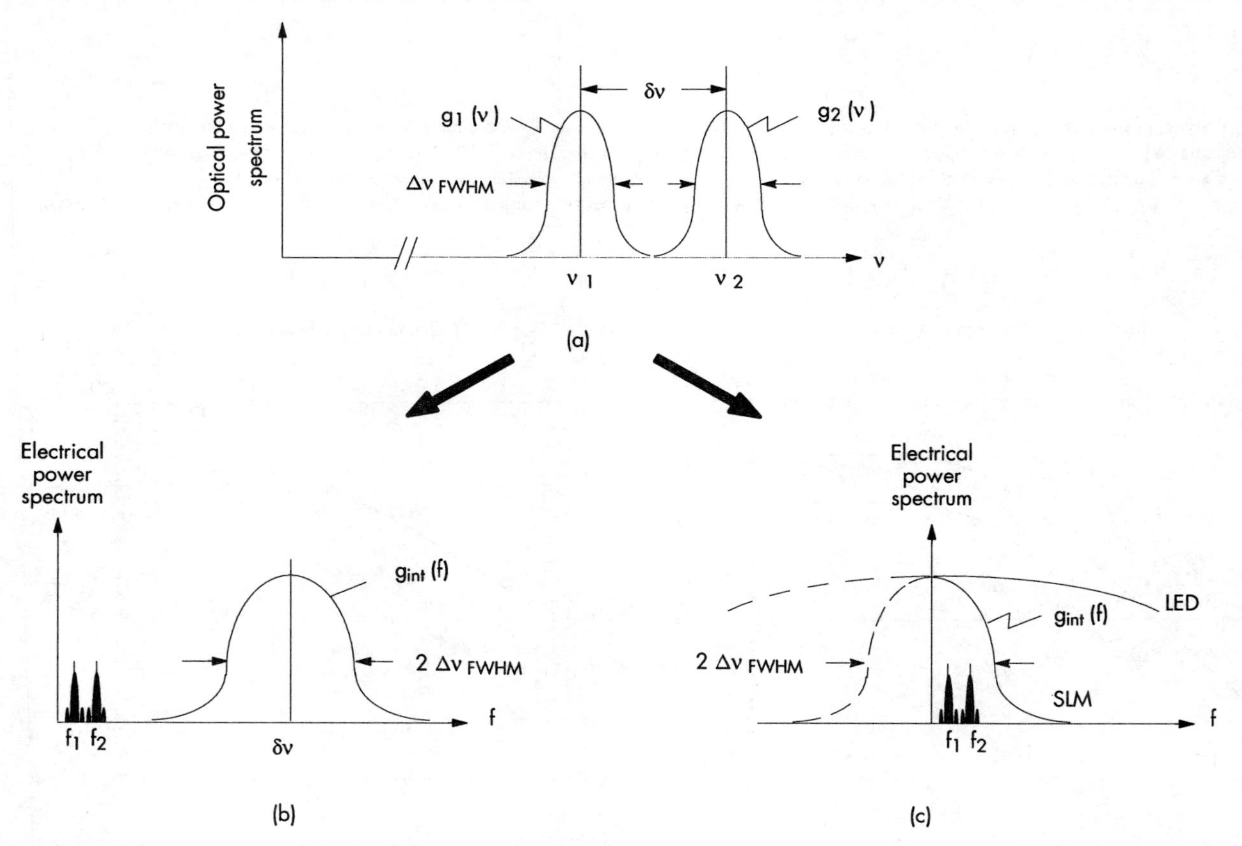

Figure 8.4 (a) Optical power spectra of two SLM lasers whose center frequencies differ by $\delta\nu$, (b) electrical power spectrum of the interference term when $\delta\nu \neq 0$, and (c) same as (b) but with $\delta\nu = 0$.

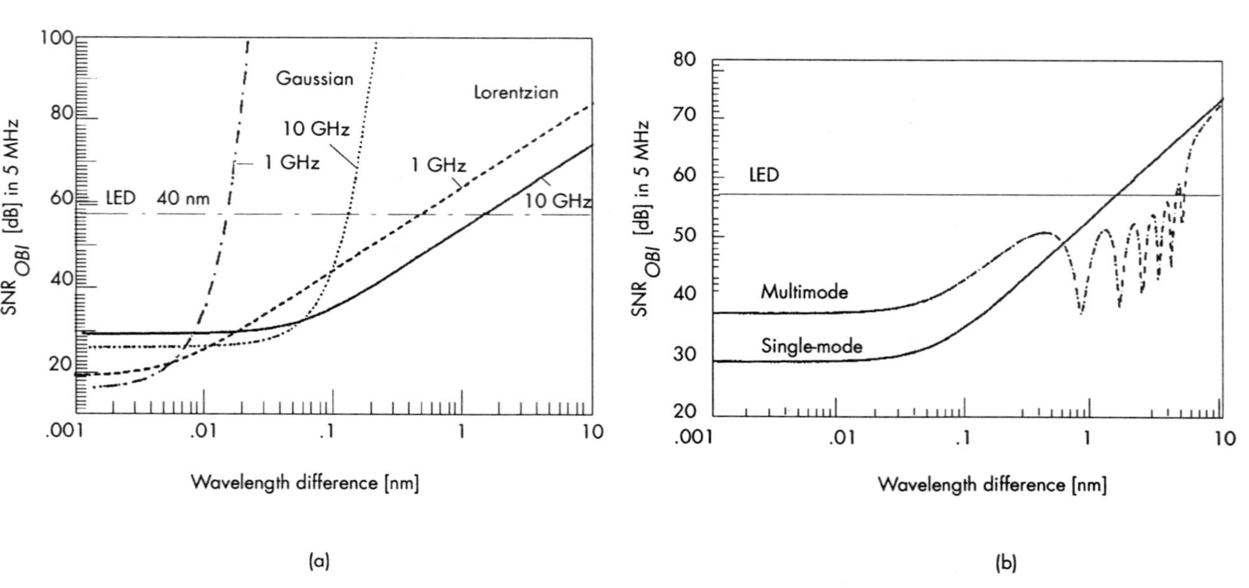

Figure 8.5 (a) The dependence of SNR$_{OBI}$ on the combined source linewidth $2\Delta\nu_{FWHM}$ and the wavelength separation $\delta\lambda$ of SLM laser sources, (b) SNR$_{OBI}$ as a function of $\delta\lambda$ for three types of light sources. The LED has a linewidth $\Delta\lambda_{FWHM} = 40$ nm, and the two laser sources have longitudinal modes with $\Delta\nu_{FWHM} = 10$ GHz. The MLM laser has seven longitudinal modes spaced by 0.8 nm. The envelope of the mode's power spectra of the longitudinal modes is Gaussian-shaped with FWHM = 5 nm. (© 1990 IEEE. [After Desem, 1990.])

This reduction depends on the source spectral shapes and is much more rapid in the case of a Gaussian line shape since it has a sharper rolloff, while Lorentzian line shapes are relatively slowly decaying functions.

Notice that the validity of (8.17) can be extended to the case of LED light sources provided that $\Delta\nu_{FWHM}$ is taken much larger than that of SLM light sources, typically of the order of 7–10 THz (e.g., for $\Delta\lambda_{FWHM} = 50$nm, $\Delta\nu_{FWHM} = 8.9$ THz at 1.5 μm). As shown in Figure 8.5(a), the SNR_{OBI} obtained with LEDs is approximately constant over a large range of $\delta\nu$ because the interference power is spread over a large frequency range and a small (quasi-constant) portion of it falls within the receiver bandwidth (Figure 8.4(c)).

Equation (8.17) can also be applied to individual modes of multilongitudinal-mode lasers: with two MLM lasers, the SNR_{OBI} can be calculated with (8.15) by taking into account all possible combinations between the individual longitudinal modes of the two lasers. Figure 8.5(b) shows the SNR_{OBI} as a function of the center wavelength difference for two MLM lasers with seven modes each. The envelope of the mode's power spectra is assumed to be a Gaussian function with FWHM of 5 nm. It can be seen that when the center wavelengths of the two lasers are identical, the SNR_{OBI} is greater than that obtained with SLM laser sources. This is because the longitudinal modes are assumed to be uncorrelated so that their individual noise contributions add on a power basis rather than in amplitude.

As the center wavelength separation of the two MLM lasers increases, the SNR_{OBI} initially increases up to a point where ripples appear. The periodicity of these ripples is equal to the wavelength separation of the multilongitudinal modes. The reason is that at these separations, some longitudinal modes from the two sources overlap and therefore the beating of these modes gives rise to a (relative) maximum interference contribution.

OBI noise with two sources has been experimentally observed and measurements agree with the theoretical estimates [Desem, 1988].

Up to now, the analysis has been restricted to OBI noise generated by only two light sources. The results can be generalized to the case of M ($M \geq 2$) simultaneous optical carriers as it would be the case for single-channel SCMA networks with more than two nodes. In this case, the OBI noise will be the result of accumulation of interference power from all possible combinations of pairs of light sources. The worst case is obtained when all sources have identical center wavelengths, as indicated by (8.17).

To simplify the analysis, let us assume that all sources have the same spectral distribution and equal power, and that they are uncorrelated. Since there are $C_M^2 = \dfrac{M(M-1)}{2}$ possible combinations organized in pairs, the SNR_{OBI} degrades according to

$$SNR_{OBI}(M, \delta\nu = 0) = \left(\frac{M(M-1)}{2}\right)^{-1} \cdot SNR_{OBI}(2, \delta\nu = 0) \tag{8.18}$$

where SNR_{OBI} (2, $\delta\nu = 0$) is given by (8.17) with $\delta\nu = 0$.

(8.18) clearly shows that the $SNR_{OBI}(M, \delta\nu = 0)$ rapidly deteriorates as the number of optical carriers, or equivalently the number of nodes in the single-channel SCMA network, increases.

As an example, consider the following representative values: $m = 0.7$, $\Delta\nu_{FWHM} = 10$ GHz and a required SNR of 16 dB. Then, from (8.17) and (8.18) and for large values of M ($M \geq 10$), the bandwidth per node will be limited by $M^2B \leq 100$ MHz. Thus, if the network must support, say 10 nodes, the bandwidth per node must be smaller than approximately 1 MHz. This bandwidth can be increased by a factor of 3 to 4 if multi-longitudinal-mode lasers are used [Desem, 1990]. Note also that in (8.18) all the lasers are assumed to have the same polarizations, and that there would be an improvement of about 3 dB in the SNR_{OBI} if they were randomly distributed.

The use of LED*s* rather than laser diodes may be attractive in this particular example. With a linewidth of $\Delta\lambda_{FWHM} = 50$ nm, (8.18) provides $M^2B \leq 86$ GHz. Therefore, the network capacity will not be limited by OBI noise but by the restricted modulation bandwidth of LED light sources.

Equation (8.18) has been obtained under the unrealistic assumption that all sources have identical wavelengths. It is more likely that the center wavelengths of the semiconductor lasers are randomly distributed over a range of wavelengths. Statistical models have been used to estimate the system performance in this more realistic situation [Desem 1990, Shakaranarayanan 1991]. In these approaches, the probability of channel outage due to the OBI effect is computed and the results show that as M increases, the network capacity increases sublinearly with M and eventually saturates.

Nevertheless, if channel outage is not allowed, an effective means to avoid the OBI effect is to divide the wavelength range in distinct wavebands and to allocate a waveband to each transmitter. This requires wavelength stabilization of each source and therefore at a first glance this scheme closely resembles WDMA. However, this particular single-channel SCMA scheme does not require optical filters since the channels are distinguishable in the RF/microwave domain (each node still uses a distinct subcarrier frequency). Moreover, as will become apparent later in this section, wavelength stability requirements are not as stringent as in WDMA and can easily be fulfilled with today's technology.

Equation (8.18) must be modified to allow for the presence of different operating wavelengths. Assume that the light sources are SLM lasers separated by a wavelength $\delta\lambda$. Then, to a first approximation, the OBI will be the result of the mixing of only the adjacent optical wavebands, so that only (M - 1) noise terms will contribute. The SNR_{OBI} then becomes

$$SNR_{OBI}(M, \delta\nu) = \frac{SNR_{OBI}(2, \delta\nu)}{M - 1} \tag{8.19}$$

To illustrate (8.19), again, consider the example where $m = 0.7$, $\Delta\nu_{FWHM} = 10$ GHz, and waveband separation of $\delta\lambda = 1$ nm ($\delta\nu = 133$ GHz at 1.5 μm). Using (8.15) and (8.17), it can be seen that the condition $SNR_{OBI} \geq 16$ dB is fulfilled when $(M - 1)B \leq 8.6$ GHz.

OBI noise measurements with four 1.5-μm DFB lasers have been reported [Wood, 1993a]. It has been shown that the noise can be modeled as white Gaussian noise, so that classical information theory can be applied in the computation of the BER obtained with a specific SNR_{OBI}. It has also been suggested that in order to reduce the level of OBI noise, a large modulation index (m up to 1.8) could be used since this increases the linewidth of the light sources [Wood, 1993b].

8.2.5 Overall SNR

Since the distinct noise sources that have just been discussed are not correlated, they can be combined in such a way that the denominators of their respective SNR given by (8.5), (8.10), (8.12) and (8.18) add. We thus obtain for the overall SNR

$$SNR_{Tot}^{-1} = SNR_{shot}^{-1} + SNR_{el}^{-1} + SNR_{RIN}^{-1} + SNR_{OBI}^{-1} \tag{8.20}$$

Equation (8.20) is illustrated in Figure 8.6. The parameters used for this particular example are indicated in the captions. Electronic noise is calculated using (8.8) such that the equivalent input noise, $\sqrt{<i_{el}^2>}$, is 0.01 $\mu A/\sqrt{Hz}$ at 100 MHz and 1 $\mu A/\sqrt{Hz}$ at 10 GHz, which are representative values for commercially available optical receivers.

It is seen that in order to satisfy $SNR_{Tot} \geq 16$dB to obtain a BER $\leq 10^{-9}$, the number of nodes must be kept below $M = 38$. The major limiting effect is the receiver electronic noise. The second limiting effect is OBI, which dominates the electronic noise for

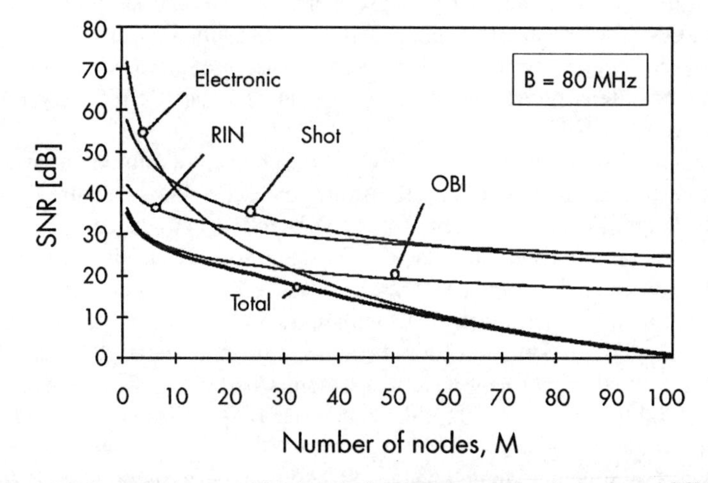

Figure 8.6 Overall SNR_{Tot} and decomposition into the four impairments for single-channel SCMA: shot noise, receiver electronic noise, RIN, and OBI. Assumptions are: B = 80 MHz, RIN = −130 dB/Hz, m = 0.7, λ = 1.55 μm.

$M \leq 33$. This limitation can be partially avoided by increasing the wavelength separation between the light. The maximum number of nodes is then limited to about $M \approx 42$, and the corresponding maximum network bandwidth is 3.4 GHz (1.7-Gbps capacity with FSK).

8.3 SINGLE-CHANNEL SCMA: EXPERIMENTAL SYSTEMS

Single-channel SCMA offers interesting possibilities in the design of broadband fiber-in-the-loop (FITL) systems. The introduction of optical fibers in the local loop, between the local switch (central office, or CO) and the subscribers, has been the subject of considerable interest by the majority of worldwide PTTs and telecommunication systems providers. The target would be fiber-to-the-home (FTTH) for everybody, to provide a broadband integrated services digital network (B-ISDN) [Van der Plas, 1992; Verbiest, 1993].

In this context, several system architectures have been proposed and experimented. Among the different proposals, passive optical networks (PONs) have proven to be potential candidates for future large scale deployment.

A PON-based FITL system is depicted in Figure 8.7. It uses a $1 \times M$ splitter to connect M subscribers to the CO. The main advantage of PONs is that they do not use active electronics except at the extremities of the network; that is, at the CO and at the customer premises (CPs) [Mestdagh, 1990].

Downstream information, from the CO to the CPs, is broadcast to all CPs. In general, this imposes constraints on the optical power budget of the system leading to a limitation on the maximum number of subscribers that can be cost-effectively connected to the same PON. The broadcast feature of PONs is well suited for distribution of services such as television, currently provided by coaxial cables of the CATV companies. The SCM technique is thus directly applicable here and has been effectively implemented [IEEE 1990, Gabla 1992].

Upstream transmission, from the CPs to the CO, needs a multiple-access scheme to avoid collision of information coming from different accessing subscribers.

In [Mestdagh, 1991], experimental 1×8 and 1×16 PON systems using single-channel SCMA for upstream transmission have been demonstrated. Both 1.3-μm LEDs and 1.3-μm MLM lasers have been used: 2-Mbps subscriber channels are transmitted by LED*s*, and broadband subscriber channels with up to 50-Mbps data rate are transmitted by MLM lasers. The FSK modulation format was used for both narrowband and broadband subscribers. To avoid impairments due to harmonic distortions (Section 8.2), especially those of LED light sources ($m = 0.8$), the subcarrier frequencies were placed within one octave of bandwidth.

Another experimental single-channel SCMA system is reported in [Wood, 1991] where four 155-Mbps data channels are transported in the upstream direction of a 1×4 PON. The subcarrier frequencies were FSK-modulated and the light sources were DFB

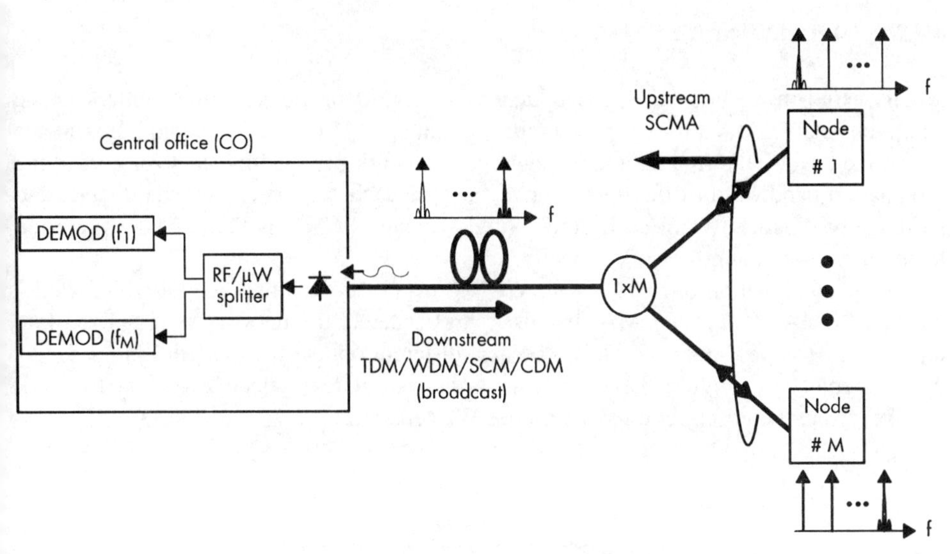

Figure 8.7 An SCMA PON-based FITL system. Downstream information may use either TDM, WDM, SCM, or CDM. Upstream information is achieved using single-channel SCMA. The nodes may be either customer premises or base stations of the mobile telephone network.

lasers operating between 1.527 and 1.545 μm. The wavelengths were stabilized and placed sufficiently far apart to avoid OBI effect.

Single-channel SCMA has also been proposed and demonstrated for the interconnection of mobile telephone base stations with the local master controller [Shiozawa, 1992]. The telephone signals received by a base station are downconverted to baseband, time-division multiplexed, and converted to a single RF/microwave signal before driving a superluminescent LED. The system can support 32 base stations with 1.5-Mbps channel capacity each. This system offers the attractive feature that it reduces the complexity of the base stations because all the switching functionalities are carried by the shared master controller.

Finally, single-channel SCMA has also been investigated for applications in computer networks [Bates 1990 & 1991]. In [Bates, 1990], 450-Mbps aggregate network throughput is achieved using low-coherence 0.8-μm self-pulsating laser diodes (those used in compact disk players). Two 150-Mbps data streams were used to BPSK-modulate two subcarriers at 300 MHz and 600 MHz, while a third 150-Mbps NRZ channel was transmitted at baseband. The network was an 8×8 multimode star. The network throughput has been increased up to 1 Gbps by using five self-pulsating laser transmitters fed with 200-Mbps *Q*PSK-modulated subcarriers at 100, 300, 500, 700, and 900 MHz [Bates, 1991]. The use of self-pulsating lasers with FWHM linewidths of ≈ 100 GHz minimizes OBI noise and modal-noise generation within multimode fibers.

8.4 MULTICHANNEL SCMA

As discussed in Section 8.2.4, the optical wavelengths of the node transmitters in the single-channel SCMA must be spaced relatively far apart (1 to 2 nm) to avoid impairments due to the OBI effect. In addition, the aggregate network bandwidth can never exceed the modulation bandwidth of the light sources because each subcarrier must be distinct and sufficiently separated from each other. Single-channel SCMA is thus not the preferred solution for maximum network capacity.

The combination of SCMA with WDMA offers an attractive alternative to single-channel SCMA or pure WDMA. In this hybrid scheme, the network is organized into clusters of M nodes, and each cluster owns a different optical wavelength λ_j ($j = 1, \ldots N$), as depicted in Figure 8.8. This allows the use of less selective optical filters at receivers than would be required with pure WDMA.

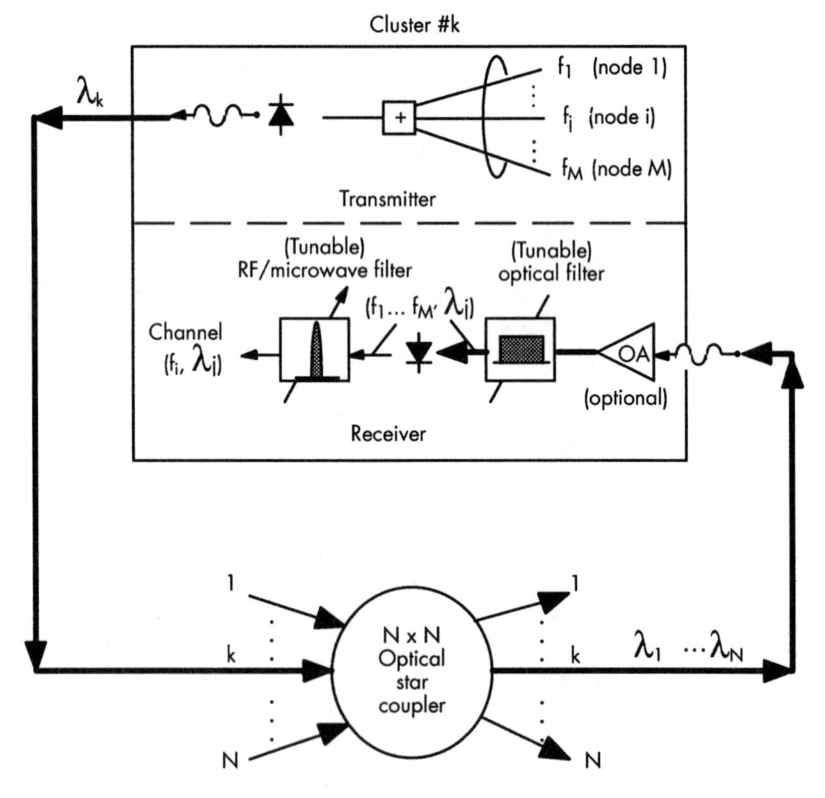

Figure 8.8 A hybrid SCMA/WDMA (or multichannel SCMA) multiaccess optical-fiber network having N wavelengths and M subcarriers per wavelength.

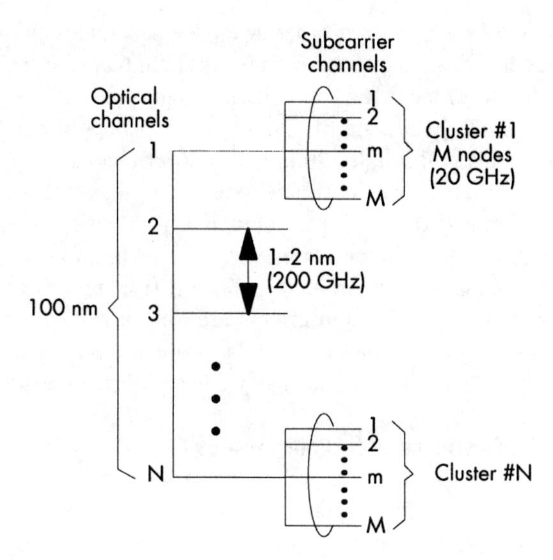

Figure 8.9 Frequency allocation in multichannel SCMA network. Each node address is identified by the pair (f_i, λ_j) corresponding to subcarrier f_i and optical wavelength λ_j.

Each node of a cluster owns a different RF/microwave transmit frequency f_i and therefore a channel on the network is identified by the pair (f_i, λ_j).

At each cluster, M modulated RF/microwave frequencies are combined together and the sum drives a single broadband laser diode operating at a prescribed wavelength λ_j. At the receive side, channel selection is performed using both (tunable) optical filters and (tunable) RF/microwave receivers. Figure 8.9 depicts the multiple-access hierarchy in a SCMA/WDMA network.

Given that each laser transmitter supports multiple RF/microwave channels, we shall call this hybrid scheme *multichannel* SCMA to distinguish it from the *single-channel* SCMA discussed in Sections 8.1 to 8.3.

8.5 MULTICHANNEL SCMA: NETWORK PERFORMANCE

The fact that each laser transmitter supports multiple modulated RF/microwave frequencies leads to new noise sources in addition to those that arise in the single-channel SCMA discussed in Section 8.2. These additional noise sources are intermodulation products (IMPs) and clipping, and will now be evaluated separately.

8.5.1 Intermodulation Products

Because a semiconductor laser is not a perfectly linear device, the various RF/microwave subcarriers are mixed within the laser cavity and new frequencies of the form $f_i \pm f_j$

(second-order IMPs) and $f_i \pm f_j \pm f_k$ (third-order IMPs) are generated. These new frequencies appear as noise if they lie within the subcarriers bandwidth. It can be shown that for digital signals, the effect of third-order IMPs can be made negligible [Olshansky, 1988]. If the subcarriers in the system are restricted to lie within one octave of bandwidth, then there is also no second-order IMPs that fall within a subcarrier channel. However, to increase the total system bandwidth, it may be desirable to use more than one octave of bandwidth, in which case second-order IMPs distort the signals and affect the system performance.

IMPs originate from two distinct nonlinearities of the laser diode. Notice that we do not consider other mechanisms that can contribute to IMPs, such as fiber dispersion and photodetector response. Fiber dispersion effects are only significant for analog signals (especially for AM-modulation format) and can be neglected for digital signals [Phillips, 1991]. Photodiodes are usually highly linear devices and their contribution to IMPs can also be neglected.

The first source of IMPs in laser diodes results from the nonlinear characteristic of the *L-I* curve. The resulting IMPs are referred to as *static* IMPs. Second-order static IMPs can be modeled by expanding the output optical power P of the laser diode in a series about the bias point (I_b, P_b) and limiting the expansion to the second-order term

$$P = P_b + P_b' \cdot (I - I_b) + \frac{P_b''}{2} \cdot (I - I_b)^2 \tag{8.21}$$

where $P_b' = \left(\frac{dP}{dI}\right)_{I=I_b}$ and $P_b'' = \left(\frac{d^2P}{dI^2}\right)_{I=I_b}$.

The drive current is the sum of M modulated RF/microwave subcarriers and has the form

$$I = I_b + \sum_{i=1}^{M} \delta I_i \cdot \cos(2\pi f_i t + \phi_i) \tag{8.22}$$

Assume that the modulating currents δI_i are adjusted so that all subcarriers have the same optical modulation index $m = \delta P_i / P_b$. Then, inserting (8.22) into (8.21) we obtain after some algebra

$$P = P_b \cdot \{1 + \sum_{i=1}^{M} m \cdot \cos(2\pi f_i t + \phi_i)$$

$$+ \sum_{i=1}^{M}\sum_{j=1}^{M} \delta m_{i-j} \cdot \cos[2\pi(f_i - f_j)t + (\phi_i - \phi_j)]$$

$$+\sum_{i=1}^{M}\sum_{j=1}^{M}\delta m_{i+j}\cdot\cos[2\pi(f_i+f_j)t+(\phi_i+\phi_j)]\} \tag{8.23}$$

where the modulation index of the second-order static IMPs, $\delta m_{i\pm j}$, due to subcarriers f_i and f_j are given by

$$\delta m_{i\pm j}=C_{2a}\cdot m^2 \text{ with } C_{2a}=\frac{P_b\cdot P_b''}{4(P_b')^2} \tag{8.24}$$

The corresponding second-order static IMP powers are given as $\delta P_{i\pm j} = P_b\delta m_{i\pm j}$.

Typical values of C_{2a} for laser diodes lie around 0.2, but can be larger if lasers with very sublinear *L-I* curve are used or if lasers are operated at too-high bias powers. For LED light sources, C_{2a} typically ranges between 0.5 and 1.0 but may also be much larger depending on the bias point.

The second source of IMPs in laser diodes results from the nonlinear photon-electron interaction mechanism in the laser cavity, and can be evaluated by the laser-rate equations. The resulting IMPs are referred to as *dynamic* IMPs. Rate-equation solutions for these dynamic IMPs have been given in [Darcie, 1986.b] and careful measurements have shown agreement with theory [Iannone, 1987]. These dynamic IMPs are given by

$$\delta P_{i\pm j}=P_b\cdot C_{2b}\left(\frac{f_i\pm f_j}{f_r}\right)\cdot m^2 \tag{8.25}$$

where

$$C_{2b}(x)=\frac{x^2}{2\sqrt{(x^2-1)^2+x^2\left(\frac{\Gamma}{2\pi f_r}\right)^2}} \tag{8.26}$$

In the above expressions f_r is the resonant frequency of the laser diode, and Γ is the damping rate which is given by the semiempirical relation $\Gamma = 0.32\times10^{-9}f_r^2$ [Olshansky, 1988].

It is seen from (8.25) and (8.26) that the dynamic second-order IMPs have the largest values when the frequency of the IMPs falls on the laser's resonant frequency f_r. At $f_{i+j}=f_r$ (worst case), the coefficient C_{2b} in (8.25) is approximately $10/f_r$ [GHz]. It is common practice to take the highest subcarrier frequency f_{max} such that $f_{max}/f_r \approx 0.8$, in which case $C_{2b} \leq 1$.

It is also seen that for both static and dynamic IMPs, the crosstalk powers increase with the modulation index as m^2. It can be shown that the second harmonic distortion

$C_{2H}(2f_i)$ is 3-dB smaller than the second-order IMP ($f_{i\pm j}$ with $i \neq j$): $C_{2H} = C_{2b} / 2$ [Daly, 1982].

Equations (8.24) and (8.25) express the powers of second-order IMPs for a specific combination of subcarrier frequencies f_i and f_j. In multichannel SCMA, a certain number of these combinations may fall within the bandwidth of a subcarrier channel. The number of combinations will depend on the details of the subcarrier frequency assignments. Assume that the subcarrier frequencies are equally spaced and located at

$$f_i = i \cdot \Delta f \text{ with } i = M_{\min}, \ldots M_{\max} \tag{8.27}$$

where Δf is the subcarrier frequency spacing. Thus, M channels lie between $M_{\min}\Delta f$ and $M_{\max}\Delta f$, as shown in Figure 8.10.

A straightforward analysis shows that for the IMPs at $f_k = f_i - f_j$, the number of combinations is given by

$$M_k^- = M_{\max} - M_{\min} - \text{k} + 1 \tag{8.28}$$

while for the IMPs at $f_k = f_i + f_j$ ($k \geq 2M_{\min}$), the number of combinations is given by

$$M_k^+ = \frac{k - 2M_{\min} + \varepsilon}{2} \tag{8.29}$$

where $\varepsilon = 1$ for k uneven and $\varepsilon = 2$ for k even.

It is seen that the maximum number of second-order IMPs occur at the edges of the band, and the number at the low-frequency edge is twice the number at the high-frequency edge.

Static and dynamic second-order IMPs add incoherently so that the total electrical noise at the receiver is proportional to the sum of the square of all the IMP combinations that fall within a given subcarrier channel. Using (8.4) with (8.24), (8.25), (8.28), and (8.29), we obtain

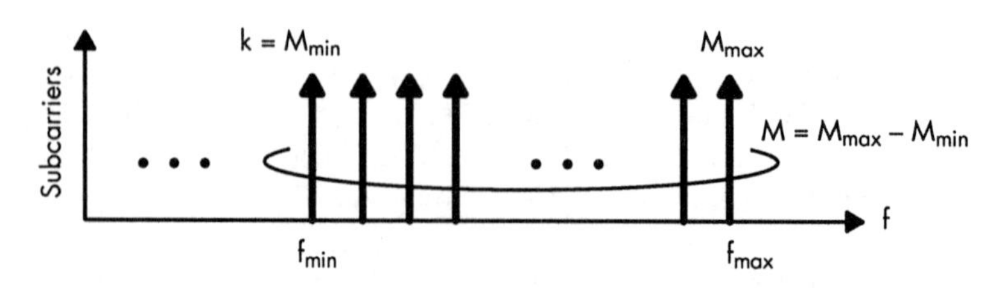

Figure 8.10 Assignment of M subcarrier frequencies, lying between $f_{\min}$ and $f_{\max}$.

$$\mathrm{SNR_{IMP_2}} = \frac{1}{(\mathrm{C_{2a}^2} + \mathrm{C_{2b}^2}) \cdot \mathrm{m^2} \cdot \mathrm{M_k}} \cdot \frac{\Delta f}{\mathrm{B}} \tag{8.30}$$

Note that for second harmonic distortion M_k is always equal to 1.

To illustrate the impact of the IMP distortion on the SNR, consider the case where $M = 20$, $\Delta f = 200$ MHz, $M_{min} = 1$ ($f_{min} = 200$ MHz), $M_{max} = 20$ ($f_{max} = 4$ GHz), $B = 100$ MHz, $f_r = 3$ GHz, and $C_{2a} = 0.2$. Then, from (8.30) the condition $\mathrm{SNR_{IMP_2}} \geq 16$ dB is fulfilled when $m \leq 8.9\%$. If we set the condition $SNR_{\mathrm{IMP_2}} \geq 30$ dB in order to avoid any IMP impairments, then the modulation index per channel must satisfy $m \leq 1.8\%$.

8.5.2 Clipping

At first glance, it might appear that using laser diodes with very linear *L-I* curve and very high resonant frequency would eliminate IMP impairments, thereby allowing higher modulation index *m* with the benefit of an improved system power budget. However, the clipping effect will set a fundamental limit on the total modulation index. This is because laser diodes are threshold devices and optical power cannot be negative. As a result, clipping occurs at the zero-optical-power level if the drive current falls below the threshold current I_{th}.

Neglecting IMP distortions ($\delta m_{i\pm j} = 0$), it is clear from (8.23) that if the modulation index per channel is constrained to $m \leq 1/M$, then no signal clipping will occur. However, this is too restrictive and would lead to a significant system power-budget limitation because the optical power per channel would have to be reduced proportionally to $1/M$. It would be more advantageous to accept some clipping by increasing *m* beyond $1/M$ and to trade off the resulting SNR loss against the SNR loss due to smaller *m*.

For small values of *M*, the allowed increase of *m* will be small. But, as *M* increases (let say, $M \geq 10$), statistical averaging will take place and the amount of clipped power is expected to grow less rapidly than the allowed increase of optical power per channel.

The computation of the frequency spectrum of the resulting clipping distortion can be made numerically. An approximate analytical expression can be obtained assuming that this spectrum is flat and of the same bandwidth as the subcarrier frequency range [Saleh 1989, Mazo 1992]. Notice that this assumption leads to the worst-case estimate of the clipping effect.

For large values of *M* and for independently modulated subcarrier channels, $P(t)$, as given by (8.23) with $\delta m_{i\pm j} = 0$ can be accurately modeled as a Gaussian random process (central-limit theorem) with a mean value of P_b and a variance of $\sigma_P^2 = MP_b^2 m^2/2$.

Therefore, the probability that at any given time the power $P(t)$ takes the value P_a is given by

$$\mathrm{Prob}\{P(t) = P_a\} = \frac{1}{\sigma_P\sqrt{2\pi}} \cdot \exp\left[-\frac{(P_a - P_b)^2}{2\sigma_P^2}\right] \tag{8.31}$$

The total mean-square value of the clipped portion of $P(t)$, which corresponds to the condition $P_a \leq 0$, is thus given by

$$<P^2_{\text{Clip}}> = \frac{1}{\sigma_P\sqrt{2\pi}} \cdot \int_{-\infty}^{0} P^2 \exp\left[-\frac{(P-P_b)^2}{2\sigma_P^2}\right] dP$$
$$\cong \sqrt{\frac{2}{\pi}} P_b^2 \mu^5 \exp\left[-\frac{1}{2\mu^2}\right] \tag{8.32}$$

where $\mu \overset{\Delta}{=} m\sqrt{M/2}$ is the total rms modulation index.

Since the mean-square optical power is $\sigma_P^2 = MP_b^2 m^2/2$, the signal-to-clipping mean-square power ratio is given as

$$\text{SNR}_{\text{Clip}} \cong \sqrt{\frac{\pi}{2}} \cdot \frac{\exp\left[\frac{1}{2\mu^2}\right]}{\mu^3} \tag{8.33}$$

For example, if we set the condition $\text{SNR}_{\text{Clip}} = 30$ dB (in this way the clipping effect becomes negligible), then (8.33) provides $\mu = 0.37$, and $m = \mu\sqrt{2}/M \cong 0.5/\sqrt{M}$.

Thus, by allowing some clipping in the transmitted signal the modulation index per channel can be increased as $1/\sqrt{M}$ rather than $1/M$ if no clipping is allowed.

Note that the clipping effect sets a fundamental limit on m, while second-order IMPs can be eliminated either by limiting the system bandwidth to within one octave or by careful assignment of the subcarrier frequencies.

8.5.3 Overall SNR

The overall SNR of multichannel SCMA is obtained by combining (8.30) and 8.(33) with (8.20).

Since in multichannel SCMA, optical filters are placed in front of each node receiver, shot noise and RIN can be neglected in comparison with other sources of noise. Moreover, if the distinct optical wavebands are sufficiently far apart, then impairments due to the OBI effect can be avoided. Therefore, the remaining sources of noise that have to be considered in the dimensioning of multichannel SCMA networks are (1) receiver electronic noise, (2) IMPs, and (3) clipping. It follows that

$$\text{SNR}_{\text{Tot}}^{-1} = \text{SNR}_{\text{el}}^{-1} + \text{SNR}_{\text{IMP}}^{-1} + \text{SNR}_{\text{Clip}}^{-1} \tag{8.34}$$

The last two terms in the right-hand side of (8.34) set a limit on the maximum modulation index per channel. For large M, the power budget of multichannel SCMA networks is thus expected to be much smaller than that of single-channel SCMA.

Suppose that IMPs have somehow been completely canceled out (e.g., perfectly linear *L-I* curve). As we saw from (8.33), in order to avoid distortions due to clipping, the modulation index per channel must satisfy $m \le 0.5/\sqrt{M}$. Putting the maximum limit value of $m = 0.5/\sqrt{M}$ into (8.9) we obtain

$$\mathrm{SNR_{Tot}} = \mathrm{SNR_{el}} = \frac{(\Re LP_b)^2}{8[K_2 + K_3 MB] \cdot M^3 B^2} \tag{8.35}$$

$\mathrm{SNR_{el}}$ can be offset with the help of optical amplifiers. Suppose that an optical preamplifier is placed in front of each node receiver and assume that the combined optical preamplifier and photodiode receiver operates in the signal ASE beat-noise limit (Chapter 5). Then, from (5.27) and with $m = 0.5/\sqrt{M}$, the total SNR becomes

$$\mathrm{SNR_{Tot}} = \frac{G \cdot L \cdot P_b}{32 \cdot h\nu \cdot n_{\mathrm{sp}} \cdot (G - 1) \cdot MB} \tag{8.36}$$

where G is the optical amplifier gain, h is the Planck's constant, ν is the optical carrier frequency, and n_{sp} is the spontaneous emission factor.

To illustrate (8.36), let us consider the case where G is large enough such that $G/(G - 1) \cong 1$, $L = L_k/N$ is the loss of the $N \times N$ star network where L_k = 5 dB is the link loss, $n_{\mathrm{sp}} = 1$ (ideal noiseless optical amplifier), λ = 1.55 μm ($h\nu$ = 0.8 eV), and there is a required $\mathrm{SNR_{Tot}}$ of 16 dB. Then (8.36) provides $NMB \le$ 20 THz. With $MB \le$ 20 GHz limited by the bandwidth of semiconductor light sources, this last inequality provides $N \le$ 1,000. Within the 150-nm usable wavelength range around 1.55 μm, the N = 1,000 channels should be spaced by only 0.15 nm. However, such a small waveband spacing would result in significant system performance degradations due to OBI effects which have been assumed negligible in (8.36). In other words, (8.36) is valid only for channel spacing of a few nanometers in order to avoid OBI effects. For example, with 2-nm waveband spacing, there will be 75 usable wavebands within the 150-nm wavelength range around 1.55 μm, resulting in an overall network bandwidth of ≈ 1.5 THz. This represents the ultimate network bandwidth achievable using multichannel SCMA in star networks.

8.6 MULTICHANNEL SCMA: EXPERIMENTAL SYSTEMS

Multichannel SCMA systems have been proposed and demonstrated for applications in two distinct areas.

First, within the framework of mobile telephony, multichannel SCMA over an eight-way-split PON network has been implemented with a capacity of six telephone channels per base station (or cluster) [Cooper, 1992]. The system has been called CTPON to refer to cordless telephony (CT) over PON. Each base station is assigned an intermediate RF/microwave frequency (IF) band and applies frequency translation to convert from the CT band (864–869 MHz) to a distinct IF band on the PON. In this way, each base station has access to the total CT band while subcarrier overlapping is avoided. In the demonstration, 1.3 μm Fabry-Perot laser diodes were used and OBI effects were negligible thanks to the narrow CT channel bandwidth (100 kHz). Provided that the IF plan extends over less than one octave of bandwidth, second-order IMPs were out of band. Compared with the similar system described in Section 8.3, the CTPON system provides the additional advantage of eliminating the need for baseband conversion and channel multiplexing/demultiplexing at each base station. However, since the CT signals are transported over the PON in their original form, the optical system must provide sufficient dynamic range because the loss incurred by the signals from the CT terminal to the base station can vary over a large range. An electrical dynamic range of 46 dB has been achieved in the demonstration.

The second application area where multichannel SCMA has been applied is the high-speed packet-switched metropolitan area lightwave network (MAN) [Liew 1989, Choy 1991 & 1993, Ramaswami 1993]. A schematic diagram of the proposed system architecture is shown in Figure 8.11. The system uses fixed-tuned wavelength optical transmitters/receivers, and fixed-tuned RF/microwave receivers. The only tunable devices in the network are the subcarrier modulators.

There are N clusters, each using a distinct wavelength for transmission. Each cluster supports M nodes. At each of the N cluster controllers, the packets from the M nodes are placed in distinct buffers for transmission and each buffer is connected to a tunable subcarrier modulator. The destination address of a packet, which is contained in the packet header, determines the subcarrier frequency of the modulator. The subcarrier frequencies are time-shared by the nodes belonging to the same cluster. A scheduler reads each packet header and selects the appropriate frequency for transmission. In the meantime, it also ensures that all subcarriers are different (otherwise a collision between two or more data packets would occur). The M modulated subcarriers are summed together and the composite signal is used to drive a single SLM laser (DFB). The output optical powers of the N lasers (N cluster controllers) are combined and split by an $N \times N$ star coupler. Thus, each cluster controller receives all the wavelengths.

At the receive side of a cluster controller, the composite optical signal is further split by a $1 \times M$ splitter and each output of the splitter is followed by a fixed-tuned optical filter (the optical filters are tuned to a wavelength λ_j which is different from that of the transmission wavelength λ_k). The M optically filtered outputs are then demodulated by M subcarrier demodulators fixed-tuned to the M distinct subcarrier frequencies used within the network.

Since each cluster controller has access to only one wavelength (λ_j) in the network, it receives directly only from the nodes that belong to the cluster controller that uses λ_j as the transmit wavelength. Full node connectivity is achieved by multihop in the optical channels on the network (Chapter 7).

To illustrate this, assume there are three cluster controllers A, B, and C that transmit

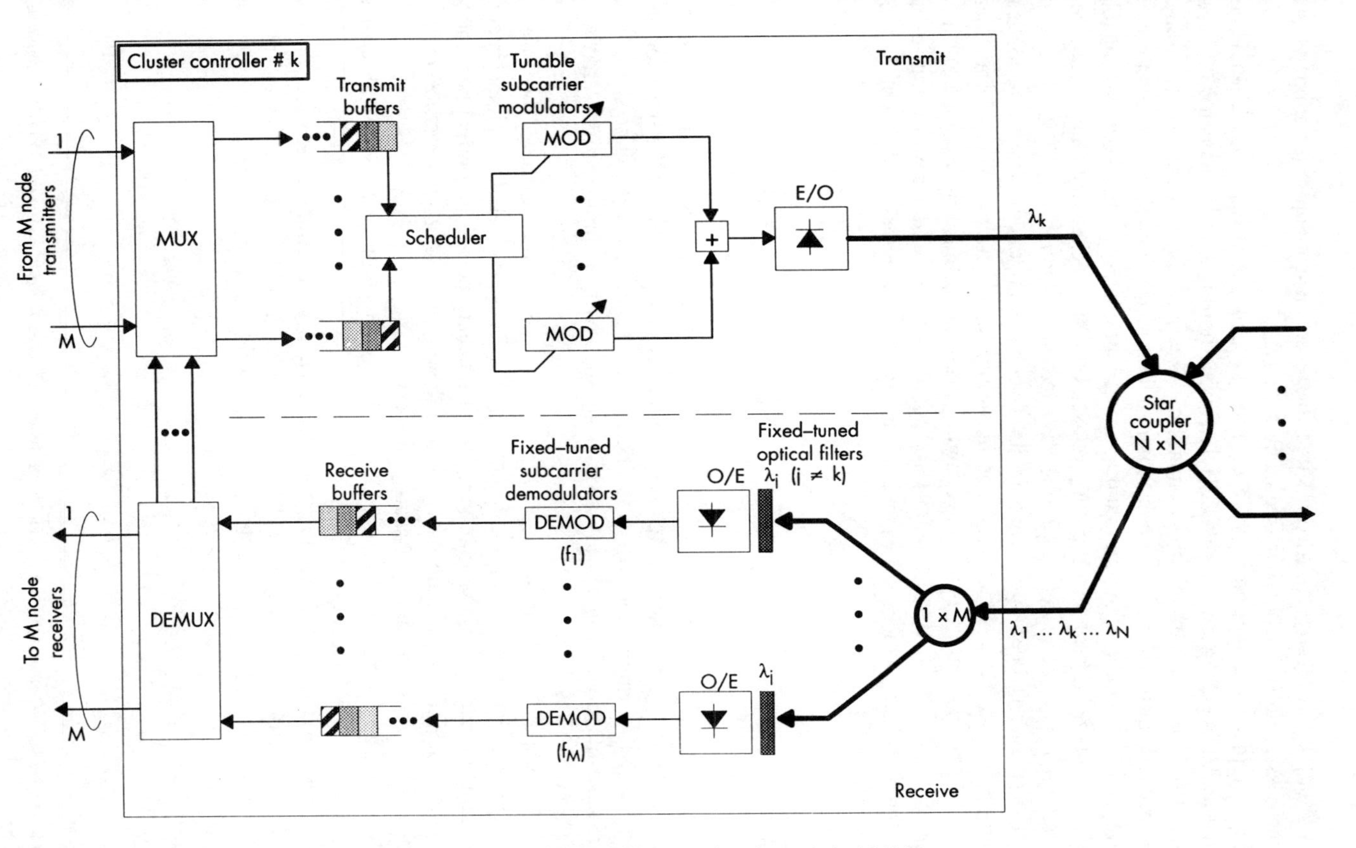

Figure 8.11 The cluster controller of a multihop multichannel SCMA high-speed packet switched MAN.

on λ_A, λ_B and λ_C, respectively. Also assume that each cluster controller supports two nodes each assigned to a subcarrier frequency f_1 or f_2.

Suppose that *B* receives on λ_A, *C* receives on λ_B, and *A* receives on λ_C. Then, nodes connected to *A* can send their packets directly (in one hop) to nodes connected to *B*, but they cannot send a packet in one hop to nodes connected to *C*. For transmission from *A* to *C*, the packets must first pass through *B*, which then forwards them in a second hop to *C*. As we saw in Chapter 7, the multihop approach eliminates the need for fast tunable optical filters and/or lasers. In the present proposal, dynamic packet switching is achieved by fast tuning (tens of nanoseconds) in the RF/microwave domain.

The system has been prototyped in its essential components (including optical preamplifiers at each cluster receiver). The feasibility has been demonstrated for a network with 32 cluster controllers (32 wavelengths) each supporting 5 nodes at a 200-Mbps data rate per node [Choy, 1993]. The average number of hops under uniform traffic has been shown to be 3.5, which corresponds to an aggregate network capacity of 7 Gbps [Ramaswami, 1993].

8.7 SUMMARY

In this chapter we have seen how SCMA can be exploited as an alternative to WDMA. SCMA is attractive, particularly in the short term, because it obviates the need for fast tunable optical transmitters or receivers. These devices are indeed replaced by their RF/microwave electronic counterparts, which are at present commercially available at relatively low cost levels.

The network capacity achievable with SCMA is, however, restricted because of several impairments typical of this multiaccess scheme. These typical SCMA impairments include the aggregate shot noise, the laser-output intensity noise, the optical beat-interference noise, the intermodulation products, and the clipping noise, and restrict the network capacity to a few gigabits per second at the very most.

Higher network capacities would require the combination of SCMA with coarse WDMA and possibly optical amplifiers.

Nevertheless, SCMA has been proven attractive in applications where maximizing the network capacity is not a major concern. In particular, SCMA is a promising technique for the interconnection of base stations for digital mobile telephony over passive optical networks.

REFERENCES

[Daly, 1982] Daly, J. C. "Fiber optic intermodulation distortion," *IEEE Trans. Selected Areas of Communication*, Vol. 30, 1982, pp. 1954–1958.

[Darcie, 1986a] Darcie, T. E., et al. "Lightwave system using microwave subcarrier multiplexing," *Electron. Letters*, Vol. 22, No. 15, July 1986, pp. 774–775.

[Darcie, 1986b] Darcie, T. E., et al. "Intermodulation and harmonic distortion in *InGaAsP* laser," *Electron. Letters*, Vol. 21, 1985, pp. 665–666; and "Erratum," Electron. Letters, Vol. 22, 1986, p. 619.

[Darcie, 1987] Darcie, T. E. "Subcarrier multiplexing for multiple access lightwave networks," *IEEE J. Light. Technology*, Vol. 5, No. 8, Aug. 1987, pp. 1103–1110.

[Iannone, 1987] Iannone, P. P., et al. "Multichannel intermodulation distortion in high-speed *GaInAsP* lasers," *Electron. Letters*, Vol. 23, 1987, pp. 1361–1362.

[Sklar, 1988] Sklar, B. *Digital Communications: fundamentals and applications*, New York, NY: Prentice Hall, 1988.

[Desem, 1988] Desem, C. "Optical interference in lightwave subcarrier multiplexing systems employing multiple optical carriers," *Electron. Letters*, Vol. 24, No. 1, Jan. 1988, pp. 50–52.

[Darcie, 1988] Darcie, T. E., et al. "Resonant *PIN-FET* receivers for lightwave subcarrier systems," *IEEE J. Light. Technology*, Vol. 6, No. 4, April 1988, pp. 582–589.

[Ohkawa, 1988] Ohkawa, N. "20*GHz* low-noise *HEMT* preamplifier for optical receivers," *Proc. European Conf. Optical Communication, ECOC'88*, Sept. 1988, pp. 404–407.

[Olshansky, 1988] Olshansky, R., et al. "High speed *InGaAsP* lasers for *SCM* optical fiber systems," *Optoelectronic-Devices and Technologies*, Vol. 3, No. 2, Dec. 1988, pp. 143–153.

[Proakis, 1989] Proakis, J. G. *Digital Communications*, Second Edition, New York, NY: McGraw-Hill, 1989.

[Yariv, 1989] Yariv, A. *Quantum Electronics*, Third Edition, New York, NY: John Wiley & Sons, 1989.

[Runge, 1989] Runge, K., et al. "4*Gb/s* subcarrier multiplexed transmission over 30*km* using *QPSK* modulation," *Proc. OFC'89*, PD-18, Feb. 1989.

[Saleh, 1989] Saleh, A.A.M. "Fundamental limit on number of channels in subcarrier-multiplexed lightwave *CATV* system," *Electron. Letters*, Vol. 25, No. 12, June 1989, pp. 776–777.

[Darcie, 1989] Darcie T. E. et al. "Wideband lightwave distribution system using subcarrier multiplexing," *IEEE. J. Light. Technology*, Vol. 7, No. 6, pp. 997–1005, June 1989.

[Nazarathy, 1989] Nazarathy, M., et al. "Spectral analysis of optical mixing measurements," *IEEE J. Light. Technology*, Vol. 7, No. 7, July 1989, pp. 1083–1096.

[Olshansky, 1989] Olshansky, R., et al. "Subcarrier multiplexed lightwave systems for broadband distribution," *IEEE J. Light. Technology*, Vol. 7, No. 9, Sept. 1989, pp. 1329–1342.

[Cunningham, 1989] Cunningham, D. G., et al. "Intensity noise in lightwave fibre systems using *LED* transmitters," *Electron. Letters*, Vol. 25, No. 22, Oct. 1989, pp. 1481–1482.

[Liew, 1989] Liew, S. C., et al. "A broad-band optical network based on hierarchical multiplexing of wavelengths and *RF* subcarriers," *IEEE J. Light. Technology*, Vol. 7, No. 11, Nov. 1989, pp. 1825–1838.

[Way, 1989] Way, W. I. "Subcarrier multiplexed lightwave system design considerations for subscriber loop applications," *IEEE J. Light. Technology*, Vol. 7, No. 11, Nov. 1989, pp. 1806–1818.

[IEEE, 1990] Special Issue "Applications of *RF* and microwave subcarriers to optical fiber transmission in present and future broadband networks," *IEEE J. Selected Areas of Communication*, Vol. JSAC-8, No. 7, Sept. 1990. See also, Special Issue "Optical fiber video delivery systems of the future," *IEEE LCS; the magazine of lightwave communication systems*, Vol. 1, No. 1, Feb. 1990.

[Hill, 1990] Hill, P. M., et al. "A 20-channel optical communication system using subcarrier multiplexing for the transmission of digital video signals," *IEEE J. Light. Technology*, Vol. 8, No. 4, April 1990, pp. 554–560.

[Desem, 1990] Desem, C. "Optical interference in subcarrier multiplexed systems with multiple optical carriers," *IEEE J. Selected Areas of Communication*, Vol. 8, No. 7, Sept. 1990, pp. 1290–1295.

[Bates, 1990] Bates, R.J.S., et al. "A 450*Mb/s* throughput subcarrier multiple-access network using *790nm* self-pulsating laser transmitters," *Proc. European Conf. Optical Communication, ECOC'90*, 1990, pp. 425–428.

[Mestdagh, 1990] Mestdagh, D.J.G., et al. "Broadband passive optical networks," *Revue HF*, Vol. 14, No. 7 & 8, 1990.

[Hein, 1990] Hein, B. "Digital optical subcarrier access networks realized with subcarrier modulation," *Proc. European Fiber-Optic Communication Conf.*, E-FOC 90, 3.6.3, 1990, pp. 369–373.

[Mestdagh, 1991] Mestdagh, D.J.G., et al. "Digital subcarrier multiple-access over passive optical networks for narrowband and broadband subscribers," *Proc. Optic-Fiber Communication Conf., OFC'91*, FE1, Feb. 1991, pp. 215. See also "Subcarrier multiple access mechanism in passive optical networks," Proc. Intl. Symp. Subscriber Loop and Services, ISSLS'91, 1991.

[Bates, 1991] Bates, R.J.S., et al. "Five-channel 1*Gb/s* aggregate throughput subcarrier multiple-access network for computer applications," *Proc. Optic-Fiber Communication Conf., OFC'91*, FE2, Feb. 1991, p. 216.

[Desem, 1991] Desem, C. "Measurement of optical interference due to multiple optical carriers in subcarrier multiplexing," *IEEE Photon. Technology Letters*, Vol. 3, No. 4, April 1991, pp. 387–389.

[Ohtsuka, 1991] Ohtsuka, H., et al. "256-*QAM* subcarrier transmission using coding and optical intensity modulation in distribution networks," *IEEE Photon. Technology Letters*, Vol. 3, No. 4, April 1991, pp. 381–383.

[Phillips, 1991] Phillips, M. R., et al. "Nonlinear distortion generated by dispersive transmission of chirped intensity-modulated signals," *IEEE Photon. Technology Letters*, Vol. 3, No. 5, May 1991, pp. 481–483.

[Shankaranarayanan, 1991] Shankaranarayanan, N. K., et al. "*WDMA*/Subcarrier-*FDMA* lightwave networks: limitations due to optical beat interference," *IEEE J. Light. Technology*, Vol. 9, No. 7, July 1991, pp. 931–943. See also, "Design of *WDMA/SFDMA* lightwave networks," *GLOBECOM'91*, 35.2.1, 1991, pp. 1251–1257.

[Choy, 1991] Choy, M. M., et al. "A 200*Mb/s* frequency-shift-keying subcarrier optical link potentially extendable to a 640-channel wavelength-division multiplexing-subcarrier multiplexing network," *CLEO'91*, CThR54, 1991, pp. 468–469.

[Wood, 1991] Wood, T. H., et al. "Demonstration of broadband-*ISDN* upgrade of Fiber-In-Loop system," *Electron. Letters*, Vol. 27, No. 24, Nov. 1991, pp. 2275–2277.

[Olshansky, 1991] Olshansky, R. "Multigigabit per second subcarrier multiplexed optical fiber ring network," *Electron. Letters*, Vol. 27, No. 23, Nov. 1991, pp. 2098–2100.

[Gitlin, 1992] Gitlin, R. D., J. F. Hayes, and S. B. Weinstein. *Data Communications Principles*, Plenum Press, 1992.

[Cooper, 1992] Cooper, A. J., et al. "*CTPON*-Cordless telephony services over a passive optical network using fibre radio techniques," *Proc. Intl. Conf. Communication, ICC'92*, 304. 4. 1, pp. 91–96, 1992.

[Shiozawa, 1992] Shiozawa, T., et al. "Upstream-*FDMA*/Downstream-*TDM* optical fiber multiaccess network," *Proc. Intl. Conf. Communication*, 304.7.1, 1992, pp. 105–109.

[Budman, 1992] Budman, A., et al. "Multigigabit optical packet switch for self-routing networks with subcarrier addressing," *Proc. Optic-Fiber Communication Conf., OFC'92*, TuO4, 1992, pp. 90–91.

[Mazo, 1992] Mazo, J. E. "Asymptotique distortion spectrum of clipped *dc*-biased Gaussian noise," *IEEE Trans. Selected Areas of Communication*, Vol. 40, No. 8, Aug. 1992, pp. 1339–1344.

[Gabla, 1992] Gabla, P. M., et al. "45*dB* power budget in a 30-channel *AM-VSB* distribution system with three cascaded erbium-doped fiber amplifiers," *Proc. European Fiber-Optic Communication Conf., EFOC/LAN'92*, 29, 1992, pp. 134–136.

[Van der Plas, 1992] Van der Plas, G., et al. "*ATM* over passive optical networks: system design and demonstration," *Proc. OE/Fibers'92 (SPIE)*, Boston, Sept. 1992, pp. 8–11.

[Blauvelt, 1992] Blauvelt, H., et al. "High-power 1,310nm DFB lasers for AM video transmission," *IEEE J. Light. Technology*, Vol. 10, No. 11, Nov. 1992, pp. 1766–1772.

[Verbiest, 1993] Verbiest, W., et al. "*FITL* and *BISDN*: a marriage with a future," *IEEE Communication Magazine*, June 1993.

[Ramaswami, 1993] Ramaswami, R., et al. "A packet-switched multihop lightwave network using subcarrier and wavelength division multiplexing," Submitted to *IEEE Trans. Selected Areas of Communication*, 1993.

[Wood, 1993a] Wood, T. H., et al. "Measurements of the effect of optical beat interference on the bit error rate of a subcarrier-based passive optical network," *Proc. Optic-Fiber Communication Conf., OFC/IOOC'93*, ThM3, Feb. 1993, pp. 231–233.

[Choy, 1993] Choy, M. M., et al. "A 200*Mb/s* packet-switched *WDM-SCM* network using fast *RF* tuning," Submitted to *IEEE J. Light. Technology*, 1993.

[Wood, 1993b] Wood, T. H., et al. "Operation of a passive optical network with subcarrier multiplexing in the presence of optical beat interference," *IEEE J. Light. Technology,* Vol. LT-11, No. 10, Oct. 1993, pp. 1632–1640.

Chapter 9
Time Division Multiaccess Networks

Time division multiple access (TDMA) is a well-known technique that has long been applied in satellite communication systems [Sklar, 1988]. More recently, it has successfully been applied in mobile communication systems such as the European GSM system (GSM stands for global system for communication) [Mouly, 1992].

Although TDMA does not really take advantage of the beneficial properties offered by optics, it has been proven to be very useful for fiber-system applications such as fiber in the loop (FITL). The local loop (i.e., the part of the public telecommunication network located between the local exchange and the subscriber homes) represents a major portion of the capital investment and maintenance expenses of telecommunication operators, and is highly cost-sensitive. With the trend towards broadband integrated services digital network (B-ISDN), a variety of FITL system architectures, which may offer cost advantages and flexibility to accommodate present and new customer demands, have been proposed. Basically, the proposed architectures fall into three major categories: the switched star, the active double star, and the passive optical network (PON), as illustrated in Figure 9.1 [Lin 1989, Kashima 1993].

Switched star networks, shown in Figure 9.1(a), run individual fiber from the central office (CO) directly to each subscriber, and each fiber carries only signals intended for that subscriber. In this architecture, there is neither sharing of physical media nor optoelectronics, resulting in a significant system cost.

The active double star architecture shown in Figure 9.1(b) reduces the amount of fiber required by transmitting a multiplexed signal to/from an active remote node located close to the subscriber premises. At the active remote node, the signal streams are multiplexed/demultiplexed, and each home is connected either with an optical drop fiber (in which case the network is referred to as fiber to the home, FTTH) or with a copper drop cable (in which case the network is referred to as fiber to the curb, FTTC).

A PON architecture, shown in Figure 9.1(c), shares optoelectronics and physical media by multiplexing signal streams destined to and coming from several homes onto a

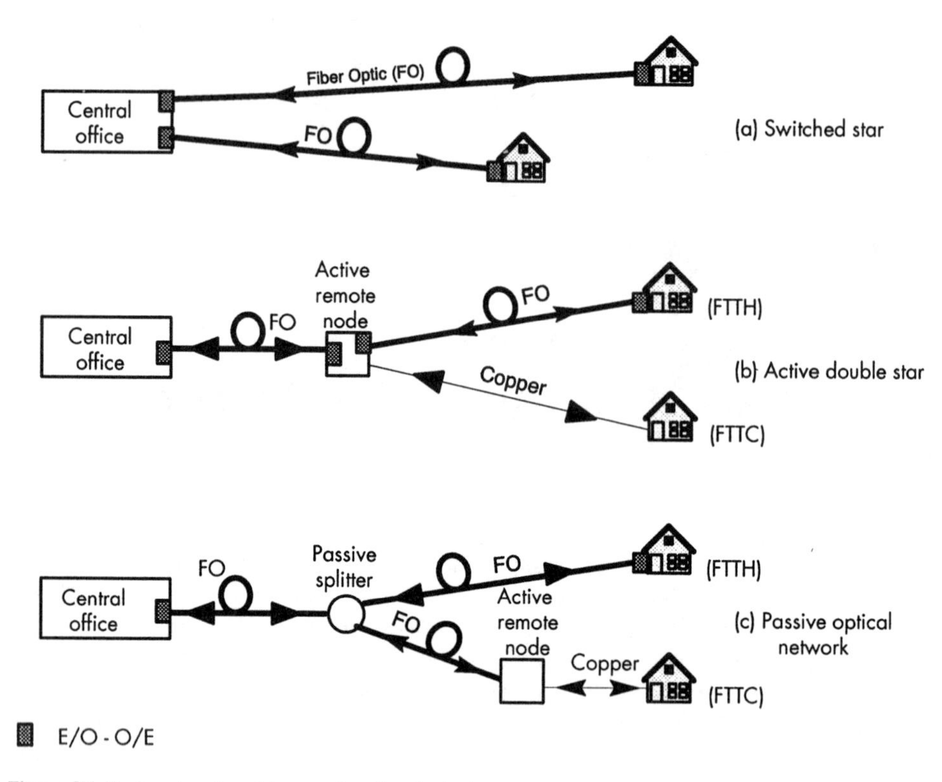

Figure 9.1 Basic network architecture for fiber-in-the-loop systems.

single fiber at the CO. Typically, a passive optical splitter is used to broadcast the multiplexed signal from the CO to the subscriber homes (a bus topology using passive optical taps or any other passive-broadcast fiber-network topology can be used as well). Upstream transmission, from the subscribers to the CO, requires a multiaccess scheme to combine the signals in a noninterfering way. TDMA is currently considered a very attractive candidate for such a purpose. In fact, the main advantages of the TDMA technique applied to FITL systems can be summarized as follows:

- The systems can be implemented with today's commercially available components.
- Overall system cost reduction can be achieved by resource sharing among a number of subscribers (i.e., nodes).

This chapter discusses TDMA fiber-optic networks with focus on PON systems. Section 9.1 introduces the basic concepts and highlights the major issues that arise when TDMA is used within fiber-optic networks. Section 9.2 is devoted to the description of three TDMA-based PON systems that have recently been demonstrated and trialed in the field. Finally, a summary is provided in Section 9.3.

9.1 BASIC CONCEPT

With TDMA, each node shares the communication resource by sending data in a synchronized way in order to avoid data collision. There are essentially two distinct ways to implement the TDMA approach: bit-based or block-based TDMA.

A bit-based TDMA data stream is created by interleaving the bits from each of the nodes. This approach only requires the storage of one bit for each node at any time and is therefore attractive with regard to the required memory space at each node. Unfortunately, the method requires that all nodes be bit-synchronized and is therefore impractical for high-bit-rate data streams usually encountered within fiber-optic systems.

The block-based TDMA approach somewhat relaxes the requirements on synchronization in that a small amount of dead time (later referred to as the guard time), which may be a few bits long, is allowed between multiplexed data blocks without adversely affecting the transport efficiency. The longer the blocks, the longer the guard time can be for the same transport efficiency.

Block-based TDMA can further be subdivided into two distinct classes: frame-based or packet-based. In frame-based TDMA, each node has assigned a fixed amount of capacity in every TDMA frame. In packet-based TDMA, each node is allowed to send a whole data packet when access is granted. Examples of block-based TDMA alternatives will be discussed in Section 9.2.

Notice that regardless of the type of TDMA used, whether bit or block based, the physical bit rate of the TDMA data stream is roughly equal to N times the data rate of the input/output data at each node, where N is the number of nodes connected to the network. In other words, even though the useful data rate per node is B (bit/sec), each node must be able to process data at a rate of $\approx NB$ (bit/sec). Since current optoelectronic transceivers operate at a maximum rate of, say, 10 Gbps [Hanenschild, 1991], the overall network capacity will also be limited to such a value at the most. Although this may seem quite restrictive, such a network capacity is large enough to accommodate most present (and possibly future) services envisaged within FITL systems.

To highlight the major issues that arise when TDMA is used within fiber networks, let us consider Figure 9.2, which represents a bus network with N nodes connected through passive taps. There is a central node controller that manages the data flow to and from the N nodes. Downstream data from the central controller (e.g., the central office) to the network nodes is broadcast, while upstream transmission is implemented through block-multiplexed TDMA.

A node i that has data destined to node j must first transmit it in the upstream direction towards the central controller, which then relays it in the downstream direction for final reception at node j.

The implementation of TDMA within such a network involves three main issues: (1) the ranging (or distance equalization), (2) the synchronization, and (3) the optical power leveling. Optical transceivers that implement these three functions are usually called burst-mode transceivers [Eldering, 1991].

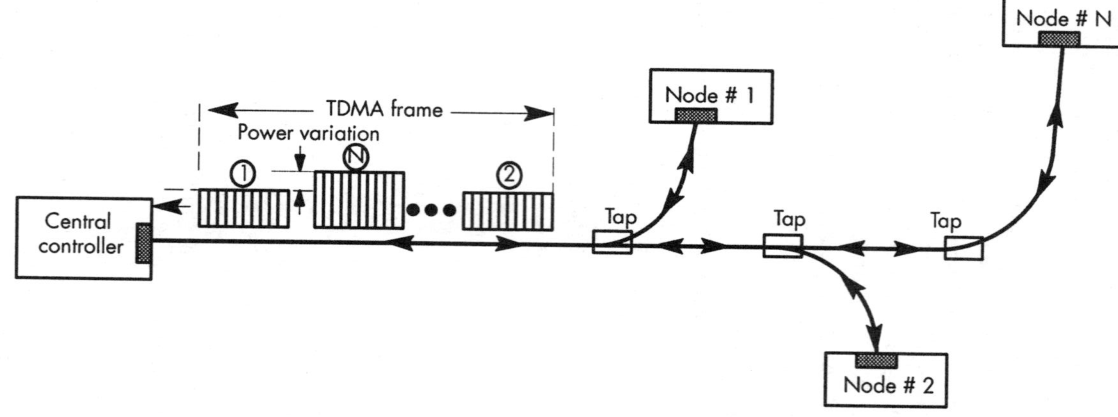

Figure 9.2 A passive optical network based on bus topology. The figure shows that bursts sent by distinct nodes arrive at the central receiver with different optical power levels and with a guard time between them (if this guard time is negative, then the bursts overlap in time, which induces errors during the detection process).

1. *The ranging*: due to the topology, the propagation distance from the nodes to the central controller will be, in general, different for each node. To form a TDMA stream without data block overlapping (i.e., without collision), some means are needed to virtually equalize the connection distance from all nodes to the central controller. This is generally achieved through what is called the ''ranging'' procedure, which measures the connection distance from each node to the central controller and determines the amount of electronic delay that must be inserted at each node: data at nodes that are the closest to the central controller will experience a large delay before being transmitted, while data at the farthest nodes will experience a small delay (which can possibly be zero for the most distant node). This way, all nodes appear at the same (virtual) propagation distance from the central controller irrespective of their physical location in the network.
2. *The synchronization*: Once the ranging has been performed, some means are needed to quickly recover the clock of each burst coming from distinct nodes. This is because the ranging cannot be achieved with infinite precision, so that guard time between bursts must be inserted to accommodate the ranging inaccuracy. Since nodes are assumed to send data at the same nominal rate, only the correct clock phase of each burst must be recovered. This can be achieved in distinct ways, using either phase locked loop (PLL) or more specific methods optimized for the intended application. Generally, the use of a PLL requires long preambles in front of each data block and therefore reduces the payload transport efficiency of the TDMA stream. Another method, based on oversampling, will be discussed in Section 9.2.3. This method requires as few as three bits of preamble to recover the correct clock phase of the TDMA bursts at 155 Mbps and is therefore very attractive in terms of transport efficiency.
3. *The optical power leveling*: the third issue concerns the variation of the received optical power at the central controller from burst to burst sent by distinct nodes. Indeed, bursts emitted by different nodes will experience different losses when propagating through the network and will impinge the central controller receiver with different optical power. The receiver thus requires a large dynamic range and should be able to set the threshold that discriminates bits ''0'' from bits ''1'' as quickly as possible [Eldering, 1993]. Reported devices for burst mode reception have shown a large dynamic range and performance up to 1 Gbps [Ota 1990, Ota 1992]. Burst mode reception incurs a power penalty with respect to continuous-mode operation usually applied in conventional optical-fiber communication systems. It has been shown that this penalty depends upon the number of preamble bits used to establish the threshold of the discriminator [Eldering, 1993]. The penalty is exactly 3 dB when a single bit preamble is used, it drops to 0.5 dB for an 8-bit preamble and becomes negligible (i.e., < 0.2 dB) for a 36-bit preamble. A simple implementation to rapidly set up the optimum receiver threshold will be described in Section 9.2.3.

9.2 DEMONSTRATED SYSTEMS

As mentioned in the introduction of this chapter, most of today's developed TDMA-based fiber-optic systems have been intended for deployment in the local loop. Pioneers in this field were researchers from British Telecom Research Laboratories who in 1987 demonstrated the feasibility to carry digital telephony over passive optical networks [Stern, 1987]. Since then, several other vendors have developed their own PON systems to support a variety of services ranging from plain old telephone service (POTS) to broadband ATM [Hawley 1991, Verbiest 1993].

The present section will briefly discuss some of these systems with emphasis on the relevant issues related to the TDMA approach. More specifically, we will focus on three distinct FITL systems; namely, the TPON, the MACNET, and the APON. Several other TDMA-PON systems are described in [Triboulet 1991, Abiven 1993, Smith 1993, Perrier 1993].

Besides these FITL systems, it is noteworthy to mention that an ATM photonic switch based on TDMA has been proposed in the literature [Tsukada, 1991]. It has been shown that this system can achieve ≈ 100 Gbps throughput by using 40-ps pulses generated by a DFB laser. Another high-capacity ultrafast TDMA optical system has been proposed and experimented in [Prucnal, 1986]. This system applies similar techniques to those used in CDMA networks (to be discussed in Chapter 10 and not be described here for the sake of space).

9.2.1 The TPON

The first system that has implemented the TDMA approach to FITL systems was the so-called TPON, which stands for telephony over passive optical network. The system, schematically shown in Figure 9.3, was demonstrated and trialed in the field by British Telecom Research Laboratories in 1989 [Rowbotham, 1991].

The system operates at a line rate of 20.48 Mbps in both directions and can support a maximum splitting factor of 128. The TDM/TDMA frames are composed of 294×64 Kbps channels, which can be configured to support digital telephony (64-Kbps speech and 8-Kbps signaling) and 144-Kbps ISDN links (18×8 Kbps) [Davies 1988, Hoppitt 1989].

Variants of the TPON system have been developed. In one of them, time compression multiplexing (TCM, or sometimes also called ping-pong) has been applied to provide duplex transmission at a single wavelength of 1.3 μm. In TCM, downstream TDM and upstream TDMA frames are transmitted alternatively, as shown in the timing chart of Figure 9.3. This makes it possible to relax the requirements on the reflectivity of the optical network as compared to the case of simultaneous full-duplex at a single wavelength [Rosher 1987, Yoshida 1989].

However, the line rate B_{TCM} at which the TCM transceivers must operate is more than twice the original source bit rate B. Indeed, referring to the time chart of Figure 9.3,

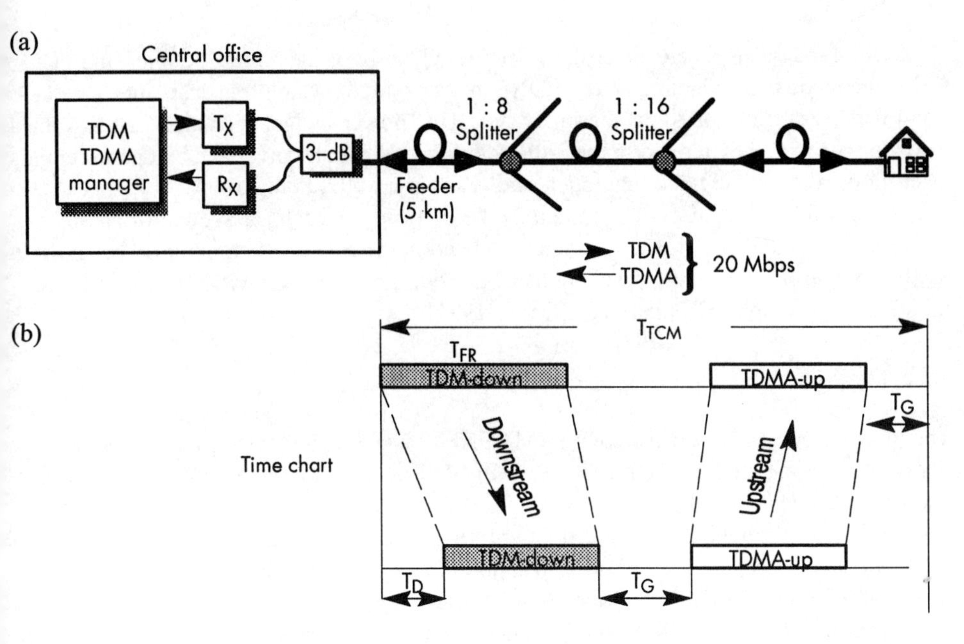

Figure 9.3 (a) Architecture of the TPON system and, (b) time chart for TCM (i.e., ping-pong) transmission.

it is seen that the burst cycle time T_{TCM}, defined as the delay between two successive frames transmitted in the same direction, is given according to

$$T_{\mathrm{TCM}} \geq 2(T_D + T_{\mathrm{FR}} + T_G) \tag{9.1}$$

where T_D is the delay time associated with the transmission time of optical signals in the fiber cable, T_{FR} is the time of the downstream or upstream TCM frames (including the time for preamble bits) and T_G is the guard time associated with the switching time between the receiving and transmitting modes of the transceivers.

Since $(T_{\mathrm{FR}} B_{\mathrm{TCM}})$ corresponds to the number of bits in the downstream or upstream frames, we have

$$T_{\mathrm{FR}} \cdot B_{\mathrm{TCM}} = T_{\mathrm{TCM}} \cdot B \tag{9.2}$$

Using (9.1) and (9.2), the operating line rate in the TCM system is obtained as:

$$B_{\mathrm{TCM}} \geq \frac{2BT_{\mathrm{TCM}}}{[T_{\mathrm{TCM}} - 2(T_D + T_G)]} \tag{9.3}$$

which shows that B_{TCM} is always greater than $2B$, even in the ideal case where $T_G = 0$.

Since the receiver sensitivity drops by roughly 3 dB as the bit rate doubles, the TCM technique leads to a reduction of the PON's power budget, which translates into a reduction of the maximum possible sharing factor. This power-budget reduction can to some extent be overcome by removing the 3-dB coupler at each transceiver and by operating the laser diode alternatively as a photodetector [Kashima 1992, Semo 1993].

A key feature of the TPON system is the inclusion in each transceiver of an optical filter that passes only the 1.3-μm wavelength band in the customer's terminal. This allows the future addition of other wavelengths to provide new services without disturbing the initial telephone and ISDN services [Cook, 1991].

9.2.2 The MACNET

The multiple access customer network (MACNET) is a TDM/TDMA-based PON system that was demonstrated at the prototype level by Telecom Australia and Standard Telephones & Cables Ltd. in 1989 [McGregor, 1989]. The system is similar to the TPON system described previously except for a splitting factor of 16 and a network capacity of 8.192 Mbps. This capacity is provided to support 20 basic accesses at 160 Kbps, 1 primary access at 2 Mbps and one broadcast channel at 2 Mbps.

The prototype MACNET, shown schematically in Figure 9.4, uses a single fiber to support both directions of transmission. Full-duplex transmission is achieved using the coarse WDM approach in the 1.3-μm fiber window; that is, 1,320 ±7 nm is used for downstream transmission and 1,280 ±7 nm for upstream transmission.

Both directions of transmission use 1-ms frames. The frame structures for the downstream and upstream directions are shown in Figure 9.4. At the beginning of each 1-ms downstream frame there is a 40-bit frame header for frame synchronization and some additional spare capacity provided for maintenance functions. The remaining space in the frame is used for actual data blocks, one block for each channel on the network separated by short blocks of the 2-Mbps broadcast channel.

The system is designed so that the maximum difference in distance between the closest and the farthest customer to the star coupler is 1 km. Hence, the loop-delay difference in the fiber drops (i.e., between the splitter and the homes) can be up to nearly 10 μs at the most. Instead of performing a fine ranging to artificially reduce this loop-delay difference, the MACNET introduces fixed guard time of 10 μs between upstream blocks to avoid collision. Upstream transmission associated with a particular channel starts immediately (i.e., the nodes do not require any delay line) after reception of the downstream block associated with that channel. Insertion of the broadcast blocks in the downstream frame provides the guard time for upstream collision-free transmission.

9.2.3 The APON

APON refers to ATM-based passive optical networks and has independently been demonstrated by several companies in the early 1990s [Ballance 1990, Mestdagh 1991,

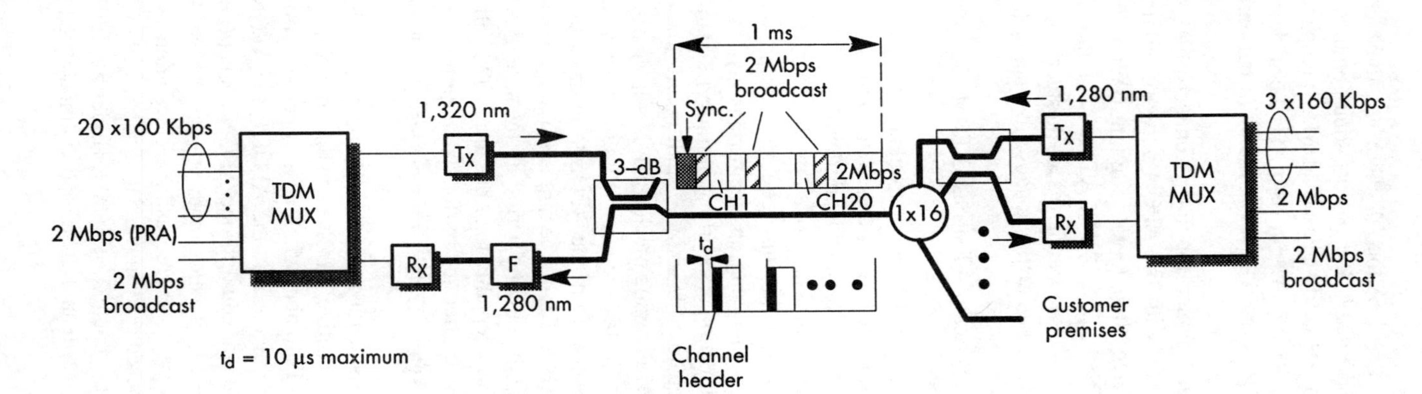

Figure 9.4 The MACNET system. (© 1989 IEEE. [After McGregor, 1989.])

Ishikura 1991, du Chaffaut 1993]. Although there are some characteristic differences between these systems (e.g., the network capacity), their common feature is that upstream transmission is performed through ATM-based TDMA (i.e., packet-based TDMA).

Figure 9.5 shows the APON system as it has been developed by Alcatel Bell [Mestdagh 1991, Van Der Plas 1992]. This system is currently being trialed in the Bermuda's islands.

The 1×16 passive star coupler topology connects the APON line termination (APOLT) at the exchange side with the APON network termination (APONT) at the subscriber side. At the exchange, the interworking unit (IWU) gives subscriber access to the different networks (PSTN, PSDN, ISDN, ATM, . . .) and service providers (video server, connectionless server, . . .) via standardized interfaces. At the subscriber side, the service unit (SU) adapts the data from different subscriber applications to the ATM format and vice versa.

The line rate of the downstream ATM-based TDM data is 622.08 Mbps while that of the upstream TDMA is 155.52 Mbps.

For both directions, ATM cells are encapsulated in so-called APON packets: an 8-bit packet overhead is added to each ATM cell to provide synchronization and network-transport-related functions such as APONT identification.

TDMA upstream transmission is controlled by the APOLT that generates grants for each APONT according to their bandwidth requests. The grants are transported in the downstream overhead octets. Grants and upstream transmission are pipelined in such a way that in the upstream direction a nearly continuous stream of APON packets is multiplexed on the feeder, separated by only a small guard time.

To limit this guard time, a ranging mechanism is implemented. This ranging can be performed by measuring the delay between transmission of a ranging signal in the APOLT and reception of the replied signal from the APONT. In order to prevent interference of the data packets from active APONT with the ranging signals, upstream transmission must be regularly interrupted during the ranging period. In the case of ATM-based TDMA, this interruption period might be unacceptably long because ATM cell accumulation at active APONTs must be avoided due to jitter and buffer size requirements. In order to minimize the interruption period, the APON system implements the ranging mechanism in two steps.

During the first step, a coarse ranging is carried out without interruption of upstream data transmission from already-active APONTs. This coarse ranging reduces the range difference between the APONTs to about 550m (i.e., one APON packet at 155 Mbps): in response to a ranging signal received from the APOLT, the APONT transmits a number of ranging pulses with pulse width equal to one APON packet (i.e., 360 Kbps) and with a low peak amplitude (e.g., −10 dB below data pulses). At the APOLT, the embedded low-frequency ranging pulses are recovered using a correlation technique. Alternatively, the ranging pulses can be imposed on a subcarrier and detected using an RF demodulator as discussed in Chapter 8 in the context of SCMA [Eldering, 1993b].

During the second step, upstream transmission is interrupted for a time correspond-

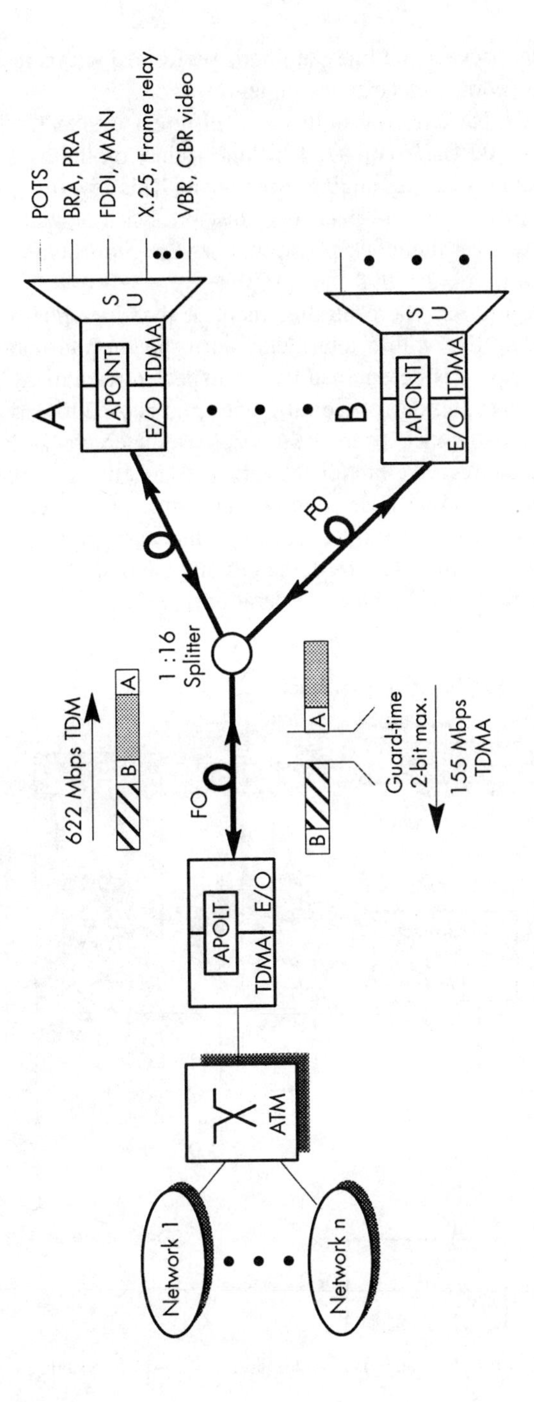

Figure 9.5 The APON system as proposed and field tested by Alcatel Bell Telephone.

ing to only two APON packets and fine ranging is performed with a resolution of 1-bit at 155 Mbps using conventional electronic counters.

Upstream APON packets, transmitted by different APONTs, interleave on the feeder and arrive at the APOLT with +/- 1 bit uncertainty on the packet position due to the unavoidable inaccuracy of the ranging mechanism. This requires bit synchronization to be performed for every received packet. A fast bit synchronization mechanism that makes it possible to recover the clock phase with a very short preamble of only a 3-bit length is implemented in the APOLT.

As shown in Figure 9.6, the central element of the clock-phase alignment (CPA) circuit is a tapped delay line with a total delay corresponding to one bit at 155 Mbps, where the number of taps n is determined by the expected resolution. This delay line is tuned to provide a precise 360-deg phase difference and is used to derive n clock signals with increasing phase difference from the crystal oscillator clock. Each of the clock signals samples three consecutive bits of the received data using D-type flip-flops. The phase recovery is then based on the detection of the center of a specific 3-bit pattern, for instance "010." Two latched priority encoders, which are triggered from the central clock, are used to detect the first "1" from the left and the first "1" from the right of the $3n$ data samples. Using a few arithmetic operations, the center of the 3-bit pattern is

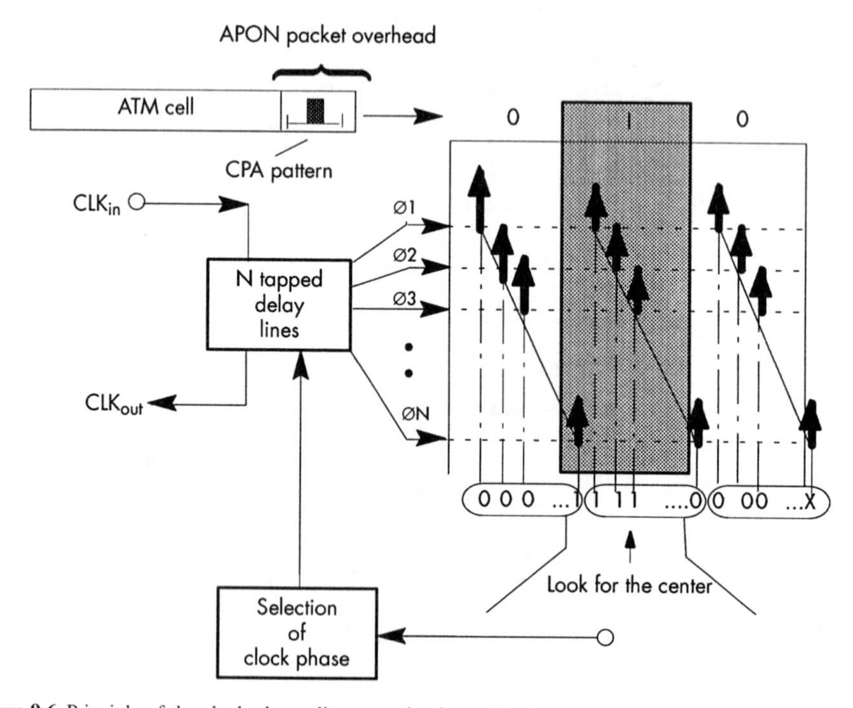

Figure 9.6 Principle of the clock phase alignment circuit used in the APON system.

calculated and the tap giving the right clock phase is selected and used for subsequent sampling of the ATM cell.

Fast gain control at the APOLT receiver is implemented with the help of the APON system protocol. Since the APOLT controls the upstream access, it knows when an APON packet from a particular APONT is expected to arrive at its receiver. Hence, the APOLT can preset the receiver gain for each expected APON packet in a synchronized way. The preset gain values are determined on the basis of high and low power measurements over the previous packets: at the end of a particular packet, the high and low power values measured are converted into digital values and saved in memory. At the beginning of the next APON packet from the same APONT, the memorized values, which are continuously updated, are used to determine the start value of the gain-control circuit. The threshold, usually set halfway between the level of bit "1" and bit "0" is calculated by taking the mean of the two memorized high and low power values. Since the APOLT knows when a packet from a particular APONT is expected to arrive, it provides this threshold value to the discriminator just before the packet arrives at the receiver.

9.3 SUMMARY

Among the various multiple-access techniques that can be applied to optical-fiber networks, TDMA is likely to be the less performant in terms of achievable network capacity. This is because TDMA is limited by the operating speed of optoelectronic transceivers, which represents less than one tenth of a percent of the huge available bandwidth of optical fibers.

Nevertheless, where minimizing cost rather than maximizing capacity is the major objective, TDMA appears to be one of the most promising techniques for short-term implementation of multiaccess optical-fiber networks. This is particularly important for FITL systems, which may represent the major capital investment of public telecommunication operators in the near future.

The implementation of optical TDMA requires some modifications of conventional transceivers that typically operate in a continuous mode. Burst-mode transceivers supporting rates of up to 1 Gbps have been demonstrated in research laboratories and several burst-mode TDMA-PON systems supporting a range of services from POTS to broadband ATM are currently being trialed in the field.

These systems may find widespread deployment in the years to come as they present several attractive features such as a low first-installation cost and maintenance expense, the capability to satisfy existing and expected future service demands, and the possibility to be upgraded by using other multiaccess techniques like WDMA and SCMA as an overlay to TDMA.

REFERENCES

[Prucnal, 1986] Prucnal, P. R., et al. "Ultra-fast all optical synchronous multiple access fiber networks," *IEEE J. Selected Areas of Communication*, Vol. JSAC-4, No. 9, Dec. 1986, pp. 1484–1493.

[Stern, 1987] Stern, J. R., et al. ''Passive optical local networks for telephony applications and beyond,'' *Electron. Letters*, Vol. 23, No. 24, Nov. 1987, pp. 1255–1257.

[Rosher, 1987] Rosher, P. A., ''Receiver penalty calculations in duplex optical systems,'' *Proc. Globecom'87*, 44.7, 1987, pp. 1550–1555.

[Sklar, 1988] Sklar, B. *Digital Communications: Fundamentals and Applications*, Englewood Cliffs, NJ: Prentice Hall, 1988.

[Davies, 1988] Davies, P. A., et al. ''System considerations for the upstream channel of a passive optical local network,'' *Proc. Sixth European Fiber-Optic Communication & Local Area Networks, EFOC/LAN'88*, 4.6, 1988, pp. 389–394.

[Lin, 1989] Lin, Y-K. M., et al. ''Fiber-based local access network architectures,'' *IEEE Communication Magazine*, Oct. 1989, pp. 64–73.

[Hoppitt, 1989] Hoppitt, C. E., et al. ''The provision of telephony over passive optical networks,'' *British Telecom Technology J.*, Vol. 7, No. 2, April 1989.

[Yoshida, 1989] Yoshida, A. ''Design considerations for optical duplex transmissions,'' *Electron. Letters*, Vol. 25, No. 25, Dec. 1989, pp. 1723–1725.

[McGregor 89] McGregor, I. M., et al. ''Implementation of a *TDM* optical network for subscriber loop applications,'' *IEEE J. Light. Technology*, Vol. LT-7, No. 11, Nov. 1989, pp. 172–178.

[Ballance, 1990] Ballance, J. W., et al. ''*ATM* access through a passive optical network'' *Electron. Letters*, Vol. 26, No. 9, April 1990, pp. 558–560.

[Ota, 1990] Ota, Y., et al. ''Burst-mode compatible optical receiver with a large dynamic range,'' *IEEE J. Light. Technology*, Vol. LT-8, No. 12, Dec. 1990, pp. 1897–1903.

[Rowbotham, 1991] Rowbotham, T. R., ''Local loop development in the U. K.,'' *IEEE Communication Magazine*, June 1993, pp. 50–59.

[Hawley, 1991] Hawley, G. T., and J. W. Shumate, Eds. *Special Issue on the 21st Century Subscriber Loop, IEEE Communication Magazine*, Vol. 29, No. 3, March 1991, pp. 24–119.

[Cook, 1991] Cook, A.R.J., et al. ''Broadband digital transmission over passive optical networks,'' *Proc. Globecom'91*, 18.1.1–5, 1991, pp. 597–601.

[Triboulet, 1991] Triboulet, M., et al. ''Experience, lessons and conclusions resulting from early trials of passive optical networks (*PON's*) and their related *O&M* systems,'' *Proc. ISSLS'91*, 1991, pp. 223–229.

[Tsukada, 1991] Tsukada, M., et al. ''Ultrafast photonis *ATM* switch based on time-division broadcast-and-select network,'' *Proc. Globecom'91*, 34.5.1–5, 1991, pp. 1230–1234.

[Mestdagh, 1991] Mestdagh, D.J.G., et al. ''*ATM* local access over passive optical networks,'' *Third IEEE Workshop on Local Optical Networks*, G.3.1–8, Tokyo, Japan, Sept. 1991.

[Hanenschild, 1991] Hanenschild, J., et al. ''Silicon bipolar chipset with maximum data rates from 10 to 23*Gbits/s* for optical communications,'' *Electron. Letters*, Vol. 27, No. 25, Dec. 1991, pp. 2383–2384.

[Eldering, 1991] Eldering, C. A., et al. ''Transmitter and receiver requirements for *TDMA* passive optical networks,'' *Third IEEE Conf. Local Optical Networks*, Tokyo, Japan, Sept. 24–25, 1991.

[Ishikura, 1991] Ishikura, A., et al. ''A cell-based multipoint *ATM* transmission system for passive double star access networks,'' *Third IEEE Workshop on Local Optical Networks*, G3.4.1–10, Tokyo, Japan, Sept. 1991.

[Ota, 1992] Ota, Y., et al. ''*DC-1Gb/s* burst-mode compatible receiver for optical bus applications,'' *IEEE J. Light. Technology*, Vol. LT-10, No. 2, Feb. 1992, pp. 244–249.

[Kashima, 1992] Kashima, N. ''Time compression of multiplex transmission system using a 1.3 μm semiconductor laser as a transmitter and a receiver,'' *IEEE Trans. Communication*, Vol. 40, 1992, p. 584.

[Van der Plas, 1992] Van der Plas, G., et al. ''*ATM* over passive optical networks: system design and demonstration,'' *SPIE's International Symposium OE/Fiber'92*, Boston, MA, Sept. 1992.

[Mouly, 1992] Mouly, M., and M-B Pautet. *The* GSM *System for Mobile Communications*, published by the authors, ISBN-2950719007, 1992.

[du Chaffaut, 1993] du Chaffaut, G., et al. ''*ATM-PON*: une famille de systemes optiques pour la distribution—l'exemple *SAMPAN*,'' *L'Echo des Recherches*, No. 154, 1993, pp. 27–38.

[Kashima, 1993] Kashima, N. *Optical Transmission for the Subscriber Loop*, Norwood, MA: Artech House, 1993.

[Semo, 1993] Semo, J., et al. "High responsivity 1.3 μm transceiver module for low-cost optical half-duplex transmission," *Electron. Letters*, Vol. 29, No. 7, April 1993, pp. 611–612.

[Verbiest, 1993] Verbiest, W., et al. "*FITL* and *B-ISDN*: a marriage with a future," *IEEE Communication Magazine*, June 1993, pp. 2–8.

[Eldering, 1993] Eldering, C. A. "Theoretical determination of sensitivity penalty for burst mode fiber optic receivers," *IEEE J. Light. Technology*, Vol. LT-11, No. 12, Dec. 1993, pp. 2145–2149.

[Abiven, 1993] Abiven, J., et al. "*TOME*: une approche originale des reseaux optiques passifs," *L'Echo des Recherches*, No. 154, 1993, pp. 19–26.

[Perrier, 1993] Perrier, P. A., "A *PON* architecture with a single optical carrier bidirectional transmissions," *Fourth Workshop on Optical Local Networks*, 1993.

[Smith, 1993] Smith, P. J., et al. "A high speed optically amplified TDMA distributive switch network," *Proc. European Conf. Optical Communication, ECOC'93*, MoC1–5, 1993, pp. 89–92.

[Eldering, 1993. b] Eldering, C. A., et al. "Out of band signalling for passive optical networks," *Intl. J. Digital & Analog Communication Systems*, Vol. 6, Jan.-March 1993, pp. 49–52.

Chapter 10
Code Division Multiaccess Networks

This last chapter deals with the spread-spectrum version of TDMA, known as code division multiple access (CDMA). The CDMA technique has been extensively studied in the context of microwave communications as it allows users to access any shared channel randomly at an arbitrary time. Its use in optical-fiber networks has attracted considerable attention since 1985 [Hui, 1985].

Optical CDMA exploits the possibility to generate ultrashort light pulses (i.e., in the picosecond or even femtosecond ranges) to encode each bit of data from the source nodes into a pulse train with a unique pattern, called the CDMA code or the address/signature sequence. The CDMA optical signal emitted by each node occupies, therefore, a bandwidth much in excess of the minimum bandwidth necessary to send the information.

Nevertheless, as we will see in this chapter, these signals can be generated purely optically by means of fiber-optic delay lines. At the receiver, optical correlation is performed with a set of parallel fiber delay lines and data is recovered using a threshold device.

By a proper choice of the CDMA codes, the signals from all network nodes can be made as mutually noninterfering as possible so that simultaneous multiaccess with no delay can be achieved without the need for a complex network protocol to coordinate the data transfer among the communicating nodes.

The major advantages of the CDMA technique with respect to the other multiple-access techniques that have been discussed in the previous chapters can be summarized as follows.

1. Compared with TDMA, CDMA does not require that all nodes be synchronized together.
2. Compared with WDMA, the need for wavelength-tunable transceivers or wavelength-stabilization schemes is eliminated. In other words, all nodes are allowed to emit at the same center wavelength.

3. Compared with SCMA, higher network capacity can be achieved because the required signal processing can be performed in the optical domain rather than electronic one.

In addition, CDMA offers the advantage of spread-spectrum methods in that it is difficult to jam or intercept a signal destined to another node because of its coded nature. This represents a particularly attractive feature where data security/privacy is of prime concern.

On the other hand, CDMA also presents some major disadvantages that have so far restricted their usefulness in practical systems.

1. The method requires light sources capable of generating ultrashort pulses in the picosecond or even femtosecond ranges. Although semiconductor lasers generating ultrashort light pulses have been demonstrated in research laboratories, such devices are still in their infancy and much effort is needed to make them usable in real-life systems.
2. Present CDMA optical encoders and decoders require electronically-controlled optical switches and fiber delay lines, resulting in rather bulky devices.

This chapter is organized as follows. In Section 10.1, we will introduce the basic concept of CDMA optical-fiber networks based on direct detection (DD). In Section 10.2, we will discuss the design issues of optical codes that essentially determine the performance of optical CDMA networks. Section 10.3 is devoted to the description of a novel CDMA optical technique based upon spectral encoding of ultrashort light pulses, referred to as coherent optical CDMA. Finally, Section 10.4 will summarize the main characteristics of optical CDMA networks.

10.1 BASIC CONCEPT

A typical fiber-optic CDMA network is represented in Figure 10.1 where the nodes are connected together through a passive $N \times N$ star coupler. At the node transmitter, an optical encoder maps each bit "1" of source information into a very high rate optical sequence of ultrashort light pulses. The bits "0" of source information are not coded; that is, they are represented by an all-zero sequence.

The coded signal is then coupled into the input single-mode fiber and broadcast to all node receivers. The optical-pulse sequence, unique to each node, represents the address code (also called the signature sequence) of the node. To send data from node j to node k, the address code of receiver k is impressed upon the data by the encoder at the jth node. The coded bit "1" of node j destined to node k can be represented by

$$c_j(t) = \sum_{n=1}^{F} c_{k,n} \cdot p(t - t_n) \tag{10.1}$$

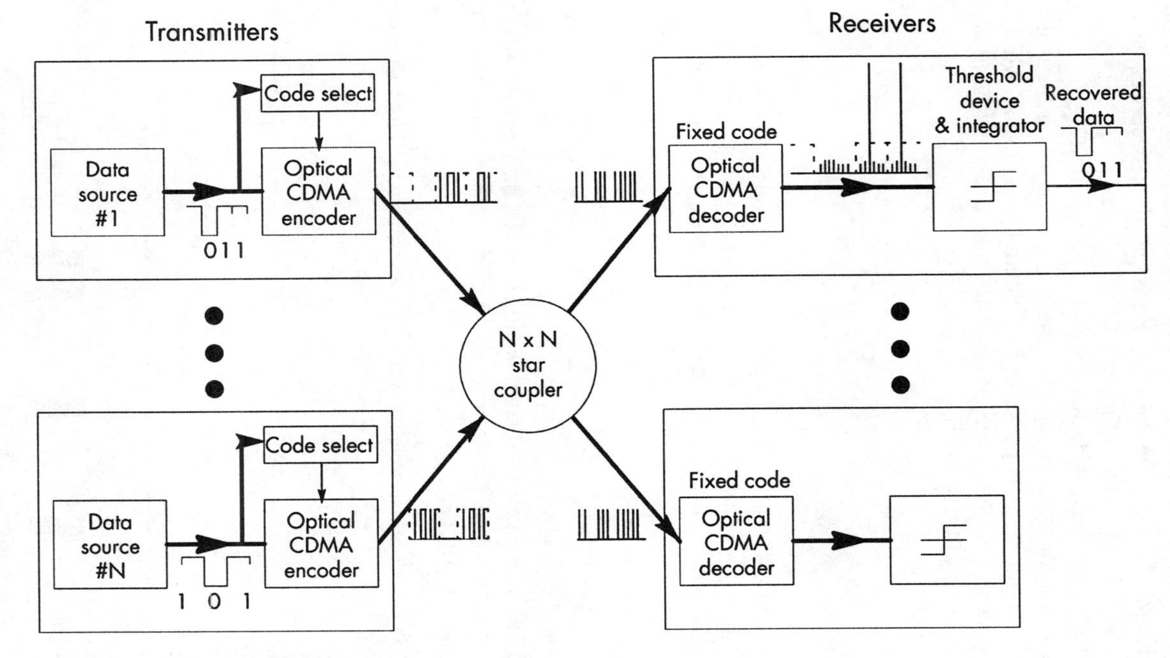

Figure 10.1 A fiber-optic CDMA network.

where $p(t)$ is an ultrashort light pulse of duration T_c and is referred to as a "*chip*." The value $c_{k,n} = 0$ or 1 for $1 < n < F$ and determines the CDMA address code of node k. F is referred to as the code length and is given by $F = T/T_c$,where T is the duration of the source bit. The sum of bits "1" in the sequence is called the code weight K. An example of a CDMA coded bit is shown in Figure 10.2.

At receiver k, the optical decoder demaps the received aggregate optical signal composed of the sum of N coded signals sent simultaneously (in the worst case) by all node transmitters. Demapping is performed by a correlation process, which compares the optical-pulse sequences to the stored-address sequence associated with node k. The received signal can be expressed as

$$r(t) = \sum_{j=1}^{N} c_j(t - t_j) \tag{10.2}$$

where $c_j(t - t_j)$ corresponds to the jth node's signal and t_j represents a random time associated with the jth signal. The intended node's receiver must be able to extract its address sequence from $r(t)$ through a correlation process whose output is given by

$$s_k(t) = \sum_{l=-\infty}^{+\infty} r(t) \cdot c_k(t - t_l) \tag{10.3}$$

where $t_l = l \cdot T_c$. To maximize the discrimination between the desired signal and all other signals (destined to other node receivers), the set of CDMA codes must verify the following correlation properties.

1. Each CDMA code must have an autocorrelation peak as large as possible; that is, for any j

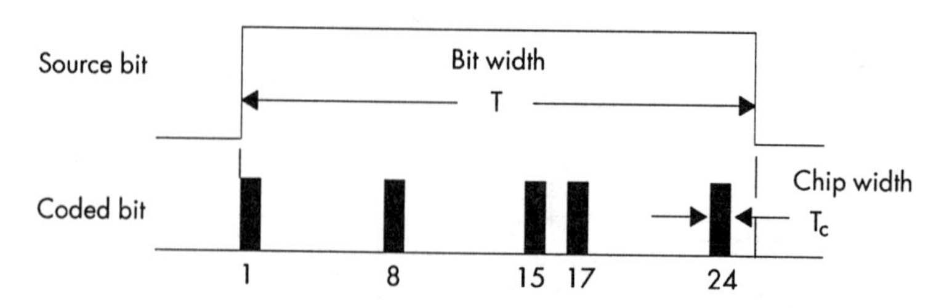

Figure 10.2 Encoding of a source bit "1" with a CDMA chip sequence. Number of chips, $F = 25$. Number of 1's, $K = 5$. (© 1987 IEEE. [After Prucnal, 1987.])

$$\sum_{n=1}^{F} c_{j,n} \cdot c_{j,n} = K \gg 0 \tag{10.4}$$

where K corresponds to the number of 1's in the CDMA code sequence and is called the code weight.

2. Each CDMA code can easily be distinguished from every other CDMA code in the set. In other words, the cross-correlation of two distinct CDMA codes must take a value as low as possible. This can be mathematically expressed as (for $j \neq k$)

$$\sum_{n=1}^{F} c_{j,n} \cdot c_{k,n+m} \leq \lambda_c \text{ for } 0 \leq m \leq F - 1 \tag{10.5}$$

where λ_c denotes the cross-correlation constraint.

Although these two conditions provide good discrimination properties among the different CDMA codes in the set, a third condition is generally required for reasons that will become clear shortly. This third condition is:

3. Each CDMA code can easily be distinguished from a shifted version of itself. In other words, it is required that all CDMA codes in the set exhibit the following autocorrelation property:

$$\sum_{n=1}^{F} c_{j,n} \cdot c_{j,n+m} \leq \lambda_a \text{ for } 1 \leq m \leq F - 1 \tag{10.6}$$

where λ_a denotes the autocorrelation constraint.

This property ensures that all pulses at the output of the decoder have amplitude less than or equal to λ_a except the dominant decoded pulse, which has an amplitude equal to K (10.4). This makes it possible to reduce the probability of jitter (and hence to reduce the BER) due to early threshold crossing caused by the sidelobes of the autocorrelation function [Santoro, 1987]. This condition may be dropped if the receiver is synchronized to the expected position of the autocorrelation peak; that is, a scheme referred to as *synchronous* CDMA [Prucnal 1986a, Wu 1992].

Note that when condition (10.6) is fulfilled, in which case the multiple-access method is referred to as an *asynchronous* CDMA, the number of sequences in the code set is reduced by a factor F - 1 as compared to the synchronous case [Prucnal 1986b]. Indeed, with synchronous CDMA, the F - 1 time-shifted versions of a given code sequence can be used, while this is not allowed for asynchronous CDMA. Therefore, from a code set standpoint, a synchronous CDMA network can support F - 1 times more nodes than an asynchronous CDMA network. However, the major drawback of synchronous CDMA as compared with asynchronous CDMA, or even with TDMA, is the tight synchronization requirement that is required. Therefore, even though the asynchronous CDMA approach supports fewer nodes, it is generally preferred to synchronous CDMA.

In general, CDMA codes are characterized by the set of the four parameters (F, K, λ_a, λ_c) and the performance of optical CDMA networks is essentially determined by the choice of these parameters. The CDMA code design issues are discussed in Section 10.2.

CDMA coders and decoders can be constructed using all-optical structures as shown in Figure 10.3(a, b).

At the transmitter, a $1 \times M$ power splitter divides the input-ultrashort light pulse into M copies of equal power. These copies are injected into a parallel set of fiber-optic delay lines and the variously delayed outputs are then combined by a $M \times 1$ coupler to form the CDMA code sequence.

At the receiver, the optical decoder consists of parallel fiber-optic delay lines, with lengths corresponding to the positions of the "1" of the CDMA code sequence. The

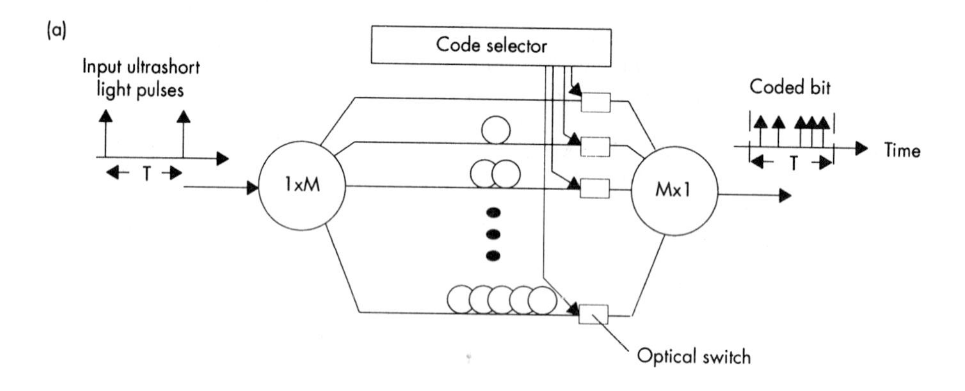

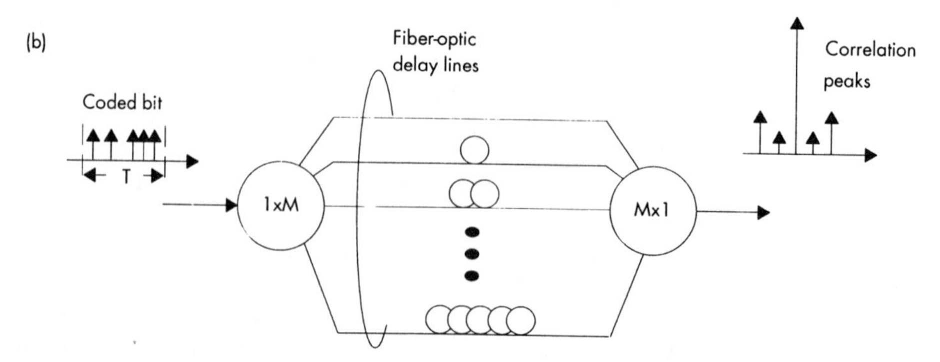

Figure 10.3 (a) CDMA optical coder and (b) decoder using parallel fiber-optic delay lines. The coder in (a) is made tunable by means of optical switches.

received optical signal is split among these delay lines and then recombined. The output of the optical decoder is detected either by an optical-threshold device (i.e., an optical bistable) [Prucnal, 1990] or by a high-speed photodetector whose electrical output is compared to a threshold level at the comparator for data recovery. When the signal level is above the threshold, a decoded bit "1" is issued. Conversely, when it is below the threshold, a decoded bit "0" is issued.

Decoding errors can occur when a node transmits a bit "0" (which corresponds to sending an all-zero sequence) and the interference due to the other $N - 1$ node's signal causes a cross-correlation peak with a level above threshold. The probability of error depends upon the threshold level (T_h), the correlation properties of the CDMA code set, and the number of interfering signals.

For example, suppose that the CDMA code set (F, K, λ_a, λ_c) has a cross-correlation constraint $\lambda_c = 1$. If the total number of interfering signals $N - 1$ is less than the code weight K, then no error will occur if one chooses a threshold level such that $T_h > N - 1$ (assuming that other noises such as quantum noise and thermal noise are neglected). If the number of interfering nodes is greater than or equal to the code weight (i.e., if $N - 1 \geq K$), then the interfering signals can lead to cross-correlation peaks that are above T_h, resulting in errors with some probability (indeed, T_h must be set lower than K for detection of the auto-correlation peak of the desired signal).

In general, the BER of CDMA systems is relatively high (typically > 10^{-6}) unless the network traffic load is kept to low values or forward-error correction schemes are used. Optical CDMA would, therefore, be desirable in a network where only a fraction of the many nodes having bursty traffic communicate at a given time. As we will see in Section 10.2, a BER << 10^{-10} can be achieved by a proper choice of the code parameters and network dimensioning.

For switched network applications, either the coder, the decoder, or both must be tunable. This tunability can be achieved by means of electro-optical switches, which control access to the different fiber-optic delay lines in the coder/decoder, as shown in Figure 10.3(a) [Zhang, 1993].

The question of where to place the tunability has already been addressed in Section 7.1.1 in the context of single-hop broadcast-and-select WDMA networks, and the same discussion applies here for CDMA networks.

To conclude this section, it is noteworthy that an alternative to direct-sequence optical CDMA networks, as discussed above, has been proposed in [Kiasaleh, 1991]. This alternative applies a frequency-hopping technique where the carrier frequency of the modulated data is shifted periodically according to a CDMA code. Frequency hopping can be performed in the optical (i.e., wavelength hopping) or electrical domain. A tunable light source is required for optical-domain implementation. The frequency-domain approach can be implemented by hopping the frequency of a microwave subcarrier and then using the SCMA technique discussed in Chapter 8. This approach has the advantage that it can be implemented with existing commercial microwave components and standard frequency-hopping transmitters and receivers.

10.2 CDMA OPTICAL CODES

During the last two decades, a considerable study effort from communication engineers and mathematicians has been devoted to the search for optimum sequences for application in CDMA-based systems such as radar systems, ranging systems, and mobile communication systems [Gold 1967, Cooper 1978, Sarwate 1980, Weber 1981, Holmes 1982]. Optimum CDMA codes have been found assuming bipolar signals (such as those that can be manipulated by conventional electronic circuits), which can take on positive (+1's) and negative (-1's) values. Though optical signals can also be processed coherently to provide bipolar signals, today's practical optical-fiber systems use direct detection and can therefore process only unipolar signals consisting of ''1''s and ''0''s. Since CDMA codes designed for bipolar signals do not necessarily maintain their desired properties for unipolar signals, new codes that are specific to (positive) optical systems must be found.

CDMA codes consisting entirely of 1's and 0's are referred to as *optical codes* and several variants have been recently proposed in the literature [Prucnal 1986, Salehi 1987, 1989a, Chung 1989, 1990, Petrovic 1990]. However, the latest search for optimum CDMA optical codes has only yielded some design guidance, and the requirements and the effects of the various code parameters on system performance have not yet been clearly established.

For all proposals, the cross-correlation constraint is set at either $\lambda_c = 1$ (which represents the minimal value for optical systems) or $\lambda_c = 2$.

An important class of CDMA optical codes is the so-called set of orthogonal optical codes (OOCs) for which the auto and cross-correlation constraints are equal to unity (i.e., $\lambda_a = \lambda_c = 1$). Figure 10.4 shows an example of two OOCs with a code length $F = 32$ and a code weight $K = 4$.

It can be shown [Salehi, 1987] that the number of network nodes that can be accommodated with OOCs is upper bounded by

$$N \leq \text{int}\left[\frac{F-1}{K(K-1)}\right] \tag{10.7}$$

where int[x] denotes the integer part of the real value x.

For a given code length F, it is seen from (10.7) that in order to support a large number of nodes N, the value of K must be made as small as possible. In other words, the OOCs must be very sparse in 1's. This general conclusion is also valid for the other optical codes proposed so far. This sparseness of 1's has two important consequences for network design:

- First, the overall network power budget may be significantly limited since the energy per bit will be low unless chips with high peak power are used.

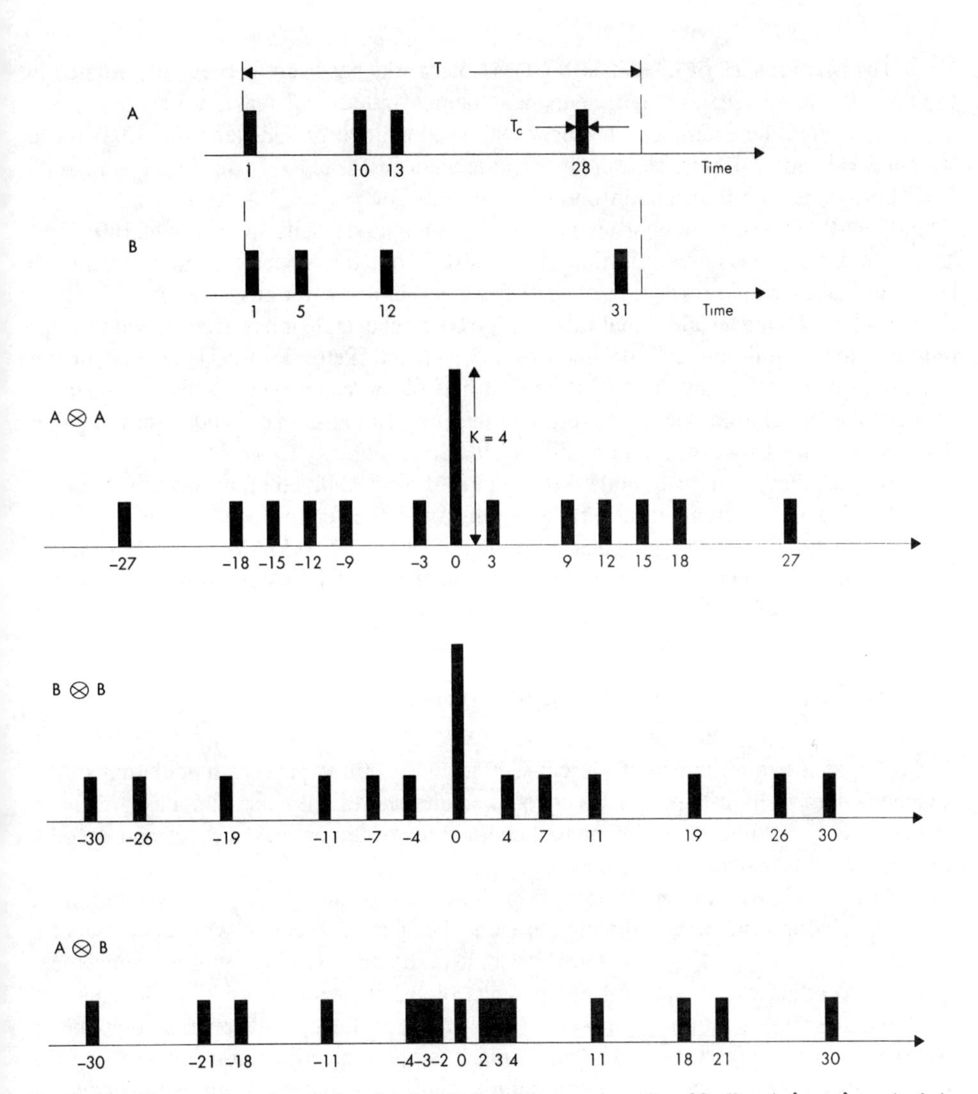

Figure 10.4 Two orthogonal optical codes A and B with parameters: $F = 32$, $K = 4$, $\lambda_a = \lambda_c = 1$, their autocorrelations (i.e., $A \otimes A$ and $B \otimes B$) and crosscorrelation (i.e., $A \otimes B = B \otimes A$).

- Second, the ratio F/K must be large, which means that the chip width must be much smaller than the width of source bits. Therefore, fiber dispersion effects may restrict the geographical coverage of CDMA networks to such sizes usually encountered in LAN applications, but not larger. This limitation can eventually be overcome by using, for example, soliton pulses that have been discussed in Section 3.5.1 (in which case the use of optical CDMA could be extended to MANs).

The performance of OOC-based CDMA networks has been theoretically studied in [Salehi, 1987] and several experiments have demonstrated their feasibility.

As a particular example, it has been estimated that with a code length $F = 6{,}000$ and weight $K = 8$, about 100 nodes can be supported at a BER << 10^{-10} [Salehi, 1989a]. With $T = 600$ ps (i.e., the information rate of each node source is $1/T \cong 1.6$ Gbps) and $T_c = T/6{,}000 = 100$ fs, such a network would have a potential capacity in excess of 100 Gbps. This would, however, be very difficult to achieve in practice since the chip interval must be controlled to within a fraction of T_c (i.e., to within a few femtoseconds).

By introducing an additional rule into the code set design, it has been shown that this tight constraint on timing can to some extent be relaxed [Petrovic, 1991]. This additional rule states that all 1's have to be in the first half of the code word so that the distance between the last chip in one code word and the first chip in the next code word is larger than the distance between any two chips inside code words.

Besides these options of code design, a novel method based upon spectral encoding and decoding of coherent ultrashort light pulses (in the femtosecond range) has been proposed to provide ultrahigh-speed all-optical CDMA networks [Weiner 1988, Salehi 1989b]. This method is referred here to as *coherent optical* CDMA and is discussed in the next section.

10.3 COHERENT OPTICAL CDMA NETWORKS

The basic idea behind coherent optical CDMA is to utilize parallel modulation in the frequency domain to reshape (i.e., to encode) an ultrashort light pulse into a low-intensity pseudonoise burst. Figure 10.5 depicts the structures of the optical encoder and decoder used in coherent optical CDMA networks.

At the node transmitter, an ultrashort light pulse of duration T_c, representing a bit "1" of source information, is directed towards the spectral encoder, which consists of a suitably configured grating and lens apparatus. The first grating of the arrangement spatially decomposes the spectral components of the incident ultrashort light pulse. A phase mask is inserted at the focal point of lens L_1, where the optical-spectral components experience maximal separation. The phase mask imposes specific phase shifts among the different spectral components of the initial ultrashort pulse and thus serves to encode its spectrum. The coded spectral components are reassembled by the second lens L_2 and the second grating.

The shape of the encoded pulse will depend upon the choice of the phase mask but, in general, the pulse will be wider than the incident ultrashort light pulse. A desired characteristic of the encoded pulses is that they have a low intensity so that they resemble a pseudonoise signal. It has been shown that a pseudorandom phase mask is almost optimal in minimizing the peak intensity of the encoded light pulses [Hajela, 1992].

Each encoded signal is injected into the $N \times N$ star coupler and broadcast to all node receivers. At the receiver, the intended signal is recovered by a spectral correlation

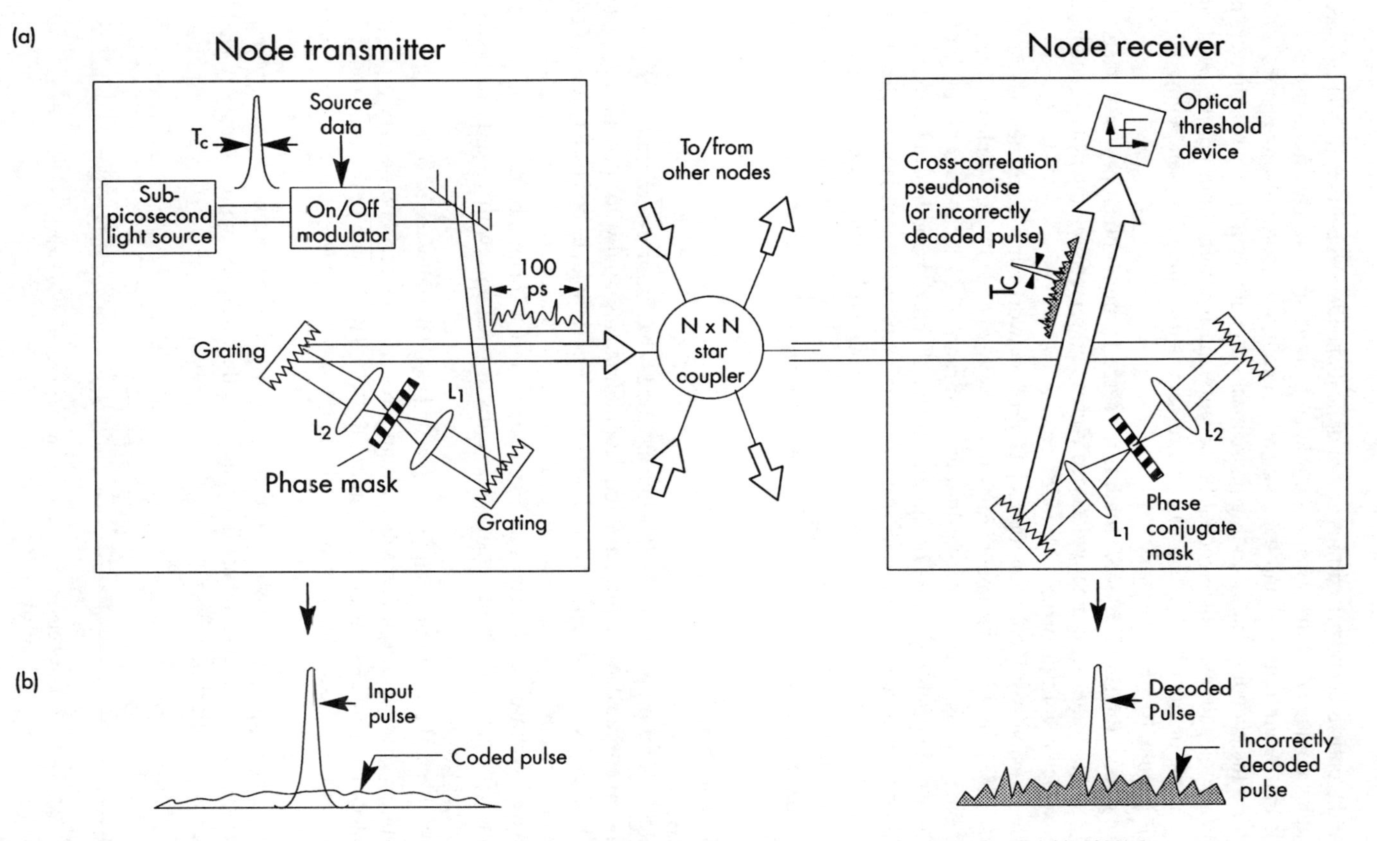

Figure 10.5 Spectral encoding and decoding of ultrashort light pulses in coherent optical CDMA networks. [After Salehi, 1989b.]

process: the optical decoder is similar to the optical encoder except that its phase mask is the complex conjugate of the encoding mask, so that the specific spectral-phase shifts of the coded pulse are removed and the original coherent ultrashort pulse is restored.

When the encoding and decoding phase masks do not match (i.e., when they are not a complex conjugate pair), the spectral phase shifts are rearranged but not removed, and the pulse at the output of the optical decoder remains a low-intensity pseudonoise burst as shown in Figure 10.5(b).

An optical threshold device, which is triggered by high optical power, detects data corresponding to intense, correctly decoded pulses and rejects low-intensity pseudonoise bursts associated with incorrectly decoded pulses.

The feasibility of coherent optical CDMA networks has been demonstrated in [Weiner, 1988]. In this experiment, 75-fs pulses were encoded by a 44-element pseudo-random binary phase mask resulting in a contrast ratio of 25:1 between the peak of the correctly decoded pulse and that of the incorrectly decoded pseudonoise bursts.

Although they are still only used in advanced optical-research laboratories, femto-second pulse-shaping techniques applied to optical networks, such as coherent CDMA networks discussed here, would suggest a new class of optical communication systems that may have several attractive features for application in the future.

10.4 SUMMARY

Since it was first proposed in 1985, optical CDMA has been theoretically and experimentally investigated by a few research laboratories essentially located in the United States. It is expected that this optical multiple-access technique, borrowed from microwave communication systems, will be suitable for use in a network environment where the nodes have asynchronous bursty traffic and where only a fraction of them communicate at a time. Optical CDMA has also some attractive features with regards to network security and privacy.

However, the technique requires advanced optical components such as ultrashort optical-pulse generators, optical switches and threshold devices, and/or very high-speed photodetectors, which excludes its commercial use today or even in the near future.

Nevertheless, continuous progress is under way to achieve the necessary break-throughs.

As often happens in advanced research, this progress may eventually become useful in as yet unexplored fiber-optic applications other than CDMA and may possibly lead to a new class of optical networks.

REFERENCES

[Gold, 1967] Gold, R. ''Optimal binary sequences for spread spectrum multiplexing,'' *IEEE Trans. Information Theory*, Vol. 13, Oct. 1967, pp. 619–621.

[Cooper, 1978] Cooper, G. R., et al. ''A spread spectrum technique for high-capacity mobile communications,'' *IEEE Trans. Vehicular Technology*, Vol. VT-27, Nov. 1978, pp. 264–275.

[Sarwate, 1980] Sarwate, D. V., et al. ''Cross-correlation properties of pseudorandom and related sequences,'' *IRE Proc.*, Vol. 68, No. 5, May 1980, p. 593–619.

[Weber, 1981] Weber, C. A., et al. ''Performance considerations of Code Division Multiple-Access systems,'' *IEEE Trans. on Vehicular Technology*, Vol. VT-30, Feb. 1981, pp. 3–10.

[Holmes, 1982] Holmes, J. K. *Coherent Spread Spectrum Systems*, New York, NY: John Wiley & Sons, 1982.

[Hui, 1985] Hui, J. Y. ''Pattern code modulation and optical decoding—A novel code-division multiplexing technique for multifiber networks,'' *IEEE J. Selected Areas of Communication*, Vol. JSAC-3, No. 6, Nov. 1985, pp. 916–927.

[Prucnal, 1986a] Prucnal, P. R., et al. ''Ultra-fast all optical synchronous multiple access fiber networks,'' *IEEE J. Selected Areas of Communication*, Vol. JSAC-4, No. 9, Dec. 1986, pp. 1484–1493.

[Prucnal, 1986b] Prucnal, P. R., et al. ''Spread spectrum fiber-optic local area network using optical processing,'' *IEEE J. Light. Technology*, Vol. LT-4, No. 5, 1986, pp. 547–554.

[Santoro, 1987] Santoro, M. A., et al. ''Asynchronous fiber optic local area network using *CDMA* and optical correlation,'' *IRE Proc.*, Vol. 75, No. 9, Sept. 1987, pp. 1336–1338.

[Salehi, 1987] Salehi, J. A., et al. ''Fundamental principles of fiber optics code division multiple access (*FO-CDMA*).'' *Proc. Globecom'87*, 46.3, 1987, pp. 1601–1609.

[Prucnal, 1987] Prucnal, P. R., et al. ''Photonic switch with optically self-routed bit switching,'' *IEEE Communication Magazine*, Vol. 25, No. 5, May 1987, pp. 50–55.

[Weiner, 1988] Weiner, A. M., et al. ''Encoding and decoding of femtosecond pulses,'' *Optics Letters*, Vol. 13, No. 4, April 1988, pp. 300–302.

[Salehi, 1989a] Salehi, J. A., et al. ''Code division multiple-access techniques in optical fiber networks—Parts I and II,'' *IEEE Trans. Communication*, Vol. 37, No. 8, Aug. 1989, pp. 824–842. See also Salehi, J. A. ''Emerging optical code-division multiple access communications systems,'' *IEEE Network*, March 1989, pp. 31–39.

[Salehi, 1989b] Salehi, J. A. ''Temporal and statistical analysis of ultrashort light pulse code-division multiple access communications network,'' *Proc. Intl. Conf. Communication, ICC'87*, 23.2, June 1989, pp. 728–733.

[Chung, 1989] Chung, F.R.K., et al. ''Optical Orthogonal Codes : Design, analysis, and applications,'' *IEEE Trans. Information Theory*, Vol. 35, No. 3, May 1989, pp. 595–604.

[Prucnal, 1990] Prucnal, P. R., et al. ''Optically-processed routing for fast packet switching,'' *IEEE Lightwave Communication Systems (IEEELCS) Magazine*, May 1990, pp. 54–67.

[Chung, 1990] Chung, H., et al. ''Optical orthogonal codes—new bounds and an optimal construction,'' *IEEE Trans. Information Theory*, Vol. 36, No. 4, 1990, pp. 866–873.

[Petrovic, 1990] Petrovic, R., et al. ''Orthogonal codes for *CDMA* optical fiber *LANs* with variable bit interval,'' *Electron. Letters*, Vol. 26, No. 10, May 1990, pp. 662–664.

[Kiasaleh, 1991] Kiasaleh, K. ''Fiber optic frequency hopping multiple access communication system,'' *IEEE Photon. Technology Letters*, Vol. 3, No. 2, Feb. 1991, pp. 173–175.

[Petrovic, 1991] Petrovic, R., et al. ''*CDMA* techniques for optical fiber *LANs*,'' *J. of Optical Communications*, Vol. 12, No. 3, 1991, pp. 101–106.

[Wu, 1992] Wu, J-H., et al. ''Synchronous fiber-optic code division multiple access networks with error control coding,'' *Electron. Letters*, Vol. 28, No. 23, Nov. 1992, pp. 2118–2120.

[Hajela, 1992] Hajela, D. J., et al. ''Limits to the encoding and bounds on the performance of coherent ultrashort light pulse code-division multiple-access systems,'' *IEEE Trans. Communication*, Vol. 40, No. 2, Feb. 1992, pp. 325–336.

[Zhang, 1993] Zhang, J-G., et al. ''Tunable prime-code encoder/decoder for all optical *CDMA* networks,'' *Electron. Letters*, Vol. 29, No. 13, June 1993, pp. 1211–1212.

Appendix A
dBm Units

In the engineering jargon, the optical power is usually expressed in terms of dBm instead of watts (or milliwatts, mW).

The relation between these units is defined by

$$P\,[\text{dBm}] = 10 \log_{10} P\,[\text{mW}] \tag{A.1}$$

The number of dBm thus scales logarithmically with the number of milliwatts. Table A.1 gives the correspondence between some often encountered power levels.

Since $10 \log_{10} 2 \cong 3$, each time the power is multiplied or divided by a factor 2, we have simply to add or subtract 3 dBm to the initial power expressed in dBm, respectively.

Table A.1
The Correspondence Between Watts and dBm

P	P
1 μW	−30 dBm
10 μW	−20 dBm
100 μW	−10 dBm
0.5 mW	−3 dBm
1 mW	0 dBm
2 mW	+3 dBm
10 mW	+10 dBm
100 mW	+20 dBm

Appendix B
The Relation Between $\Delta\lambda$ and $\Delta\nu$

Given the center wavelength λ and the deviation $\Delta\lambda$ around λ, the corresponding deviation $\Delta\nu$ in frequency can be obtained as follows. Let us start with the fundamental relation

$$\lambda \cdot \nu = c \tag{B.1}$$

which relates the wavelength λ to the carrier frequency ν, where c is the light velocity in vacuum.
By differentiating (B.1), we have immediately

$$\frac{\Delta\nu}{\nu} = -\frac{\Delta\lambda}{\lambda} \tag{B.2}$$

Replacing ν by c/λ in (B.2), we obtain

$$\Delta\nu = -\frac{c}{\lambda^2}\Delta\lambda \tag{B.3}$$

For example, $\Delta\lambda = 1$ nm around $\lambda = 1.3$ μm corresponds to $\Delta\nu = -180$ GHz. Thus, the 1.3-μm fiber window, which spans from 1,260 nm to 1,360 nm ($|\Delta\lambda| = 100$ nm) has a bandwidth of approximately 18,000 GHz. Similarly, the 1.5-μm fiber window, with $|\Delta\nu| = 150$ nm around 1.5 μm, has a bandwidth of approximately $|\Delta\nu| = 20{,}000$ GHz. The total usable fiber bandwidth (where the loss is less than about 0.5 dB) is thus approximately 40,000 GHz.

Appendix C
Dispersion Parameters

The factors D(λ) in (3.38) and β_0'' in (3.52) are two distinct forms used to express the same effect of dispersion. The mathematical relation between these two factors can be obtained as follows.

Since $\beta(\omega) = \dfrac{\omega N(\omega)}{c}$, where $N(\omega)$ is the effective refractive index that takes both material and waveguide dispersions into account, we have

$$\frac{\partial \beta}{\partial \omega} = \frac{1}{c}\left(N + \omega \frac{\partial N}{\partial \omega}\right) \tag{C.1}$$

and $$\frac{\partial^2 \beta}{\partial \omega^2} = \frac{1}{c}\left(2\frac{\partial N}{\partial \omega} + \omega \frac{\partial^2 N}{\partial \omega^2}\right) \tag{C.2}$$

Using $\lambda\omega = 2\pi c$ [Eq.(B.1)], we also have

$$\frac{\partial}{\partial \omega} = \left(\frac{\partial \lambda}{\partial \omega}\right)\frac{\partial}{\partial \lambda} = -\frac{\lambda^2}{2\pi c}\frac{\partial}{\partial \lambda} \tag{C.3}$$

and $$\frac{\partial^2}{\partial \omega^2} = \frac{\lambda^3}{(2\pi c)^2}\left(2\frac{\partial}{\partial \lambda} + \lambda\frac{\partial^2}{\partial \lambda^2}\right) \tag{C.4}$$

Inserting (C.3) and (C.4) into (C.1) and (C.2), we finally obtain

$$\frac{\partial\beta}{\partial\omega} = \frac{1}{v_g} = \frac{N}{c}\left(1 - \frac{\lambda}{N}\frac{\partial N}{\partial\lambda}\right) \tag{C.5}$$

and $$\frac{\partial^2\beta}{\partial\omega^2} = \beta'' = \frac{\lambda^2}{2\pi c}\left(\frac{\lambda}{c}\frac{\partial^2 N}{\partial\lambda^2}\right) = -\frac{\lambda^2}{2\pi c}D(\lambda) \tag{C.6}$$

Appendix D
Shot Noise

The derivation of (4.23) is somewhat qualitative and is based on the remarkable property of the Poisson probability distribution, which states that the variance is equal to the mean. In this appendix, a rigorous derivation of (4.23) is presented with the advantage being to bring out the physical meaning of the shot-noise current.

What we wish to determine is the mean-square current fluctuation of the shot-noise current, σ^2_{shot}, at the receiver output with postdetection bandwidth Δf.

We start from the observation that (4.18) results from the fact that photons within the laser beam are distributed uniformly over the time. In other words, the a priori probability that a given photon will impinge the photodetector in any time interval is distributed uniformly over that interval.

Therefore, an observation during the time interval T of the electrical current $i(t)$ generated by the incident photons will yield

$$i(t) = e \cdot \sum_{\alpha=1}^{N} \delta(t - t_\alpha) \tag{D.1}$$

where N is the total number of photons during the interval T, and t_α are random times uniformly distributed over T (we assume an ideal photodetector such that $\eta = 1$).

Equation (D.1) can be expanded in a Fourier series such as

$$i(t) = \sum_{\mu=-\infty}^{+\infty} I_\mu \cdot e^{i(2\pi/T)\mu t} \tag{D.2}$$

The dc component of $i(t)$, I_μ, is given by (4.19) in which <n> is substituted by N; that is, $I_{\mu=0} = \bar{I} = eN/T$. All other coefficients I_μ represent random fluctuations of the current around its dc average value and must be regarded as random noise sources.

With (D.1), these coefficients are given by

$$I_\mu = \frac{1}{T} \cdot \int_{-T/2}^{T/2} i(t) \cdot e^{-i(2\pi/T)\mu T}\, dt = \frac{e}{T} \cdot \sum_{\alpha=1}^{N} e^{-i(2\pi/T)\mu t_\alpha} \tag{D.3}$$

Since the current $i(t)$ is real, we have $i(t) = i^*(t)$, and therefore

$$<[i(t) - \bar{I}]^2> = \sigma_{shot}^2 = < \sum_{\substack{\mu=-\infty \\ \mu,\sigma\neq 0}}^{+\infty} \sum_{\sigma=-\infty}^{+\infty} I_\mu I_\sigma^* \cdot e^{i(2\pi/T)(\mu-\sigma)t}> \tag{D.4}$$

Because the times t_α are random and uniformly distributed over T, one can assume that the phases of the complex coefficients I_μ are random so that the average over an ensemble of a very large number of physically identical phenomena yields

$$\sigma_{shot}^2 = \sum_{\substack{\mu=-\infty \\ \mu\neq 0}}^{+\infty} <|I_\mu|^2> \tag{D.5}$$

Using (D.3), we have

$$<|I_\mu|^2> = \frac{e^2}{T^2} \cdot \sum_{\alpha=1}^{N} \sum_{\beta=1}^{N} <|e^{i(2\pi/T)\mu(t_\beta - t_\alpha)}|> = N \cdot \frac{e^2}{T^2} \tag{D.6}$$

The summation in (D.5) must be restricted to the bandwidth Δf of the receiver since out-of-band frequencies do not pass through the receiver and consequently do not contribute to σ_{shot}^2. Assuming that the receiver passes a band of frequency Δf from $f_n = \frac{n}{T}$ to $f_m = \frac{m}{T}$, one has

$$\sigma_{shot}^2 = 2 \cdot \sum_{\mu=n}^{m} N \frac{e^2}{T^2} = 2N \frac{e^2}{T^2} (m - n) \tag{D.7}$$

The factor 2 comes from the fact that $i(t)$ is real, which yields to $I_{-\mu} = I_\mu^*$, so that the summation over negative indices can be included in the sum over positive indices.

With

$$\Delta f = \frac{m - n}{T} \text{ and } \bar{I} = \frac{eN}{T} \quad \text{(D.8)}$$

Equation (D.7) becomes

$$\sigma_{\text{shot}}^2 = 2e\bar{I}\Delta f \quad \text{(D.9)}$$

which is equivalent to (4.23).

Notice that (D.1) is an approximation since it uses "δ" functions to represent the current resulting from a single electron emission within the photodetector. This is justified as long as f_m, the highest frequency in the band Δf, satisfies $f_m << 1/t_e$ where t_e is the transit time of the electrons as they pass from the cathode to anode of the photodetector. When $f_m > 1/t_e$ one has to replace the "δ" function in (B.1) by a function $h(t—t_\alpha)$, which describes the form of the current generated by one electron. It can be shown that, even in this case, (D.9) still applies.

To end this appendix, we must notice that one can show that the shot noise (or quantum noise) is due to Heisenberg's uncertainty principle, which cannot be violated. This uncertainty principle states that the in-phase amplitude noise and the quadrature-phase amplitude noise cannot be simultaneously reduced below the quantum noise. However, one of these two amplitudes can be reduced below the quantum-noise level while the other increased above the quantum-noise level. This leads to new optical signals having quantum-noise imbalance in the two noise amplitudes, which are referred to as *squeezed states of light*. These new states of light lead to a modification of the shot-noise limit for coherent detection receivers. However, for current practical optical communication systems, the above is valid.

Appendix E
Preamplifier Noise

Equation (4.33) is a general expression for the mean-square noise current introduced by the preamplifier. For FET preamplifiers, it has been shown that [Personick, 1973 in Chapter 4]

$$<i^2_{\text{amp}}> = \left(\frac{4kT}{R_L} + 2eI_{\text{gate}}\right)I_2B + \frac{4kT\Gamma}{g_m}(2\pi C_T)^2(f_cI_fB^2 + I_3B^3) \tag{E.1}$$

where B is the bit rate ($\Delta f = B/2$ for NRZ, $\Delta f = B$ for RZ), R_L is the load resistor (or feedback resistor for a transimpedance front-end), I_{gate} is the FET gate leakage current, g_m is the FET transconductance, C_T is the total input capacitance, f_c is the FET $1/f$ noise corner frequency, and Γ is the FET channel-noise factor. Table E.1 gives the ranges of typical FET parameters for the three FET types discussed in Section 4.2.2.2.

I_2, I_3, and I_f are known as the *Personick integrals*. They depend only on the pulse shape entering and leaving the fiber and the type of line coding used. Table E.2 gives values of the Personick integrals for rectangular input pulses and raised-cosine output pulses.

For bipolar preamplifiers, (E.1) is modified according to

$$<i^2_{\text{amp}}> = \left(\frac{4kT}{R_L} + 2eI_b\right)I_2B + \left[\frac{2eI_c}{g_m^2}(2\pi C_T)^2 + 4kTr_{bb'}(2\pi(C_d + C_S))^2\right]I_3B^3 \tag{E.2}$$

where I_b and I_c are respectively the base and collector bias currents, $r_{\text{bb'}}$ is the base-spreading resistance, C_d is the detector capacitance, and C_s is the stray capacitance. Table E.3 gives typical values of these parameters.

Table E.1
Parameters of Field-Effect-Transistors (FETs)

	GaAs MESFET	*Si MOSFET*	*Si JFET*
g_m [mS]	15–50	20–40	5–10
C_T[pF]	0.3–0.7	0.6–1.2	3–6
Γ	1.1–1.75	1.5–3.0	0.7
I_{gate} [nA]	1–1,000	0	0.01–0.1
f_c [MHz]	10–100	1–10	< 0.1

Table E.2
Personick Integrals for NRZ and RZ Formats

	NRZ	*RZ*
I_2	0.562	0.403
I_3	0.0868	0.0361
I_f	0.184	0.0984

Table E.3
Bipolar Junction Transistor-
Preamplifier Parameters

I_b [mA]	$\approx$ 0.1
$C_d + C_S$ [pF]	0.2
C_T [pF]	1.0
$r_{bb'}$ [Ω]	20

Appendix F
Physical Constants in MKSA System of Units

c	Light velocity in vacuum	$2.997\ 925\ 10^{8}$ m/s
h	Planck's constant	$6.626\ 20\ 10^{-34}$ J·s
k	Boltzmann's constant	$1.380\ 54\ 10^{-23}$ J/K
e	Electron's charge	$1.602\ 10\ 10^{-19}$ C
μ_0	Vacuum magnetic permeability = $4\pi \cdot 10^{-7}$ H/m	$1.2566\ 10^{-6}$ H/m
ε_0	Vacuum permitivity ($\varepsilon_0\mu_0 = 1/c^2$)	$8.854\ 10^{-12}$ F/m

List of Acronyms

AGC	automatic gain control
AM	amplitude modulation
APD	avalanche photodiode
ASE	amplified spontaneous emission
ASK	amplitude shift keying
ATM	asynchronous transfer mode
BER	bit error rate
B-ISDN	broadband integrated services digital network
BPF	bandpass filter
BPSK	binary phase shift keying
CATV	community antenna (or cable television)
CCITT	Committe Consultatif International des Telegraphe et Telephone
CDMA	code division multiple access
CNR	carrier-to-noise ratio
CO	central office
COH	coherent
CP	customer premises
CPFSK	continuous phase shift keying
CSMA/CD	carrier sense multiple access with collision detection
CSO	composite second order
CTB	composite triple beat
DBR	distributed Bragg reflector
DD	direct detection
DFB	distributed feedback (laser)
DPSK	differential phase shift keying
DQDB	distributed queue dual bus
EDFA	erbium-doped fiber amplifier

ELED	edge emitting light-emitting diode
FDDI	fiber distributed data interface
FDM	frequency division multiplexing
FDMA	frequency division multiple access
FPF	Fabry-Perot filter
FPM	four-photon mixing
FSK	frequency shift keying
FWHM	full-width at half maximum
GRIN	graded index lens
GSA	ground state absorption
IF	intermediate frequency
IM/DD	intensity modulation / direct detection
IMP	intermodulation product
ITU	international telecommunication union
LAN	local area network
LD	laser diode
LED	light-emitting diode
LO	local oscillator
LPF	low-pass filter
MAN	metropolitan area network
MLM	multilongitudinal mode
MQW	multiple quantum well
MZF	Mach-Zender filter
N-ISDN	narrowband-integrated services digital network
OOC	optical orthogonal code
OOK	on-off keying
PIN	positive-intrinsic-negative (photodiode)
PLL	phase locked loop
POLSK	polarization shift keying
PON	passive optical network
POTS	plain old telephone service
QAM	quadrature amplitude modulation
QPSK	quadrature phase shift keying
RF	radio frequency
RIN	relative intensity noise
RMS	root-mean-square
SAW	surface acoustic wave
SBS	stimulated Brillouin scattering
SCM	subcarrier multiplexing
SCMA	subcarrier multiple access
SLD	superluminescent light-emitting diode
SLM	single longitudinal mode

SMSR	side-mode suppression ratio
SNR	signal-to-noise ratio
SPM	self-phase modulation
SQW	single-quantum well
SRS	stimulated Raman scattering
TDM	time division multiplexing
TDMA	time division multiple access
WAN	wide area network
WDM	wavelength division multiplexing
WDMA	wavelength division multiple access
XPM	cross-phase modulation

About the Author

Denis J.G. Mestdagh is a senior research engineer and project leader at Alcatel Bell Telephone in Antwerp, Belgium. He earned his M.S. degree in physics engineering (1983) and his Ph.D. (1988) in nonlinear fiber optics at the Free University of Brussels. He is a member of IEEE/LEOS and the New York Academy of Sciences. His most recent research has revolved around fiber-optic networks and digital signal processing in telecommunication systems.

Index

The Artech House Optoelectronics Library

Brian Culshaw, Alan Rogers, and Henry Taylor, *Series Editors*

Acousto-Optic Signal Processing: Fundamentals and Applications, Pankaj Das

Amorphous and Microcrystalline Semiconductor Devices, Optoelectronic Devices, Jerzy Kanicki, editor

Bistabilities and Nonlinearities in Laser Diodes, Hitoshi Kawaguchi

Electro-Optical Systems Performance Modeling, Gary Waldman and John Wootton

The Fiber-Optic Gyroscope, Hervé Lefèvre

Field Theory of Acousto-Optic Signal Processing Devices, Craig Scott

Fundamentals of Multiaccess Optical Fiber Networks, Denis J. G. Mestdagh

Germanate Glasses: Structure, Spectroscopy, and Properties, Alfred Margaryan and Michael A. Piliavin

High-Power Optically Activated Solid-State Switches, Arye Rosen and Fred Zutavern, editors

Highly Coherent Semiconductor Lasers, Motoichi Ohtsu

Iddq Testing for CMOS VLSI, Rochit Rajsuman

Introduction to Electro-Optical Imaging and Tracking Systems, Khalil Seyrafi and S. A. Hovanessian

Introduction to Glass Integrated Optics, S. Iraj Najafi

Introduction to Radiometry and Photometry, William Ross McCluney

Optical Control of Microwave Devices, Rainee N. Simons

Optical Document Security, Rudolf L. van Renesse

Optical Fiber Amplifiers: Design and System Applications, Anders Bjarklev

Optical Fiber Sensors, Volume I: Principles and Components, John Dakin and Brian Culshaw, editors

Optical Fiber Sensors, Volume II: Systems and Applicatons, John Dakin and Brian Culshaw, editors

Optical Interconnection: Foundations and Applications, Christopher Tocci and H. John Caulfield

Optical Network Theory, Yitzhak Weissman

Optical Transmission for the Subscriber Loop, Norio Kashima

Principles of Modern Optical Systems, Volumes I and II, I. Andonovic and D. Uttamchandani, editors

Reliability and Degradation of LEDs and Semiconductor Lasers, Mitsuo Fukuda

Semiconductor Raman Laser, Ken Suto and Jun-ichi Nishizawa

Semiconductors for Solar Cells, Hans Joachim Möller

Single-Mode Optical Fiber Measurements: Characterization and Sensing, Giovanni Cancellieri

For further information on these and other Artech House titles, contact:

Artech House
685 Canton Street
Norwood, MA 02062
617-769-9750
Fax: 617-769-6334
Telex: 951-659
email: artech@world.std.com

Artech House
Portland House, Stag Place
London SW1E 5XA England
+44 (0) 71-973-8077
Fax: +44 (0) 71-630-0166
Telex: 951-659
email: bookco@artech.demon.co.uk